Springer Series in
Surface Sciences

Editor: Gerhard Ertl

35

Springer Series in Surface Sciences

Editors: G. Ertl, R. Gomer and D. L. Mills   Managing Editor: H.K.V. Lotsch

W. Eberhardt (Ed.)

Applications of Synchrotron Radiation

High-Resolution Studies of Molecules and Molecular Adsorbates on Surfaces

With Contributions by
A.M. Bradshaw W. Eberhardt H.-J. Freund
F.M. Hoffmann H. Kuhlenbeck N. Mårtensson
D. Menzel A. Nilsson G.P. Williams
D.P. Woodruff W. Wurth

With 170 Figures

Springer-Verlag
Berlin Heidelberg New York London Paris
Tokyo Hong Kong Barcelona Budapest

Professor Dr. Wolfgang Eberhardt

Institut für Festkörperforschung,
KFA Jülich,
52425 Jülich, Germany

Series Editors

Professor Dr. Gerhard Ertl

Fritz-Haber-Institut der Max-Planck-Gesellschaft, Faradayweg 4–6,
14195 Berlin, Germany

Professor Robert Gomer, Ph.D.

The James Franck Institute, The University of Chicago, 5640 Ellis Avenue,
Chicago, IL 60637, USA

Professor Douglas L. Mills, Ph.D.

Department of Physics, University of California,
Irvine, CA 92717, USA

Managing Editor: Dr. Helmut K.V. Lotsch

Springer-Verlag, Tiergartenstrasse 17,
69121 Heidelberg, Germany

ISBN-13:978-3-642-79026-3 e-ISBN-13:978-3-642-79024-9
DOI: 10.1007/978-3-642-79024-9

Library of Congress Cataloging-in-Publication Data. Eberhardt, W. (Wolfgang), 1950- Applications of synchrotron radiation: high-resolution studies of molecules and molecular adsorbates on surfaces/W.W. Eberhardt. p. cm. — (Springer series in surface sciences; 35) Includes bibliographical references and index. ISBN-13:978-3-642-79026-3 1. Surfaces (Physics) 2. Synchrotron radiation—Industrial applications. 3. Surface chemistry. 4. Molecular dynamics. 5. Adsorption. I. Title. II. Series. QC173.4.S94E24 1994 530.4'17—dc20 94-12947

Softcover reprint of the hardcover 1st edition 1995

Typesetting: Macmillan India Ltd., Bangalore-25

SPIN: 10092384 54/3140/SPS - 5 4 3 2 1 0 - Printed on acid-free paper

Preface

Both, the fundamental interest in understanding the chemisorption bond and the dynamics of molecular interactions on and with surfaces as well as the obvious connection to the technological application in catalysis have provided an important stimulus for the research on molecules and molecular adsorption systems. A major impact in the understanding of model chemisorption systems on single-crystal surfaces originated from the application of synchrotron radiation based experimental techniques like angle-resolved photoemission, high-resolution studies of core-electron excitations, photoemission and core-hole decay processes, and infrared reflection absorption spectroscopy. These experiments have revealed new insight into the static and dynamic electronic interactions and simultaneously also provided new and more accurate information about the chemisorption geometries.

As the third-generation synchrotron-radiation facilities, offering vastly improved source characteristics for these kind of studies, are coming online at many places around the globe it seems appropriate to step back and evaluate these experimental techniques and to discuss their potential for future applications. Foremost among these capabilities are improvements in spectral and spatial resolution and in intensity for the study of low-yield secondary processes. The trend to explore the dynamics in the electronic interactions as well as the dynamically introduced changes of the molecular geometry, ranging from the excitation of vibrational modes to desorption and dissociation, is obvious in some of the most recent work. Together with spatially-resolved "spectromicroscopy" experiments the exploration of the dynamics seems to be one of the 'new horizons' in this field of research.

This quasi-monograph provides a survey of various experimental techniques, as they are applied to the study of molecules and molecular adsorption systems. The extension of the application of these experimental techniques to other systems like semiconductor surfaces and heterostructures or epitaxial growth of metallic thin films and overlayers is obvious and straightforward. Therefore this book should also be of interest to the general community of researchers and students working with synchrotron radiation in the area of materials science and information technology.

Jülich, September 1994 *Wolfgang Eberhardt*

Contents

Contributors

A.M. Bradshaw

Fritz Haber Institute, Faradayweg 4-6,
14195 Berlin, Germany

W. Eberhardt

Institut für Festkörperforschung, KFA Jülich,
Postfach 1913, 52425 Jülich, Germany

H.-J. Freund

Physikalische Chemie, Ruhr-Universität,
Universitätsstr. 150, 44801 Bochum, Germany

F.M. Hoffmann

Exxon Research and Engineering,
Annandale, NJ 08801, USA

H. Kuhlenbeck

Physikalische Chemie, Ruhr-Universität,
Universitätsstr. 150, 44801 Bochum, Germany

N. Mårtensson

Department of Physics, Uppsala University, Box 530,
75121 Uppsala, Sweden

D.Menzel

Physik Department E20, Technische Universität München,
85748 Garching, Germany

A. Nilsson

Department of Physics, Uppsala University, Box 530,
75121 Uppsala, Sweden

G.P. Williams

Brookhaven National Laboratory, Upton, NY 11973, USA

D.P. Woodruff

Physics Department, University of Warwick,
Coventry CV4 7AL, UK

W. Wurth

Physik Department E20, Technische Universität München,
85748 Garching, Germany

1. Introduction

W. Eberhardt

The application of synchrotron radiation in physics, chemistry, materials science, and biology by now has matured from an exotic experimental field to a well-established area of science. It all started with a few scientists using the radiation emitted, as an unwanted by-product, by large electron accelerators for their own scientific investigations. These electron accelerators turned out to be very powerful sources of electromagnetic radiation, since the rf acceleration power is almost exclusively converted into emission of synchrotron radiation. Actually, the power converted into synchrotron radiation limits the maximum beam energy attainable in any of these machines. Later on, scientists were allowed to establish their own laboratories, tapping into the vacuum environment of the accelerator and guiding the radiation through 'beam lines' into their experiments. Finally, dedicated synchrotron radiation sources were established, designed and constructed exclusively for the production and application of synchrotron radiation in science. By now some of these dedicated facilities have been in existence in excess of more than 15 years. At this point in time the next quantum leap is just ahead, as new synchrotron radiation sources are being commissioned world-wide, where undulators and wigglers are used as the major sources for the radiation instead of the bending magnets. These insertion-device sources have unique radiation characteristics, promising a gain in brightness of the emitted radiation by several orders of magnitude. This improvement in quality will lead to new experiments not possible at present in essentially all the fields of science with synchrotron radiation.

Synchrotron radiation offers some unique properties, which are not matched by any other source of electromagnetic radiation. The qualities making synchrotron radiation generated by bending magnets such an excellent probe can be briefly summarized as follows. First, the spectrum extends *continuously in wavelength* from the infrared region of the electromagnetic spectrum up to hard X-rays. Second, the radiation is *linearly polarized* in the plane of the electron orbit and *elliptically polarized* above and below this plane. Synchrotron radiation is *highly collimated* and originates from a *small source*, both qualities result in the exceptional *brightness* of this type of radiation. Additionally, it has a well defined *time structure* with pulse lengths from 10 to 100 ps. Moreover, it can be calculated exactly from first principles and thus may serve as a radiation standard.

Springer Series in Surface Sciences, Vol. 35
Applications of Synchrotron Radiation Ed.: W. Eberhardt

The undulator spectrum differs to the extend that it is characterized by a coherent superposition of the radiation generated by the various poles of the insertion device. Thus, the power of the coherent part of the radiation is vastly improved, the spectral distribution exhibits more or less pronounced peaks and minima due to the interference effects, and the brightness of the radiation is increased by orders of magnitude. Additionally, the polarization of the undulator radiation is changed to exhibit more complex characteristics.

We here do not want to elaborate any further on the properties of synchrotron radiation generated by bending magnets, insertion devices or Free Electron Lasers. The interested reader may be referred to the extensive reviews [1.1–9] that have been published on this subject.

The spectroscopy of molecules and molecular adsorbates on surfaces is certainly one area of science, where synchrotron radiation related studies have, in the past, made an impact on the understanding of these systems. With the new high-brightness synchrotron radiation sources ahead this will certainly be a continuing field of very active research. Therefore, at this break in time, as all over the world scientists are getting ready to explore the possibilities opened up for research with the next-generation synchrotron-radiation sources, it seems an appropriate idea to step back and evaluate the various techniques that have been developed for this research and to critically assess their potential. In this sense this compilation of reviews is intended as a comprehensive summary and snap-shot of the state of the field for the active research scientist as well a fresh graduate student, who would like to get acquainted with this kind of research.

In the context of this book we have concentrated on the study of molecular systems, in gas phase as well as in the chemisorbed state. Synchrotron radiation is, of course, much more versatile in its potential for application in scientific research. However, to give examples of all of these is beyond the scope of this book. Nevertheless, this collection of review should be of interest to researchers active in other areas of science as well, since most of the techniques presented here can and have been readily applied to the study of other systems. For example, angle-resolved photoemission has been established as a very powerful tool in the determination of the electronic structure of bulk solids and surfaces as well as in research on thin epitaxial layer systems. These latter systems are also of technological interest, as far as semiconductor heterostructures or thin magnetic films are concerned. As a second example, photoelectron diffraction as a technique has been successfully applied to the study of the epitaxial growth of metal and semiconductor layers and made important contributions in the understanding of the growth mechanism of these systems.

Moreover, molecular adsorption systems are not only of basic scientific interest, but also bear some technological relevance, since they allow to study the basic phenomena underlying catalytic reactions. Most of the systems studied and presented here are, at best, to be considered 'model catalyst' systems and thus, are far from being used in any technical applications. Nevertheless, the implication is that an enhancement of the understanding of the basic phenomena will lead to some technological advances in the future. One of the basic phenomena under investigation is the development of the chemical bond of a species with the surface and the resulting changes in the electronic structure. Also, the molecular interactions in the chemisorbed state between co-adsorbates are at the heart of the mechanism of catalysis.

Investigations of molecular chemisorption processes have initially been devoted to understand the static properties of the system. Foremost among these is the quantitative determination of the atomic structure of the surface chemisorption complex together with its electronic structure. Comparison of the electronic structure of the chemisorbed state with the isolated gas-phase species allows to deduce the electronic interactions occurring upon the formation of the chemisorption bond. In addition to these static properties, it has to be realised that the dynamic properties are equally important for the understanding of catalysis and chemical reactions on surfaces. The study of the kinetics and dynamics of the chemisorption systems is more the area of conventional surface science techniques and the roam of laboratory experiments. Nevertheless, the synchrotron radiation techniques reviewed here are all currently being extended to the study of various dynamical properties or non-stationary phenomena. This is presently the forefront of the development of the experimental techniques.

The first chapter of this book, co-authored by H. Kuhlenbeck and H.J. Freund, reviews the experimental determination of the *valence electronic structure* of chemisorbed molecular systems. The experimental technique used for these studies is *angle-resolved photoemission*, which is able to unfold its full potential, only when used in conjunction with a variable energy photon source such as synchrotron radiation. In the chemisorbed state, the electronic structure of the adsorbed molecules is modified by the formation of the chemical bond with the substrate. In the simplest picture this results in bonding and antibonding electronic orbitals formed by hybridization of the molecular orbitals with substrate states of the appropriate symmetry. The bonding states are stabilized with respect to the corresponding states of the free molecule and thus generally are observed at higher binding energies. Since the molecular states often have higher ionization potentials than the states of the substrate they are interacting with, the character of the bonding states is largely reflected by the original states of the molecule. The corresponding antibonding states, on the other hand, are often only partially occupied and display largely substrate character. These states are, in general, energetically degenerate with the substrate continuum. Therefore, the identification of the antibonding states is somewhat more difficult and involves a detailed knowledge of the substrate electronic structure.

Apart from the stabilizing effect on some of the molecular orbitals, which appears as a shift to higher binding energy, the opposite shift is observed for the orbitals not involved in the chemisorption bond due to the presence of metallic screening in the substrate. These screening effects generally lower the measured binding energy of the electrons since the metallic charge reacts to the hole generated in the photoemission process by forming an image charge in the substrate. Thus, the Coulomb attraction of the 'photohole' is reduced and the photoelectron is emitted with a larger kinetic energy.

For two-dimensional, ordered adsorbate systems band formation is observed in the electronic states due to hybridization between the electronic states of neighbouring adsorbate molecules. This is caused either by the direct overlap or indirectly by coupling through substrate states. Consequently these effects lead to the appearance of dispersion and a two-dimensional band structure is observed for these systems, which can be mapped using angle resolved photoemission.

Core-level photoemission from adsorbed molecular systems is described by N. Mårtensson and A. Nilsson in the second chapter. Even though the core electrons do not directly interact with the substrate, the binding energy of the core electrons in a molecular system is changed by the interaction with the substrate, reflecting both charge transfer between the chemisorbed molecule and the substrate and screening processes. The same concept can be applied to the substrate core level photoemission as well. Since the surface electronic structure of the substrate in general differs from the electronic structure of the bulk crystal, the surface atoms also exhibit different core-level binding energies. Consequently, in a chemisorption system the surface atoms involved in the bonding with the adsorbate reflect upon these interactions through a change in the core-level binding energy. A careful analysis of these shifts leads to a model of the structural co-ordination of the adsorbate-surface complex.

Additionally, with the high resolution and excellent statistics that recently has become available for these studies, the line shape of the core-level emission may be analyzed in great detail. The core-level line shape viewed in high resolution reflects the simultaneous excitation of vibrational quanta in the photoemission process. Thus the dynamics of the nuclear rearrangement following the emission of a core electron is probed. Different adsorption sites and geometries are apparent not only by different binding energies, but also by a change in the observed vibrational profile.

Dynamical charge rearrangements are also reflected by the shake-up structure observed as satellites in the photoemission signal. The shake-up processes correspond to the creation of excited valence-electron configurations simultaneous to the emission of the core electron. The energy position and intensity of the shake-up states is related to the strength of the coupling between the chemisorbed molecule and the substrate.

Core-level photoemission can also be used for a quantitative structural analysis of a chemisorbed system. *Photoelectron diffraction* is observed as a result of the scattering processes of the outgoing electron wave at atomic neighbours with a well defined registry relative to the atom the core electron is emitted from. These type of studies are described by A.M. Bradshaw and D.P. Woodruff in the third chapter. Directional and energy-dependent intensity modulations of the core-level photoemission signal result from the coherent superposition of scattering contributions in the final-state wave function of the photoemitted core electron. The continuously variable excitation energy at a synchrotron allows one to detect these 'interference patterns' at a fixed emission geometry, often chosen to coincide with the direction of the surface normal, as the kinetic energy and thus the wavelength of the outgoing photoelectron is varied. In the most simplistic analysis of these effects, the molecular orientation can be deduced on the basis of an intuitive forward scattering model, which also allows for a qualitative determination of the adsorption site. At the heart of this technique, however, is the quantitative determination of the atomic co-ordinates of the chemisorption complex, which is achieved by a comparison of the interference pattern with scattering calculations.

Dynamic information is contained in these interference patterns since the nuclear motion of the adsorbates results in an exponential damping factor in the interference pattern. Moreover, the sensitivity to the various modes of the adsorbate vibrations changes depending on the direction of emission. Therefore, specific vibrational modes can be probed by selecting the emission geometry most sensitive to that particular mode.

Core-electron excitations and the detection of the corresponding *core-electron absorption spectra* of chemisorbed atoms and molecules are described in the following chapter written by W. Wurth and D. Menzel. The structures at the onset of the core-electron excitations reflect the transitions of core electrons into the unoccupied electronic states. These are generally divided into transitions into molecular orbitals, Rydberg orbitals, and into the continuum states leading to photoemission of the core-electron. A comparison of the absorption spectra of the isolated molecule and of the same molecule in a chemisorbed state reflects the changes in the electronic structure in the unoccupied states. These changes depend critically on the strength of the adsorbate substrate interactions. Consequently, for physisorbed systems the spectra are much closer in appearance to the gas-phase spectra, whereas in chemisorbed systems the energy positions and line shapes are modified by the interaction with the substrate.

High-resolution studies reveal vibrational substructure for some of these adsorbate systems, which is also modified in the chemisorbed state. Moreover, small shifts and changes in the energy of the resonances can be attributed to adsorption in different sites on a substrate. Finally, the electronic states populated in the decay of these core excited states can be explored through the electron-emission spectra. Some of these electronic configurations are highly unstable, such that desorption of the chemisorbed molecule or of some fragment may be observed.

The electronic processes following the excitation of a core vacancy in a molecular system are described in more detail in the following chapter on *deexcitation electron spectroscopy* authored by W. Eberhardt. The creation of a core-electron excited state is generally followed by an electronic relaxation process, leading predominantly to the emission of additional electrons. Through the use of synchrotron radiation specific core-excited states can be created and studied selectively. The high definition and variability in the creation of core-excited states allows to study the electronic interactions in various electronic configurations with unprecedented definition and detail. Thus, these type of studies serve as a very stringent test for modern theoretical concepts for the description of correlated electronic states in molecular systems.

The spectra of these electronic decay processes of core excited states in molecules are governed by the Auger process and consequently display a distinctive local character, compared to the more global process of photoemission. Thus the core hole deexcitation reflects upon local dynamic charge rearrangement and screening occurring at a specific atomic site in the molecular system upon the creation of the core excited states. In chemisorbed systems, first of all, additional dynamical charge transfer processes in the core-excited state are possible through the interaction with the substrate, and secondly, the modification of the total electronic structure also results in specific changes of the decay processes.

High-resolution studies of the electronic core-hole decay processes give a glimpse at the evolution of the nuclear wave functions of the molecular systems on the timescale of a few fs. This corresponds to the lifetime of the core-excited electronic state, which is sufficiently short to maintain coherence among the vibrational wave functions populated in the excitation process. Accordingly, the excitation and decay cannot be viewed as independent processes. Therefore, the population of the final-state vibrational sublevels is modulated by interference effects resulting in vibrational profiles vastly different than expected from simple Franck-Condon overlap factors.

Finally, in the chapter written by F.M. Hoffmann and G.P. Williams *infrared absorption spectroscopy* of molecular adsorbates on surfaces is described. This is one of the new areas of the application of synchrotron radiation, which is advancing rapidly. The stability and brightness are the special features of synchrotron radiation which result in a superior signal-to-noise ratio for this type of spectroscopy. Thus not only the internal vibrational modes of the chemisorbed molecules can be probed with a very high sensitivity and resolution, but also the much weaker, lower-energy modes of the motion relative to the substrate can readily be investigated. A detailed analysis of the vibrational modes of the chemisorbed complex allows to deduce the adsorption site. Also inter-molecular and adsorbate substrate interactions are reflected by shifts of the vibrational frequencies. Fano type lineshapes of the infrared absorption lines result from coupling between various vibrational and electronic modes.

In summary, all standard synchrotron radiation related techniques used to investigate molecules in the chemisorbed state and the dynamics of vibrational and electronic excitation and relaxation mechanisms have been described in this book. There is only one exception to this, the area of surface EXAFS was omitted intentionally, since a comprehensive review, authored by J. Stöhr, has just appeared in print [1.10].

References

1.1 J.D. Jackson: *Classical Electrodynamics*, (J. Wiley, New York 1967) p. 464
1.2 J. Schwinger: Phys. Rev. **70**, 798 (1946) and Phys. Rev. **75**, 1912 (1949)
1.3 R.P. Goodwin: *Springer Tracts Mod. Phys.* **51**, 1 (Springer, Berlin, Heidelberg 1969)
1.4 *Synchrotron Radiation*, ed. by C. Kunz, Topics Curr. Phys., Vol. 10, (Springer Berlin, Heidelberg 1979)
1.5 *Synchrotron Radiation Research*, ed. by H. Winick, S. Doniach, (Plenum, New York 1980)
1.6 *Handbook on Synchrotron Radiation*, ed. by E.E. Koch (North Holland, Amsterdam 1983)
1.7 'An ALS Handbook', Lawrence Berkeley Laboratory, Pub-643 (1988)
1.8 J.M.J. Madey: J. Appl. Phys. **42**, 5 (1971)
1.9 H. Wiedemann: *Particle Accelerator Physics, Basic Principles and Linear Beam Dynamics* (Springer, Berlin, Heidelberg (1993)
1.10 J. Stöhr: *NEXAFS Spectroscopy* Springer Ser. Surf. Sci. Vol. 25 (Springer Berlin, Heidelberg 1993)

2. Band-Structure Determination of Adsorbates

H.-J. Freund and *H. Kuhlenbeck*

We have chosen several examples from the literature to discuss the various aspects of band-structure formation in ordered overlayers. The best suited method to study band-structure formation is Angle-Resolved Ultraviolet Photoelectron Spectroscopy (ARUPS), and in the present chapter exclusively ARUPS data are discussed. The most complete data sets exist for CO adsorbates and we shall outline the various details on the basis of these results. Then, we proceed to discuss hydrocarbon layers of molecules of increasing size, i.e. ethylene and benzene. There are only very few examples of coadsorbates where band structure studies have been undertaken. We shall briefly mention some aspects in the last section of this chapter.

2.1 Historical Introduction

In Langmuir's picture of adsorption [2.1] the adsorbed particles occupy the lattice points of a two-dimensional substrate with equal probability and with hard wall potentials between the adsorbed particles, preventing double or multiple occupancy of any particular site and a well defined adsorption energy typical of the site. As a result saturation would be characterized by $\theta = 1$ and the formation of a true 1×1 absorbate layer. Obviously, the formation of ordered layers with coverages far below $\theta = 1$ are more the exception than the rule and are a direct consequence of the existence of interaction potentials [2.2]. Such an interaction potential is exemplified in Fig. 2.1 for the system CO/Pd(100) reported by *Tracy* and *Palmberg* [2.3] in 1969, compared with a CO–CO interaction potential in the gas phase. Intermolecular interaction potentials may be either attractive or repulsive and may be classified into direct and indirect interactions. Direct interactions involve dipole-dipole (multipole-multipole) and orbital overlap interactions, and are often repulsive. On the other hand, indirect interactions mediated through the metal surface may be either attractive or repulsive depending on distance and surface sites, i.e., the kind of charge modification of the electronic structure of the substrate by the adsorbate [2.4–6]. The interplay of the intermolecular interaction potential and the adsorption energy of the isolated molecule with the clean substrate finally determines the observed structure of an ordered layer. In other words, the

Springer Series in Surface Sciences, Vol. 35
Applications of Synchrotron Radiation Ed.: W. Eberhardt

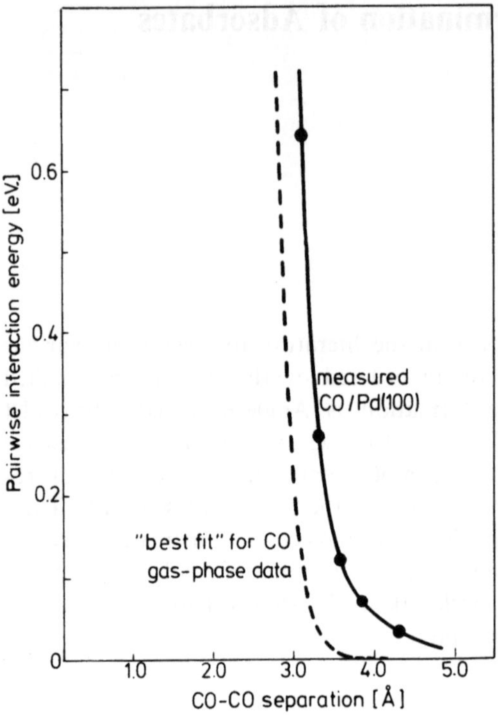

Fig. 2.1. Pairwise interaction energies as a function of intermolecular separation. A CO adsorbate is compared with the gas phase

structure of the ordered layer depends on the heat of adsorption and the coverage. Temperature (T)-coverage (θ) diagrams are usually called *phase diagrams* and have been determined as early as 1972 [2.7]. Figure 2.2 depicts one of the rare examples of a phase diagram of a molecular adsorbate, i.e. CO/Cu(100) [2.8]. CO bound carbon-end down on Cu(100) occupies only on-top sites. In this case two ordered structures, denoted by I and II_+ in Fig. 2.2 occur at $\theta = 1/2$ and $\theta = 4/7$, respectively. Phase I is the well known $c(2 \times 2)$ structure while phase II_+ consists of stripes of $c(2 \times 2)$ structure of width n = 3 separated by domain walls. The main part of the phase diagram is filled by a disordered phase. It is the well-ordered structure that plays the important role in this chapter as we shall see below.

A way to get insight into intermolecular interactions within adsorbates is to study their electronic structure. Order is particularly important because in the case of two-dimensionally well ordered overlayers the wave vector parallel to the surface ($k_\parallel$) is a good quantum number, and therefore $k_\parallel$ may be observed experimentally. Consequently, it is possible to carry out angle-dependent inelastic electron spectroscopic measurements to determine the $k_\parallel$ dependence of the particular elementary excitation. If we use, for example, electron energy loss spectroscopy (EELS) we get the phonon dispersion relations [2.9] or the plasmon dispersion relations, etc. [2.10]. If we perform angle-resolved photoelectron spectroscopic measurements we get the 'band structure' of the ordered

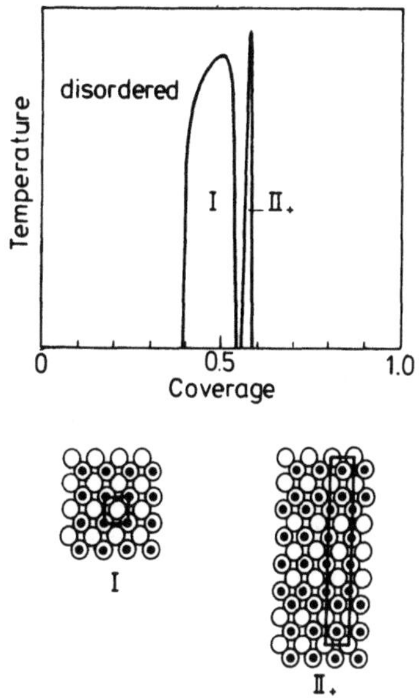

Fig. 2.2. Temperature (T)-Coverage (θ) phase diagram for the system CO/Cu(100). The two areas of ordered structures are indicated and structural models are shown

layers, or better the dispersion relations of the ion states of the system [2.11]. Experimentally, molecular band dispersions were first observed by *Horn* et al. for CO overlayers [2.12]. The experimentally determined band structure can be connected with the intermolecular interaction potential through theoretical modelling. Briefly, in the simplest form of the band-structure formulation, the tight-binding approximation, the bandwidth is determined by the strength of the intermolecular interaction as probed by the particular molecular orbital (ion state) under consideration [2.13]. If it is possible to identify a particular ion state of a molecule in ordered layers of different coverage densities, i.e. varying intermolecular distance, then we are in a position to determine ion state specific intermolecular interaction parameters as a function of intermolecular distance [2.14]. The last point is particularly interesting: Take diatomic molecules with localized orbitals. Assume the molecules interact on one end only with the surface and form an ordered layer. Ionization of an orbital localized on the far end of the molecules away from the surface most likely will lead to a dispersion relation predominantly determined by the direct, or through space interaction [2.15]. If on the other hand we choose to ionize an orbital localized on the near end of the molecules close to the surface, we are going to measure a dispersion relation determined by direct as well as indirect, 'through bond', interaction of the molecules [2.16]. It is therefore possible to use measured ion state specific dispersion relations to study intermolecular interaction mechanisms. It must be understood, however, that experimentally determined band dispersions are excited state properties and intermolecular potentials are usually considered to be ground state properties.

2.2 CO Adsorbates

2.2.1 Symmetry Considerations and CO-Substrate Interactions

CO forms ordered overlayers on most transition metal single-crystal surfaces. Before we turn to the analysis of the denser layers it is important to understand the interaction of an isolated molecule with the clean surface.

With the exception of very few examples, e.g. CO on Cr(110) [2.17], CO on Fe(100) [2.18] and CO on Mo(100) [2.19], where the molecule is tilted, CO is bound vertically towards transition metals if the adsorption energy is characteristic for a chemisorption system [2.20]. Figure 2.3 exhibits two spectra for the chemisorption system CO/Pd(111) for which an adsorption enthalpy of 142 kJ mol^{-1} has been determined. The spectra are taken at different electron-emission angles [2.21]. The top spectrum is taken in the so called *'forbidden'* geometry, the spectrum at the bottom in the so-called *'allowed' geometry* (see

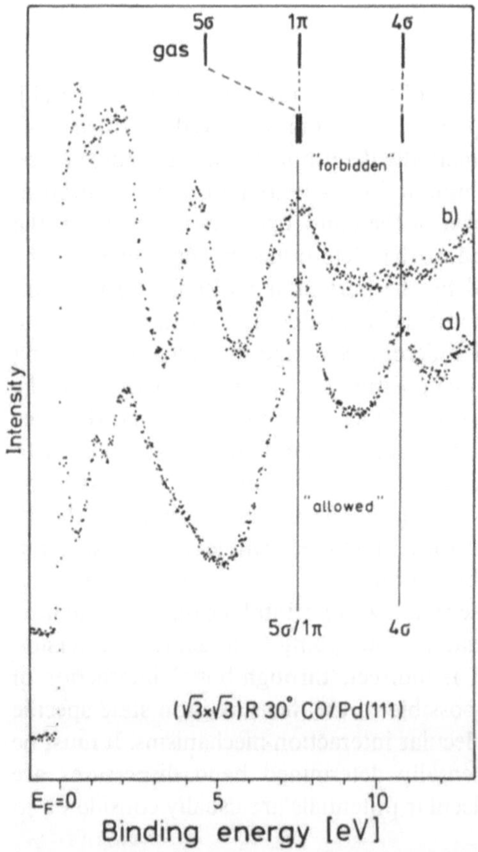

Fig. 2.3. Electron distribution curves in the so called "allowed" and "forbidden" geometries for CO/Pd(111). The positions of the two adsorbate induced features are compared with the positions in the free molecules, where the 4σ-levels have been aligned

below) [2.22]. Characteristic intensity variations are found. First, the bottom spectrum shows two bands, which are typical for a CO chemisorption system and which are shifted by about 2 eV to lower binding energies with respect to the gas phase (see below) [2.22]. The position of the ion states observed in the gas phase as marked in Fig. 2.3 have been aligned artificially to the position to the 4σ ion state in the adsorbate. The intensity variations between the two electron-distribution curves can be used to show that these two CO-induced bands are really caused by the three CO-ion states (5σ, 1π, 4σ) and that the CO molecules are oriented with their axes parallel to the surface normal [2.20, 22]. In order to recall the essentials we shall quickly step through the arguments [2.23, 24] in the following using the results in Fig. 2.3 as an example.

The basis for the discussion is Fermi's Golden Rule:

$$I \propto |\langle \varphi_f | p | \varphi_i \rangle|^2 \tag{2.1}$$

First, it has to be remembered that φ_f is the final state after electron excitation consisting of the ion state ${}^{N-1}\Psi_{e,E}$ and the emitted electron $\Phi_e(n)$, and that φ_i represents the neutral ground state of the system. Since p is a one electron operator the matrix element can be rewritten as [2.25]:

$$I \propto |\sum \langle \Phi_e(n) | p_n | \Phi_k(n) \rangle \langle {}^{N-1}\Psi_{e,E} | a_k \varphi_i \rangle|^2 \tag{2.2}$$

where a_k and $\Phi_k(n)$ are the annihilation operator and the one-electron wave function of the electron that is being emitted, respectively. These $\Phi_k(n)$ are called initial states in the following.

The first matrix element determines the angular distribution pattern, the second matrix element defines the absolute value and contains the internal degrees of freedom of the system, e.g., the line widths [2.25]. The sum takes all possible ion states ${}^{N-1}\Psi_{e,E}$ into account and explains the existence of satellite structure [2.25]. Since we are interested in ARUPS, much of the discussion will concentrate on the first matrix element. Second, in order to evaluate whether this matrix element is non zero, and thus leads us to expect a finite photoelectron current in a specific direction in space, symmetry arguments can be used. In principle the space group of the adsorbate under consideration has to be chosen, and then we have to classify the wave functions according to their irreducible representations. Often, it is sufficient to consider one specific symmetry operation belonging to the point group, instead of all possible symmetry operations, in order to predict the angular variations of electron emission. In the present case we refer to one of the mirror planes of the Pd(111) surface. If we classify the wave functions of the electron Φ_e and Φ_k, as well as the momentum operator p to be 'even' or 'odd' with respect to this mirror plane, we are in the position to differentiate between 'even' and 'odd' initial states by choosing certain light polarizations and detecting the angular-distribution pattern as long as spin-orbit interaction is not important [2.26]. For the above given situation the light incidence angle is 90° with respect to the surface plane and the light polarization direction is within the mirror plane. This corresponds to even symmetry of the momentum operator with respect to the mirror plane. Therefore, initial states with even symmetry will emit in the direction of the mirror plane because the final states have to be even in order for the matrix element not to vanish. In principle, one would expect a finite emission probability along the whole mirror plane. In the present case, however, one has to take into account the cylindrical symmetry of a CO molecule, bonded linearly towards the metal surface. Even initial states of a cylindrical molecule cannot emit in the direction given by the plane perpendicular to the mirror plane which contains the light polarization vector. This latter property can easily be understood if we remember that any plane in a cylindrical system containing the cylindrical axis is a symmetry plane. Combining this property with the fact that the momentum operator is odd with respect to this second plane (perpendicular to the light incidence) means that there cannot be any emission of even states in this direction. Therefore, in order to fulfill both conditions simultaneously, we do not expect an intense CO-4σ emission along the surface normal for light polarized in the surface plane. 4σ emission along the surface normal can only be achieved by using a light polarization perpendicular to the surface plane because in this case the momentum operator is even with respect to any plane perpendicular to the surface plane. If we combine the considerations so far, we verify the above angular distribution pattern.

As a consequence of the outlined behaviour of even initial states, we expect a complementary behaviour of odd initial states. This is exactly what is observed experimentally and is shown as a set of electron distribution curves - which is the usual way to look at ARUP-spectra-in Fig. 2.3. In this figure the complementary behaviour of σ and π emissions, which was first observed by *Lapeyre* and co-workers [2.27, 28] and then by *Plummer* and coworkers [2.22], is obvious: If we record a spectrum perpendicular (so-called *forbidden geometry*) to the incidence plane we do not observe emission in the region of the 4σ level but only in the region of the 5σ/1π levels. Note, that the 1π-ion states of CO has two degenerate components, one of which is always even with respect to a certain mirror plane. Thus, we expect to see the one odd component of the 1π-ion state. A spectrum recorded with the analyzer placed within the incidence plane, the so called *allowed geometry*, shows all states with even symmetry. From Fig. 2.3 it is clear that the feature at 8 e V below the Fermi energy contains two states, i.e. the 1π- and the 5σ-ion states. Their energies are, in contrast to the gas phase, nearly (within a few tenths of 1 e V) degenerate in the adsorbate [2.22]. This is due to the donation of the 5σ carbon lone pair into empty metal levels, thus stabilizing the CO 5σ level with respect to the 1π level which is not as intimately involved in the molecule-metal interaction for linear metal-molecule bonding [2.29, 30], a situation similar to that proposed in the introduction.

The adsorbate induced features exhibit, in addition to the described angular dependence [2.31], characteristic photon-energy dependencies, which, when recorded in an angle-dependent fashion, can be used to get further information about adsorbate site geometry [2.22]. Figure 2.4 shows a plot of the intensity of the 4σ-ion state as a function of photon energy. The data have been recorded for the system CO/Co(0001) [2.15] for three different electron emission angles. The

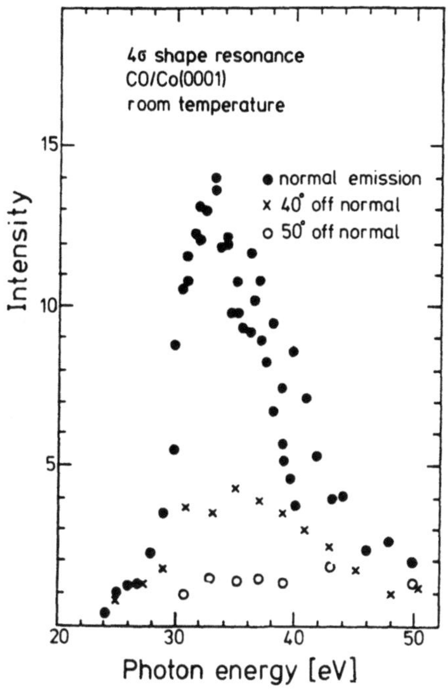

Fig. 2.4. 4σ shape resonances at three electron emission angles for the system CO/Co(0001)

observed resonance feature is caused by the so called *shape resonance*, which is well known from CO gas phase studies [2.32, 33]. It can be traced back to a molecular final state of σ symmetry in the ionization continuum, quasi-bound by a centrifugal barrier in the molecular potentials. Its symmetry confines the electron emission direction to the molecular axis, and directs the 4σ emission out of the oxygen end of the molecule. This means that for the case of a molecule oriented along the surface normal, carbon-end bound to the surface, the resonance should peak along the surface normal. Experimentally, we find in Fig. 2.4 the expected behaviour, i.e., a pronounced attenuation of the resonance intensity for off-normal emission, which corroborates the assumed adsorbate orientation. This is true basically for all CO chemisorbates on transition-metal surfaces.

2.2.2 Ordered Overlayers

a) Chemisorbed Layers

The next step is to consider an ordered overlayer. As an introduction to the quasi two-dimensional band structure of molecular overlayers we discuss the band dispersions and the symmetry properties of a hexagonal overlayer [15] of CO molecules on a (111) surface. If the intermolecular interaction potential is large enough to demand consideration of the two-dimensional crystal periodicity, the overlap of adsorbate wave functions is sufficient to produce two-dimensional Bloch states Ψ_k and to make a band description of the electronic structure more appropriate [2.12]. Then the one-electron wave function at a lattice site R_1 is related to the wave function at site R_2 by

$$\Psi_k(R_1) = \Psi_k(R_2)\exp[ik_\parallel \cdot (R_1 - R_2)] \qquad (2.3)$$

where $\exp[ik_\parallel \cdot (R_1 - R_2)]$ gives the phase difference between the two sites for a state specified by the two-dimensional wave vector $k_\parallel$. Before we consider the changes in the band structure introduced by the substrate we will briefly discuss the qualitative behaviour of the band dispersion of a hypothetical support free molecular layer [2.34], as shown in Fig. 2.5. We can illustrate the qualitative features of the dispersion by plotting schematically the real parts of a tight-binding wave function in real space for values of $k_\parallel$ corresponding to high symmetry points in reciprocal space. Figure 2.6 depicts the real- and reciprocal-space unit cells for the hexagonal structure. The real and reciprocal lattices have two different types of mirror planes, one along the $\bar{\Gamma}$-$\bar{M}$-$\bar{\Gamma}$ line (in reciprocal

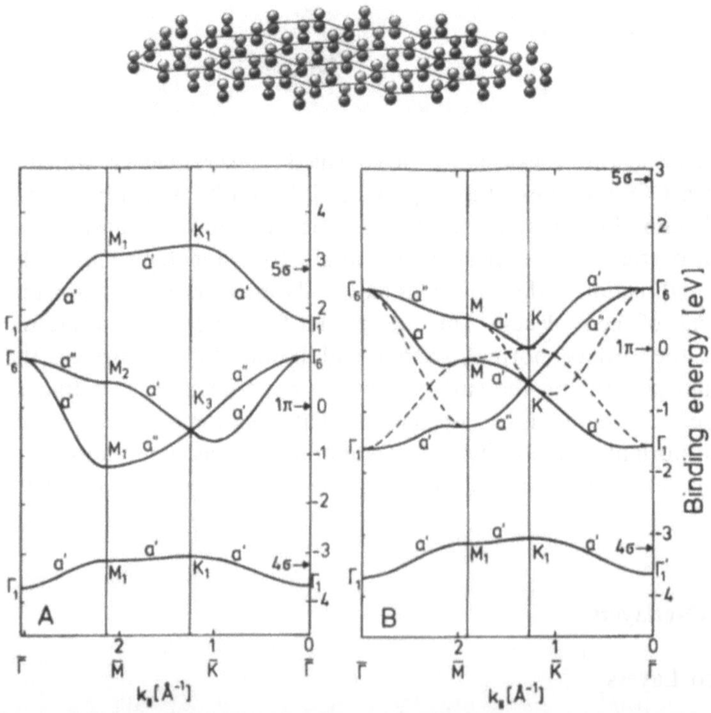

Fig. 2.5. Calculated band structure of a free unsupported hexagonal CO layer. In the left panel the bands are centered at the positions of the free molecule. In the right panel the 5σ-level has been shifted to center the band of the 1π-band center in order to demonstrate possible hybridization effects

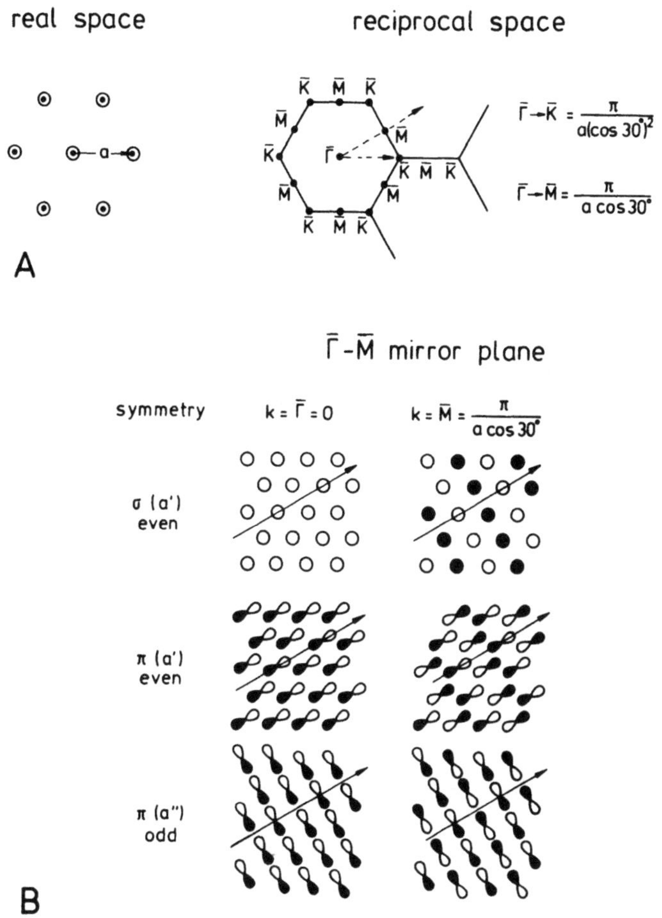

Fig. 2.6A, B. Schematic two-dimensional wave functions of a hexagonal CO layer along the $\bar{\Gamma} \to \bar{M}$ direction **A** real and reciprocal space representation of the hexagonal layer **B** wave functions

space), and the other one along the line $\bar{\Gamma}$-$\bar{K}$-$\bar{M}$-$\bar{K}$. The wave functions along these lines will be even (a') or odd (a''). Figure 2.6 illustrates the phases of a σ-state and the two π-states at $\bar{\Gamma}$ and $\bar{M}$. At $\bar{\Gamma}$ ($k_{\|} = 0$) all the wave functions at the different lattice sites are in phase. This results in a strongly bonding configuration for the σ-state (top row), but an anti-bonding configuration for both π-states because the individual π-functions change sign about the molecular axis. The π-functions have been chosen so that one is even and one is odd with respect to the $\bar{\Gamma}$-$\bar{M}$ mirror plane. Because at $\bar{\Gamma}$ the wave functions transform according to C_{6v}, the two π-components are degenerate. Therefore, at $\bar{\Gamma}$ we have a strong bonding σ-band and a degenerate antibonding π-band. Along the mirror plane $\bar{\Gamma}$-$\bar{M}$ $k_{\|}$ increases from 0 at $\bar{\Gamma}$ to $\pi/a \cos 30°$ at the zone boundary $\bar{M}$. The second column shows the wave functions at $\bar{M}$ where the arrow indicates the direction of $k_{\parallel}$. All rows of atoms parallel to $k_{\parallel}$ have the same phase but the phases of neighbouring rows differ by π. The result for the σ-states is that each atom is surrounded by four atoms of opposite phase (antibonding) and two bonding atoms. The σ-bands therefore disperse upward from $\bar{\Gamma}$ to $\bar{M}$. In contrast, the even π-state is strongly bonding since each lobe of the molecular π orbitals sees only bonding nearest neighbours. The even π-band disperses downward from $\bar{\Gamma}$ or $\bar{M}$ with the largest difference in the π-band energy. The odd π-state at $\bar{M}$ is just slightly more bonding than the π-state at $\bar{\Gamma}$ since the overlap of the lobes in a line perpendicular to $k_{\parallel}$ is antibonding but the overlap between the lines of atoms is bonding. This means that at $\bar{M}$ the two π derived bands are no longer degenerate as a consequence of the lower symmetry of the $\bar{M}$ point. Thus, we have explained the qualitative features of the dispersions in the $\bar{\Gamma}$ of $\bar{M}$ direction shown in Fig. 2.5. Analogously, the dispersions in the other directions of high symmetry in the surface Brillouin zone can be explained, and we refer to the literature for a more detailed discussion [2.15].

Such dispersions can be determined via ARUPS, as shown in Fig. 2.7, where spectra of a $(2\sqrt{3} \times 2\sqrt{3})R30°$ CO/Co(0001) overlayer for different values of the wave vector $k_{\parallel}$ are plotted [2.15]. The wave vector is varied by varying the photon energy and the emission angle θ within the direction of the considered surface azimuth as noted in the figure. The connection between $k_{\parallel}$ and the varied quantities is given by

$$k_{\parallel} = (2m_e \hbar^{-2} E_{\text{kin}})^{1/2} \sin\theta \tag{2.4}$$

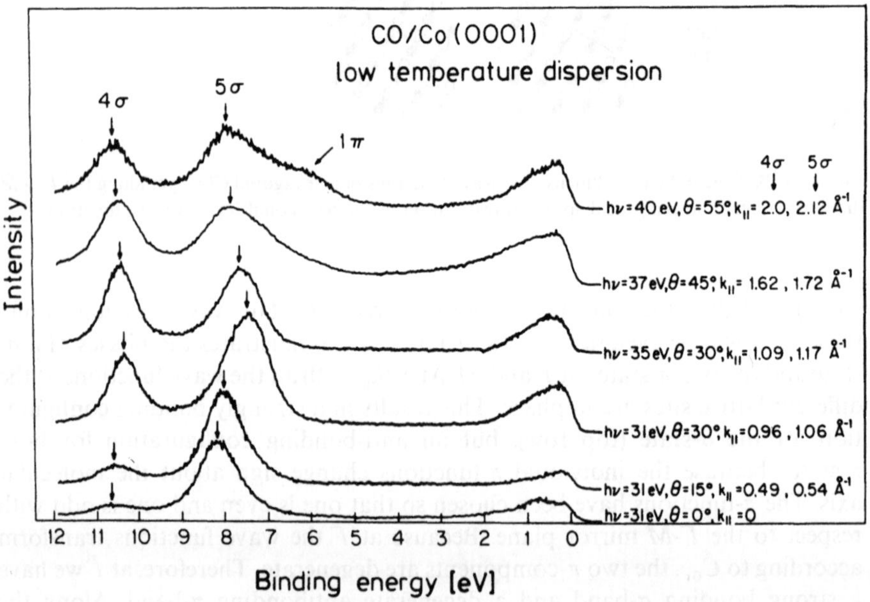

Fig. 2.7. Electron distribution curves for the system $(2\sqrt{3} \times 2\sqrt{3})R30°$ CO/Co(0001) taken at different excitation and electron emission conditions

where $k_{\parallel}$ is the wave vector parallel to the surface, which is the conserved quantity, E_{kin} is the kinetic energy of the escaping electron with respect to the vacuum level. Figure 2.8 displays a comparison of calculated dispersions for the 4σ derived band with measured 4σ dispersions of CO/Co(0001) adsorbates in $\bar{\Gamma}$-$\bar{M}$ direction of the hexagonal $(\sqrt{3}\times\sqrt{3})\,R30°$ and $(2\sqrt{3}\times 2\sqrt{3})\,R30°$ layers (partially based on Fig. 2.7). We have artificially set the lengths of the two Brillouin zones equal for a more convenient comparison. Due to the smaller CO–CO distance in the $(2\sqrt{3}\times 2\sqrt{3})R30°$ layer, the overlap of the 4σ CO wave functions increases, and concomitantly, the *bandwidth* increases. Figure 2.8 illustrates that the increase in bandwidth can be quantitatively reproduced by simple tight-binding calculations (full lines) in the case of 4σ derived bands. In the present case the comparison can be made directly because the number of nearest neighbours is the same in both systems. If, on the other hand, we want to compare dispersions in hexagonal and quadratic systems, the observed bandwidths have to be corrected for the different numbers of nearest neighbours. Such a correction is straightforward on the basis of simple tight-binding considerations. The results of such a comparison for several different adsorbate systems is exhibited in Fig. 2.9 [2.12, 15, 16, 34–42]. The data points follow an exponential dependence on the nearest-neighbour distance with a characteristic decay length of 1.23 Å if we disregard the CO–K co-adsorbate for the moment. This strongly supports the conclusions that the 4σ-dispersion is caused by direct CO–CO overlap. Intuitively, this is reasonable, because the 4σ CO level is not

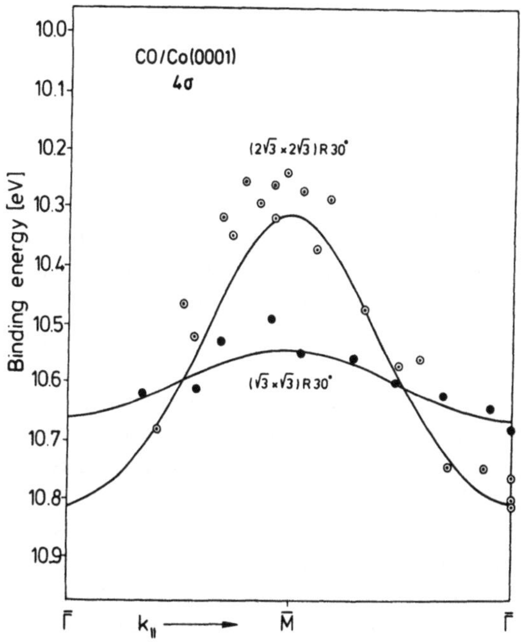

Fig. 2.8. Comparison of experimental and theoretical dispersions for two CO overlayers on Co(0001) of varying CO density

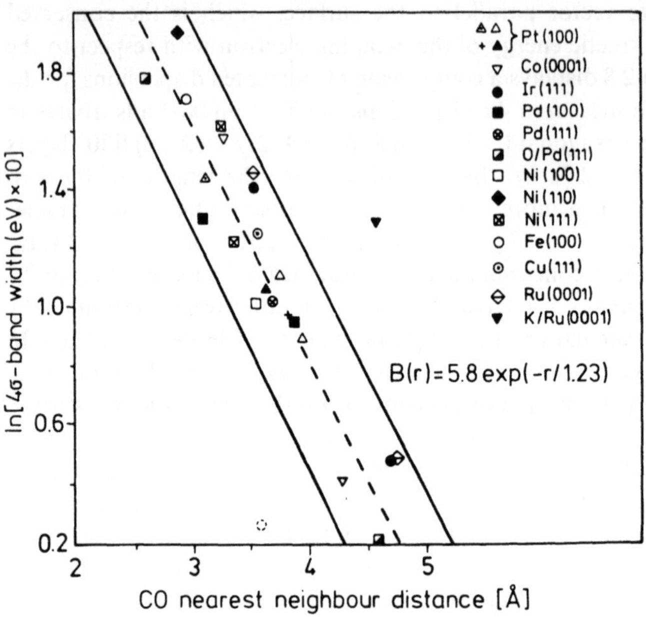

Fig. 2.9. Observed 4σ-bandwidth as a function of intermolecular separation. Six nearest neighbours are used as reference

strongly involved in the metal substrate bonding. At the same time we expect a completely different behaviour for the 5σ-level, because in this case the interaction with the substrate should have a marked influence on the measured dispersion. There is no such simple exponential dependence of the observed bandwidth on the CO–CO distances as for the 4σ-level [2.16]. A similar plot for the 5σ-level exhibits no particular functional dependence, which may be an expression of the participation of indirect through substrate interactions in the intermolecular interaction. Care has to be exercised not to jump to this conclusion prematurely, because, due to the stabilization of the 5σ-level in the region of the 1π-level we expect strong $5\sigma/1\pi$ hybridization effects which have to be taken into account in the prediction of band dispersions [2.15]. For the case of the hexagonal CO layer this situation is depicted in Fig. 2.5 on the right-hand side.

In spite of the vast knowledge on CO adsorbates on solid surface there are only very few examples where detailed and rather complete experimental information on the band structure has been collected. One example is the $CO(2\times1)p2mg/Ni(110)$ system [2.37, 2.46–48] which we want to consider in greater detail in the following. Exposure of a clean Ni(110) surface to CO at $T\sim120$ K leads to the formation of a dense CO overlayers with coverage $\theta=1.0$ which gives rise to a (2×1) LEED (Low Energy Electron Diffraction) pattern [2.49]. This LEED pattern is characterized by spot extinctions along the

[1$\bar{1}$0] direction indicating that the layer has *p*1*g*1 or *p2mg* symmetry. X-ray Photoelectron Diffraction (XPD) [2.50] and Electron Stimulated Desorption Ion Angular Distribution (ESDIAD) [2.51] as well as ARUPS studies showed that the structure actually exhibits *p2mg* symmetry; a structure model is depicted in Fig. 2.10. The CO molecules are canonically bound carbon-end down and tilted along the [001] azimuth in order to avoid the close intermolecular contact along [1$\bar{1}$0] with a separation below 3 Å which would occur if the molecules were perpendicularly oriented [2.49–54].

Figure 2.11 illustrates the full band structure of the system. The occupied part was determined by measuring the binding energies of the CO induced features as a function of the electron-emission angle along two azimuths with respect to the Ni(110) substrate, i.e. the [1$\bar{1}$0] and the [001] directions. Together with the experimental data [2.37, 2.46–48] we have plotted the result of a band-structure calculation where the position of the lowest occupied $4\sigma^+$ -valence orbital (see below) has been shifted to agree with experiments and the tilt angle has been used as a parameter. The best fit resulted for a tilt angle of 17° with respect to the surface normal. On the right-hand side of the collected dispersion data we show a set of photoelectron spectra at the Γ point (normal emission). Clearly, the number of outer valence features which are found at binding

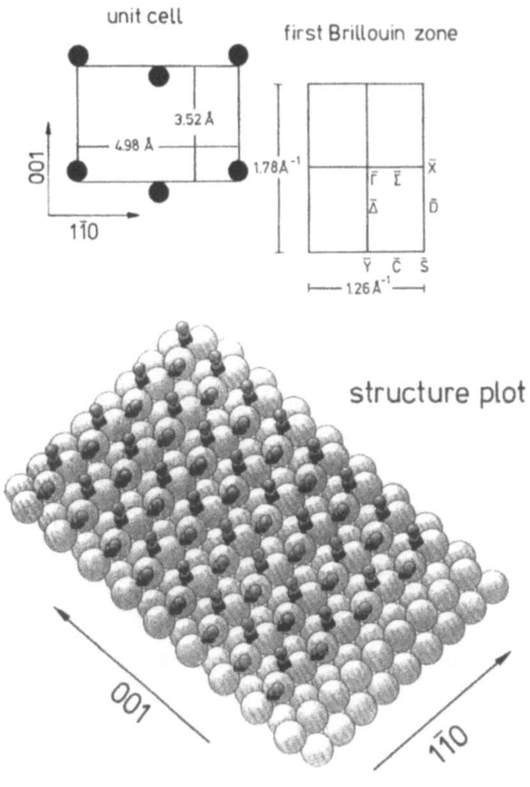

Fig. 2.10. Schematic representation of the CO(2 × 1)*p2mg*/Ni(110) structure, the unit cell and the surface Brillouin zone

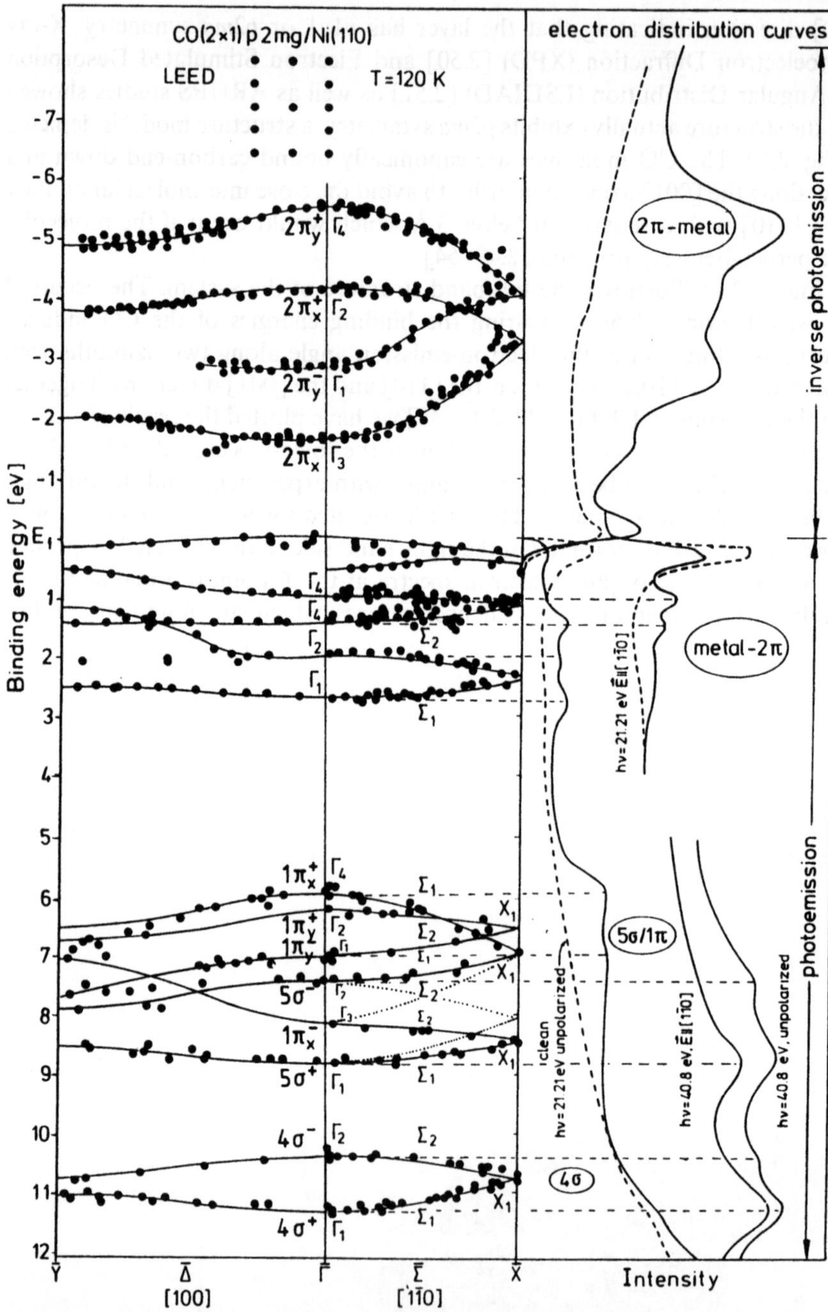

Fig. 2.11. Occupied and unoccupied band structure for CO*p*2*mg*/Ni(110). In the right panel some typical electron distribution curves are shown

energies above 6 eV is larger than four, which would be the maximum number of features for a single molecule within the unit cell (5σ, 4σ and two 1π components). This indicates that the unit cell contains more than one molecule [2.37]. The region of the energetically well separated 4σ-emissions reveals a splitting into two features consistent with two molecules in the unit cell.

For a description of the band structure the wave functions of a single molecule are therefore not appropriate. However, linear combinations

$$\chi_1 = c_1|CO^1\rangle + c_2|CO^2\rangle \qquad \text{symmetric}$$

and

$$\chi_2 = c_1|CO^1\rangle - c_2|CO^2\rangle \qquad \text{asymmetric} \tag{2.5}$$

of molecular CO wave functions located at the positions of the two molecules within the unit cell allow one to construct two-dimensional adlayer wave functions that transform according to *p2mg* symmetry and which give rise to bands (E vs. $k_{\parallel}$) easy to label by their irreducible representations at the points of high symmetry. Figure 2.12 schematically displays two-dimensional adlayer wave functions at $\bar{\Gamma}$, $\bar{X}$, and $\bar{Y}$ based on σ molecular wave functions (Fig. 2.12a). The unit cell is indicated, χ_1 and χ_2 are plotted. Consider Fig. 2.12a to represent the 4σ-orbitals of CO. Unlike the case of one molecule per unit cell leading to a single band per non-degenerate Molecular Orbital (MO) in the Brillouin zone, we have two bands in the present case, namely, one associated with the symmetric, the other one with the asymmetric combination of molecular orbitals. In the band structure the bands are labelled + and −, respectively. At $\bar{\Gamma}$ the splitting of the two 4σ derived bands is about 0.85 eV, caused by the strong lateral interaction. In the symmetric combination all MOs have the same phase, i.e, the two-dimensional wave function is strongly bonding which leads to a stabilization on a binding energy scale compared to a laterally non-interacting adsorbate. The asymmetric combination at $\bar{\Gamma}$ is antibonding with respect to the direction of the glide plane $[1\bar{1}0]$, but bonding with respect to the [001] direction. The interaction is stronger along the close packed $[1\bar{1}0]$ direction leading to a net destabilization on a binding energy scale. If we follow the bands in $k_{\parallel}$-space along $\bar{\Gamma} - \bar{X}$ (the $\bar{\Sigma}$ direction), the + band increases, while the − band decreases in energy until they are degenerate at $\bar{X}$. Figure 2.12a allows us to visualize the situation: At $\bar{X}$ the phase between adjacent unit cells changes by 180° along $[1\bar{1}0]$. Consequently, for the symmetric combination this yields a two-dimensional wave function with phase changes only between rows in [001] direction. For the asymmetric combination a similar wave function results which is evident from Fig. 2.12a. In fact, these wave functions are degenerate. As was shown by *Hund* [2.55] the bands have to be degenerate on the entire line $\bar{X} - \bar{S}$, which is perpendicular to the glide plane (Fig. 2.10). Clearly, the wave function plots indicate that the energy position of the bands at $\bar{X}$ has to be intermediate between the energies at $\bar{\Gamma}$ since there is increasing antibonding character for the symmetric combination while there is loss of antibonding character for the asymmetric combination. At $\bar{Y}$ the wave functions and thus the

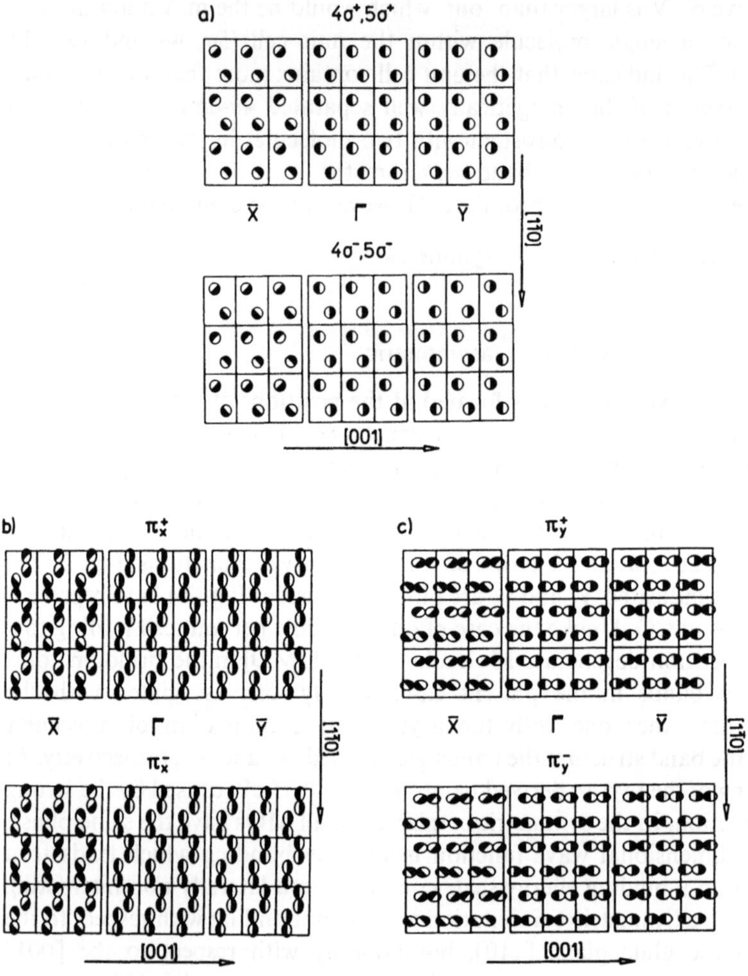

Fig. 2.12a-c. Schematic representation of the σ (**a**) and π (**b, c**) derived wave functions of the CO overlayer for different *k*-points. Each orbital lobe is half hatched. The direction of the line separating open and hatched part of the lobe marks the phase of the respective molecular wave function. In the zero phase situation the direction of the separating line points along the [1$\bar{1}$0] direction and the left part of the lobe is hatched

bands are non-degenerate energetically, since the spatial separation between sites of equal phase is larger for the $-$ than for the $+$ band. The splitting at $\bar{Y}$ is 0.3eV, which should be compared with 0.85 eV at $\bar{\Gamma}$. In order to specify the character of the wave functions within the band-structure plot the irreducible representations of the wave functions are indicated.

Next, we consider the region of the band structure of occupied levels at lower binding energies, namely, the region of the 5σ and 1π molecular orbitals. Due to the low symmetry of the overlayer 5σ and 1π derived adlayer bands are allowed to hybridize. The solid lines in this region (Fig. 2.11) refer to the bands after hybridization has been taken into account, while the dotted lines refer to the non-hybridized bands. We start the discussion by first considering the band dispersion neglecting hybridization. Clearly, the dispersion of the nonhybridized 5σ-bands should be similar to the 4σ-dispersion. In fact, this is our result. The splittings calculated for various symmetry points, however, differ considerably. This is expected since the spatial extent of the 5σ molecular orbitals is different from the 4σ molecular orbitals. This aspect has been discussed in detail by *Greuter* et al. [2.15]. Compared with the 4σ-bands the 5σ-bands show a splitting of 1.35 e V at $\bar{\Gamma}$ and 0.63 e V at $\bar{Y}$ since the 5σ molecular orbitals are more diffuse than the 4σ molecular orbitals.

The dispersions associated with the nonhybridized 1π molecular orbitals are due to their twofold degeneracy on the molecular level, slightly more complicated than the σ-band dispersions. Since the global symmetry of the adlayer is only twofold, the two 1π-components cannot be degenerate and, due to the formation of symmetric and asymmetric combinations, give rise to four bands at $\bar{\Gamma}$. In order to label the bands, we have chosen x to denote the component in $[1\bar{1}0]$, y to denote the component in [001] direction. The wave functions belonging to the x components are exhibited in Fig. 2.12b. Again, symmetric and asymmetric phase relations are identical to those for the σ-bands shown in Fig. 2.12a. Unlike the σ-bands, however, the antibonding combination is symmetric (+) and the bonding combination is asymmetric (−) in the 1π-case. As outlined above, the + or − sign refers to phase factors within the unit cell, while bonding and antibonding refers to the interaction of the wave functions within the unit cell: An in-phase combination of two π functions is antibonding and thus energetically destabilized, the out-of-phase combination of two π functions is bonding and thus energetically stabilized. In other word, π bands associated with phase factors + and − are energetically reversed with respect to the σ-bands. The splitting between + and − bands differs for the x and y components, in agreement with expectations since the lateral interactions along $[1\bar{1}0]$ are much stronger than along [001]. The magnitudes of the splitting are 2.1 eV and 0.78 eV at $\bar{\Gamma}$, respectively. At $\bar{\Gamma}$, $1\pi_x^+$, $1\pi_x^-$, $1\pi_y^+$ and $1\pi_y^-$ all belong to different irreducible representations. At $\bar{X}$ the bands are pairwise degenerated for the same reason we used for the σ-bands. The bonding character of the wave functions at $\bar{X}$ is clearly higher for the x-component than for the y-component as is evident from Fig. 2.12b and 2.12c. Along the $\bar{\Sigma}$ line there are only two irreducible representations. The bands are labelled according to *Litvin's* compatibility relations [2.56]. The $1\pi_y^-$-band has the same symmetry as the $1\pi_x^{-1}$ band along $\bar{\Sigma}$, leading to a very small gap around the crossing point. The gap is very small since the interaction is 'weak' for topological reasons. Perpendicular to the $\bar{\Sigma}$ direction, namely in the $\bar{\Delta}$ direction, the π-bands, like the σ-bands, are not degenerate at the zone boundary. Along the $\bar{\Sigma}$ direction the two $1\pi_y$-bands cross and, having the same symmetry, are allowed to hybridize. Figure 2.12c shows on the right hand side that the $1\pi_y^-$-band is, in addition to its bonding character in [001], antibonding along $[1\bar{1}0]$. The $1\pi_y^+$-band is bonding in both directions. σ/π-hybridization can only occur if the bands for π and σ parentage transform according to the same irreducible representations. Along $\bar{\Gamma}$-$\bar{X}$, i.e. in the $\bar{\Sigma}$-direction, the $1\pi_x^-$-band hybridizes with the $5\sigma^-$-band shifting the energetic position of the point of degeneracy. Note that the degeneracy at $\bar{X}$ is not lifted by σ/π-hybridization. However, it is evident that the hybridized band structure is considerably different from the nonhybridized one. At $\bar{\Gamma}$, for instance, the σ/π-hybridization leads to reversal of the $1\pi_y^-$ and $1\pi_y^+$ binding energies.

So far we have discussed the development of the band structure without explicit reference to the experimental observations. The basis for an experimental analysis is Table 2.1 where the dipole selection rules for the *p2mg* system at $\bar{\Gamma}$ are summarized. At $\bar{\Gamma}$ the symmetry is C_{2v} and the symmetry selection rules shown in Table 2.1 may be applied. Because the final state of the electron must be totally symmetric (Γ_1) there is evidence for a Γ_2- and a Γ_1-state in the range of the 4σ-ionization if we compare the spectra recorded with polarized (electric field vector $\boldsymbol{E}$ parallel to the $[1\bar{1}0]$ direction) and unpolarized light in Fig. 2.11. The symmetry restrictions for the $4\sigma^-$- and $4\sigma^+$-bands hold in the same way for the $5\sigma^-$- and $5\sigma^+$-bands. The comparison of the same two spectra supports this conclusion. If the photon energy is smaller, i.e. around 20 eV, the cross section of the π derived components becomes more prominent. The π derived Γ_4-band at lowest binding energy can be well used to demonstrate that the symmetry of the ordered overlayers is *p2mg* and not *p1g1*. According to the *p2mg* selection rules Γ_4-bands can only be observed with light polarized along the $[1\bar{1}0]$ azimuth whereas in *p1g1* these states can also be ionized with light polarized perpendicular to the surface (*z*-direction). The result of such a test is shown in Fig. 2.13 [2.57]. For the upper spectrum light polarized predominantly along $[1\bar{1}0]$ was used whereas in the lower spectrum the component of the electric field vector along $[1\bar{1}0]$ was zero. If the adsorbate symmetry were *p1g1* the $1\pi_x^+$-state, which transforms according to Γ_4, should also be visible in the lower spectrum. Since this is not the case we conclude that the adsorbate symmetry is *p2mg* and not *p1g1*.

The splitting of the 1π-levels has been discussed above and the experimental data basically corroborate the conclusions as is evident from Fig. 2.11. The

Table 2.1. Dipole selection rules for CO(2 × 1)*p2mg*

Electric field vector E	Initial state bands		
Orientation	Symmetry	Molecular assignment	Symmetry
$\boldsymbol{E}\|\|[110]$: *z*-polarization	Γ_1	$4\sigma^+, 5\sigma^+, 1\pi_y^-$	Γ_1
$\boldsymbol{E}\|\|[110]$: *y*-polarization	Γ_2	$4\sigma^-, 5\sigma^-, 1\pi_y^+$	Γ_2
	Γ_3	$1\pi_x^-$	Γ_3
$\boldsymbol{E}\|\|[110]$: *x*-polarization	Γ_4	$1\pi_x^+$	Γ_4

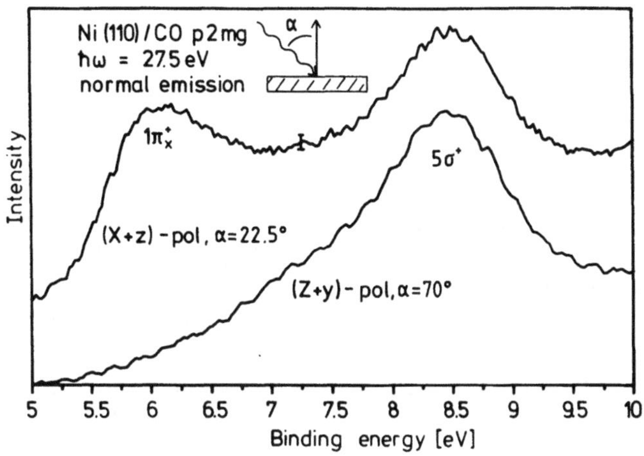

Fig. 2.13. Electron distribution curves in the range of the $5\sigma/1\pi$-levels using X and Z polarized light

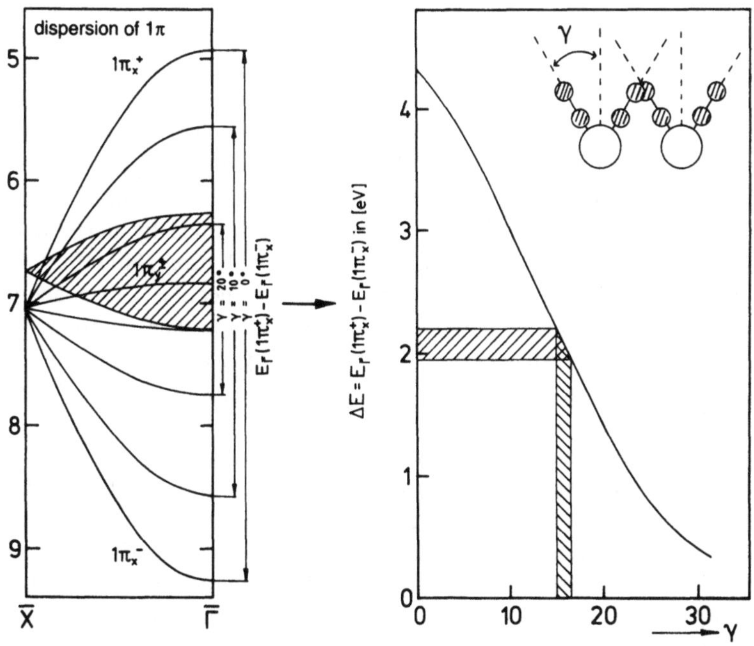

Fig. 2.14. Calculated 1π-bandwidth as a function of the tilt angle of the molecular axis (γ) with respect to the surface normal. In the right hand panel the observed splitting is used to deduce the tilt angle

strong splitting of the $1\pi_x$ components can be used -and has been used- to estimate the tilt angle of the CO molecules. The procedure can be visualized with the help of Fig. 2.14. In the left panel we show the dispersion of the $1\pi_y$- and the $1\pi_x$-bands as a function of the tilt angle γ. The $1\pi_y$-bands disperse within the hatched area for the considered angular variations. We use the $1\pi_x$ bands as an example because for these bands the angular variations are largest so that the fitting procedure is rather precise. For $\gamma = 0°$ the splitting is calculated to be ≈ 4 eV between the bands belonging to the bonding and the antibonding combination. Note that for $\gamma = 0°$ the glide plane disappears and we reach a $p(1 \times 1)$ structure. In this case the unit cell contains only one molecule, and therefore, instead of having a splitting at Γ into two bands, the size of the Brillouin zone doubles. However, for comparison we have folded back the bands into a reduced zone, identical to the one of the *p2mg* structure. Clearly, as soon as we allow the CO molecules to avoid each other by tilting the splitting decreases dramatically, until we reach agreement with the experimentally observed splitting of 2.2 eV at $\gamma = (17 \pm 2)°$. This is graphically shown in the panel on the right-hand side of Fig. 2.14. This strong variation of the $1\pi_x$-splitting is a consequence of the strong variation of intermolecular overlap as a function of γ. The splitting of the 1π-levels in the range of occupied orbitals should be compared with the corresponding splitting of the 2π-orbitals in the unoccupied range (Fig. 2.11). The experimental results have been taken from inverse photoemission studies of *Memmel* et al. [2.47, 48]. The observed splitting for the both the $2\pi_x$ (3.8 eV) and the $2\pi_y$ (1.2 eV) components is considerably larger than observed for the 1π-levels. This may be caused again by substrate mediated interaction due to 2π-backdonation but an additional role may be played by the 2π-wave functions themselves which are expected to be considerably more diffuse as compared with the 1π-levels. This point can be supported through a very nice series of experiments performed by *Bertel* and coworkers [2.48] on three isomorphic structures, i.e., CO *p2mg*/Ni(110), CO *p2mg*/Pd(110) and CO *p2mg*/Pt(110) where the intermolecular CO spacing varies and also the substrate mediated interaction changes. The comparison between the Ni and the Pd systems reveals that the splitting in the 2π-manifold is slightly smaller for Pd due to the increase in CO–CO separation. For Pt the situation is more complicated probably due to coupling of the unoccupied CO levels with the metal unoccupied states in this energetic range.

A direct observation of the substrate-mediated interaction is possible for the system CO *p2mg* on Ni(110) by following the CO-induced features in the region of the metal-band structure, i.e. between the Fermi energy and a binding energy of 3 eV [2.46]. In this energy range Fig. 2.11 shows bands which exhibit dispersions of up to 0.5–1.0 eV indicating a substantial contribution via this type of interaction. Figure 2.15 shows how these features develop as a function of coverage. The states that give rise to these features are located in the surface region because they do not change their binding energy when the photon energy changes as clearly demonstrated in Fig. 2.16. At $\bar{\Gamma}$ the symmetry of the states involved can be deduced from polarization-dependent spectra (Fig. 2.17) by using Table 2.2 where the symmetry selection rules for CO-2π-derived and Ni-derived states are given under the assumption that the CO molecules are bridge-bonded [2.15–53]. The situation would change a little when the molecules were terminally bonded as suggested by *Voigtländer* et al. [2.58] because

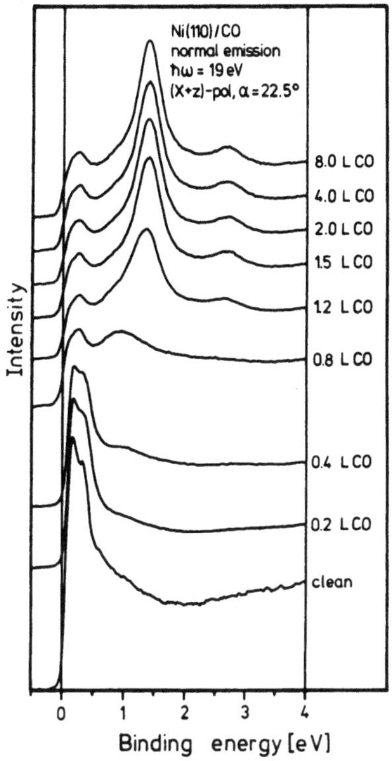

Fig. 2.15. Adsorbate induced features in the region of metal states in the system CO/Ni(110) as a function of coverage

the lateral position of the CO molecules determined the position of the mirror plane in *p2mg* symmetry. Figure 2.18 depicts the band dispersions as determined experimentally (Fig. 2.11) superimposed onto the projected band structure of the substrate along the two azimuths, i.e. [001] and [1$\bar{1}$0] [2.59]. In the projected band structure odd and even bands are differentiated. Within the projected band structure of odd symmetry we find gaps of noticeable size and some of the observed bands change their character from surface resonances to surface states if they exist in a gap. However, a detailed analysis of the observed peak intensities and widths reveals that these properties do not change when the bands disperse into a gap. The solid lines through some of the data points are included to guide the eye and are not due to explicit calculations. The four experimental points at $\bar{\Gamma}$ in Fig. 2.18 are the π-d features shown in Fig. 2.17. The polarization dependence of these states as obtained from Fig. 2.17 can be used to identify the character of the bands of $\bar{\Gamma}$. The 2.7 eV state is excited by Z-polarized light only and thus it transforms according to Γ_1. Neglecting the $d_{x^2-y^2}$ and d_{xy}-derived bands, this feature can be identified from Table 2.2 to be the $2\pi_y^- - d_{z^2}^+ - s^+$ hybrid band. In this case the amount of s-character in the wave function cannot be ignored since the $d_{z^2} - s$ mixing is known to be strong [2.60].

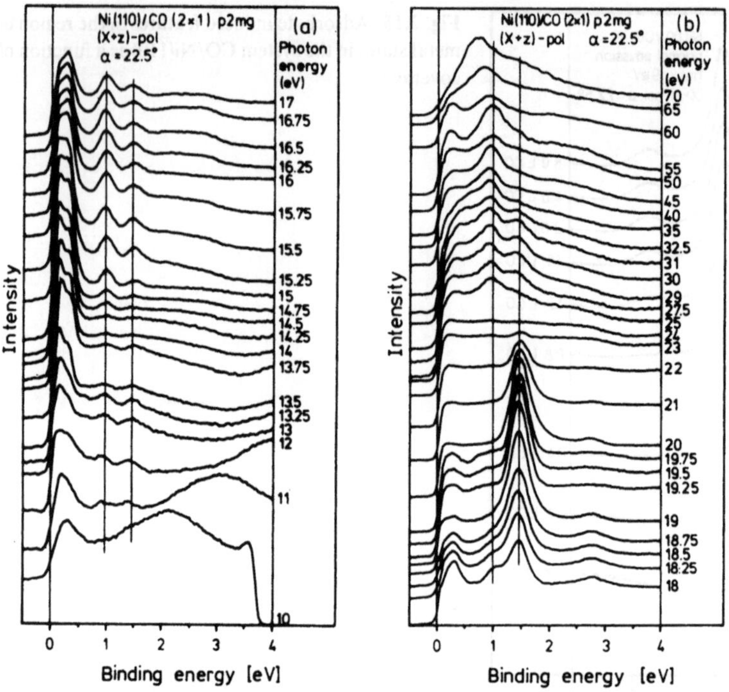

Fig. 2.16. Energy dependence of CO induced features in the region of metal states for the system CO*p2mg*/Ni(110). The spectra are taken in normal emission

In a similar fashion the 2.0 eV feature can be assigned. This state belongs to the irreducible representation Γ_2 and thus has to be one of the hybrid bands $2\pi_y^+ - d_{yz}^+$ and $2\pi_y^+ - d_{xz}^-$. To further identify this band its dispersion as shown by the solid lines in Fig. 2.18 can be considered. If this band was mainly d_{xz}^- in character at $\bar{\Gamma}$, their should not be much dispersion along $\bar{\Delta}$ because the interaction of the d_{xz}-derived bands is only weak along [001]. Figure 2.18 reveals that the dispersion bandwidth along $\bar{\Delta}$ is 0.75 eV, which is the greatest bandwidth of all observed π-d-bands. From this we assign the main character of this band to be d_{yz}^+ since the d_{yz}-orbitals interact significantly along [001], giving rise to a much greater dispersion bandwidth compared to the d_{xz}-derived bands. The remaining two π-d features at 1.0 and 1.5 eV in Fig. 2.18 are excited only by x-polarized light, indicating Γ_4 symmetry. According to Table 2.2 the metal part of their wave functions have to have d_{xz}^+- or d_{yz}^--character, respectively. An analysis like that done for the 2.0 eV-feature shows that the main character of the 1.0 eV-feature has to be d_{yz}^-, whereas the 1.5 eV-feature has to be d_{xz}^+ in character.

There are two additional bands at a binding energy of approximately 2.2 e V around $\bar{Y}$, belong to the irreducible representations Δ_1 and Δ_2, respectively. Since the Δ_1 band dispersing from 2 e V at $\bar{\Gamma}$ to 1.25 e V at $\bar{Y}$ in the first Surface

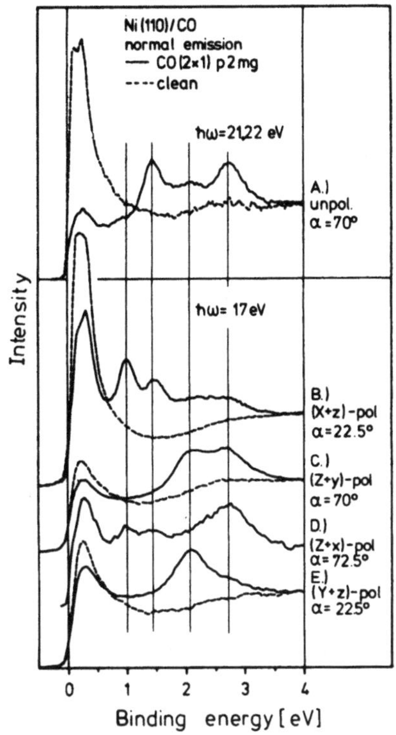

Fig. 2.17. CO induced features in the range of metal states as a functional of light polarization

Table 2.2. Symmetry selection rules for bridge-bonded CO

Representation	CO 2π	Ni valence states	Polarization
Γ_1	$2\pi_y^-$	$s^+, d_{z^2}^+, d_{xy}^-, d_{x^2-y^2}^+$	Z
Γ_2	$2\pi_y^+$	d_{xz}^-, d_{yz}^+	Y
Γ_3	$2\pi_x^-$	$s^-, d_{z^2}^-, d_{xy}^+, d_{x^2-y^2}^-$	
Γ_4	$2\pi_x^+$	d_{xz}^+, d_{yz}^-	X
Σ_1	$2\pi_y^-, 2\pi_x^+$	$s^+, d_{z^2}^+, d_{xy}^-, d_{x^2-y^2}^+, d_{xz}^+, d_{yz}^-$	Z, X
Σ_2	$2\pi_y^+, 2\pi_x^-$	$s^-, d_{z^2}^-, d_{xy}^+, d_{x^2-y^2}^-, d_{xz}^-, d_{yz}^+$	Y
Δ_1	$2\pi_y^-, 2\pi_y^+$	$s^+, d_{z^2}^+, d_{xy}^-, d_{x^2-y^2}^+, d_{xz}^-, d_{yz}^+$	Z, Y
Δ_2	$2\pi_x^-, 2\pi_x^+$	$s^-, d_{z^2}^-, d_{xy}^+, d_{x^2-y^2}^-, d_{xz}^+, d_{yz}^-$	X

Brillouin Zone (SBZ) will, of course, also exist in the second Brillouin zone, the two experimental data points (Fig. 2.18) making up the Δ_1 band in the second SBZ around 2.2 eV may belong to this band. The Δ_2 band near $\bar{Y}$ at 2.2 eV must transform according to either Γ_3 or Γ_4 at $\bar{\Gamma}$. Table 2.2 shows that there are only two bands belonging to Γ_4 at $\bar{\Gamma}$. These are the two bands in the binding-energy range from 0.75 to 1.5 eV along $\bar{\Delta}$, and thus the Δ_2 band at 2.2 eV must belong to the representation Γ_3 at $\bar{\Gamma}$. The symmetry-selection rules in Table 2.2

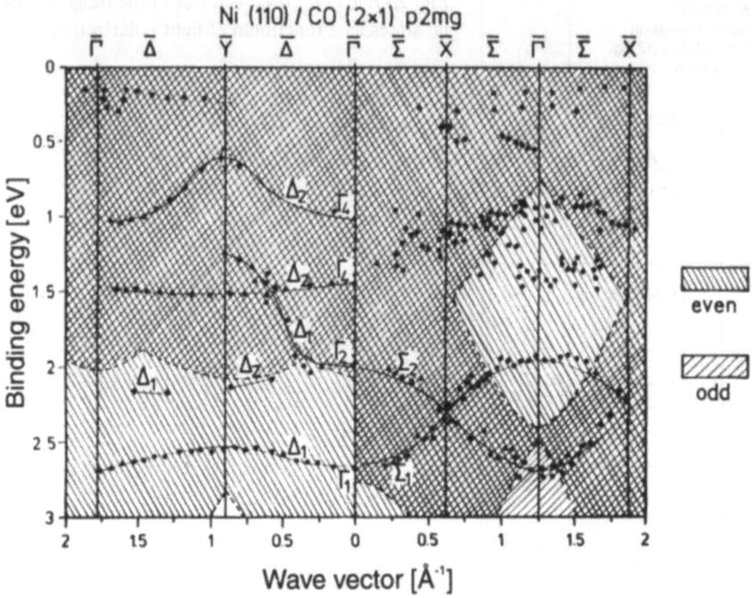

Fig. 2.18. Measured dispersions (●) of the 2π-d wavefunctions plotted on top of the calculated bulk band structure projected onto the [110] surface. The projection has been done separately for even and odd bands

demonstrate that such a state cannot be exited by either polarization of the light and so this band could not be observed in the normal emission spectra.

The experimental points within 0.5 eV below E_F could not be assigned along either direction. Because the main part of the substrate emission occurs in this energetic range, these features may be due to surface umklapp processes. However, emission from hybrid bands in this energy range cannot be conclusively excluded yet [2.46].

The range from 1.0 to 1.5 eV binding energy contains many data points along $\bar{\Sigma}$ that could not be assigned to definite bands. In normal emission ($\bar{\Gamma}$) in this range only the two Γ_4 features are observed. Since the symmetry allows only twofold-degenerate bands at $\bar{X}$ which have to belong to different representations along $\bar{\Sigma}$ and at $\bar{\Gamma}$, the two Γ_4 bands cannot be degenerate at $\bar{X}$. Thus there must be two additional bands in this energetic range transforming according to either Γ_2 or Γ_3 at $\bar{\Gamma}$. The spectra show no additional emission in this range at $\bar{\Gamma}$ so that these bands will most probably belong to Γ_3 and thus will not be excitable in normal emission. In contrast to the situation near to the Fermi energy, the dispersion of the bands in the energetic range from 2 to 2.7 eV is rather clear. These bands are degenerate at $\bar{X}$ and their periodicity is just as demanded by the symmetry of the system. At $\bar{X}$ certain symmetry operations transform each of the degenerate wave functions into the other one. The band starting at 2.0 eV from $\bar{\Gamma}$ was assigned to have $2\pi_y^+$-d_{yz}^+-character at $\bar{\Gamma}$, whereas the band starting from 2.7 eV was assigned to have $2\pi_y^- \text{-} d_{z^2}^+ \text{-} s^+$ character. The transformation of $2\pi_y^+$ into $2\pi_y^-$ is a symmetry operation of $\bar{X}$, whereas the transformation of the metal part $d_{z^2}^+ \text{-} s^+$ into d_{yz}^+, and vice versa, of course, cannot be done by a symmetry operation. Thus the degeneracy of these bands apparently seems not to be possible by symmetry. But, as is well known, the composition of the bands is a function of the wave vector $k_\parallel$ and thus the compositions of the bands at $\bar{X}$ will differ from their composition at $\bar{\Gamma}$. To discuss this in more detail, assume, for instance, that the band starting at 2.0 eV from $\bar{\Gamma}$ ($2\pi_y^+ \text{-} d_{yz}^+$) begins to pick up $d_{z^2}^- \text{-} s^-$ character when dispersing along $\bar{\Sigma}$. This is possible since $2\pi_y^+$, d_{yz}^+, and $d_{z^2}^- \text{-} s^-$ all belong to the same irreducible representation along $\bar{\Sigma}$. If the second band starts to pick up d_{yz}^--character, we get two bands at $\bar{X}$, the first one being composed of $2\pi_y^+$, d_{yz}^+, and $d_{z^2}^- \text{-} s^-$, and the second consisting of $2\pi_y^-$, d_{yz}^-, and $d_{z^2}^+ \text{-} s^+$ wave functions. These bands are allowed to be degenerate at $\bar{X}$ if they contain corresponding metal or CO wave functions of equal amounts. This means that if, for instance, the first band contains 10% d_{yz}^+, the second one has to contain 10% d_{yz}^-. Figure 3.19 shows a plot of the wave functions of these bands at $\bar{Y}$, $\bar{\Gamma}$, and $\bar{X}$, assuming that the composition of the bands at $\bar{Y}$ does not differ from their composition at $\bar{\Gamma}$. It should be noted that the qualitative composition of the bands as shown in

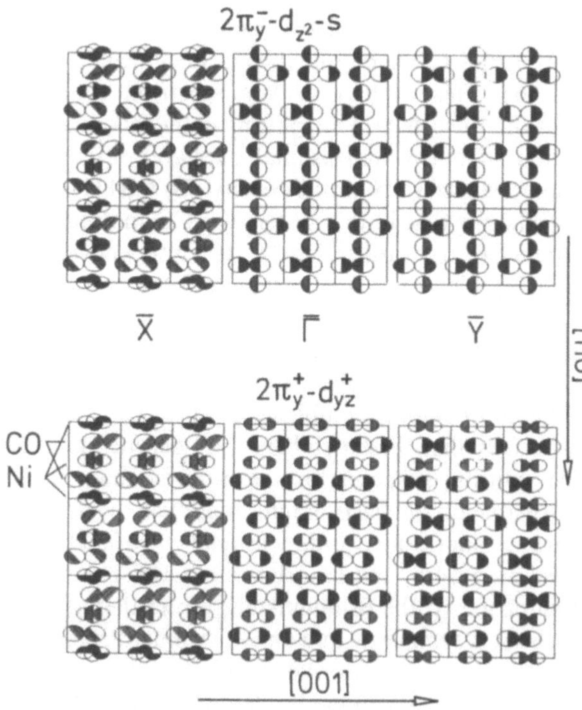

Fig. 2.19. Schematic representations of the 2π-d wave functions at different k-points. (See comments in the caption of Fig. 2.12). The positions of CO molecules and Ni atoms are marked

Fig. 2.19 at $\bar{X}$ and $\bar{Y}$ is just an arbitrary choice. There is no specific physical reason to state that this must be the right composition although it may be the correct one.

As stated above, the bands in the range from 1.0 to 1.5 eV below E_F will most probably be degenerate at $\bar{X}$ with bands belonging to Γ_3 at $\bar{\Gamma}$. For these bands the same problems must be discussed; there must also be a change in the amount of the metal character of the wave functions along $\bar{\Sigma}$.

Summarizing this section we have a full data set of the band structure of a molecular adsorbate available but a full understanding of all observed details is still not in sight although one has come some way.

b) Weakly Chemisorbed and Physisorbed Layers

The examples for dispersions in quasi two-dimensional systems were chosen so far from the many examples of strongly chemisorbed systems. One question is what happens to the dispersion when weakly chemisorbed or physisorbed systems are considered. The latter case is easy.

Figure 2.20 demonstrates how the spectra change when we gradually go from the chemisorbed CO adsorbate via the weakly chemisorbed molecule and the physisorbed molecule finally to the condensed phase [2.12, 2.61–64]. From the top of the bottom the heat of adsorption (E_{ad}) decreases from 142 kJ mol^{-1} to 19 kJ mol^{-1} [2.65–69]. This is accompanied by changes in the adsorbate spectra. In CO/Ag(111) at $T = 20$ K CO is physisorbed as documented by the small $E_{ad} = 19$ kJ mol^{-1} [2.66]. This explains why a spectrum similar to condensed CO is observed for this adsorbate. The splitting in the 5σ-emission is connected with the formation of a two-dimensional overlayer. Figure 2.21 exhibits the dispersion measured via ARUPS for the system CO/Ag(111) [2.62]. We know from a symmetry analysis that in this system the CO molecules are oriented with their axes parallel to the surface [2.63]. It is also known from LEED studies that CO and N_2 physisorbed on graphite form herringbone structures, as shown in the inset in fig. 2.21 [2.70]. Such structures again belong to nonsymmorphic space groups with two molecules in the unit cell. This is the reason why the molecular ionization bands appear to be split in two components, i.e., a bonding and an antibonding combination at $\bar{\Gamma}$ [2.71]. From symmetry considerations it is clear that these two bands are degenerate at the zone boundary [2.55]. The splitting is larger for the 5σ-level than for the 1π-level, which is not unreasonable on the basis of intermolecular overlap considerations. Figure 2.22 shows schematically two-dimensional wave functions of a herringbone structure based on 5σ, 4σ-, and 1π-orbitals of CO for the center and the border of the Brillouin zone. At the Brillouin-zone center there are two split components for each orbital, a symmetric and an antisymmetric combination, i.e, two components for the 5σ- and 4σ-orbitals and four components for the 1π-orbital. Upon going from the center to the border of the Brillouin zone, the components become pairwise degenerate for similar reasons as discussed for the system CO(2 × 1) *p2mg*/Ni(110) [2.37]. The smaller splitting of the π-levels

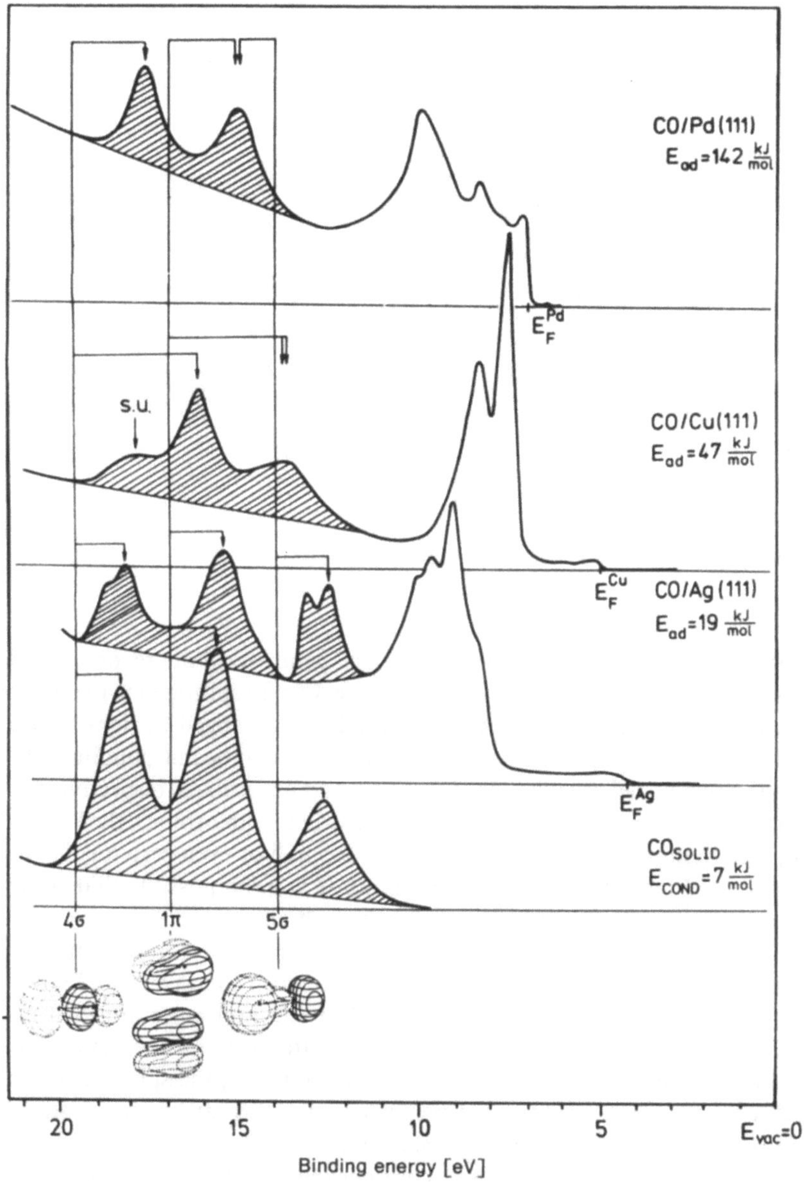

Fig. 2.20. Electron distribution curves of CO adsorbates of varying CO-substrate bond strength. For reference the gas phase values of the outer valence ionizations of CO are given

compared with the 5σ-levels can be qualitatively deduced if we estimate the overlap between neighboring molecules. Clearly, even though the radial extent of the 1π-orbitals is largest, phases are such that the total overlap cancels to a large degree. For the 5σ-orbital this is not the case as it is not for the 4σ-orbital

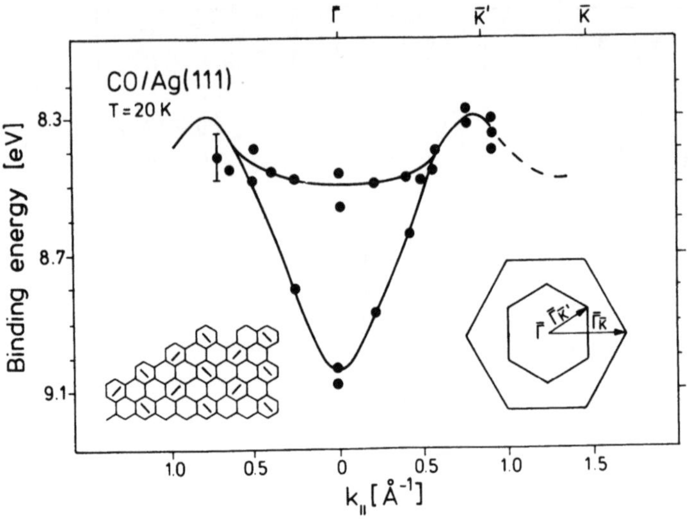

Fig. 2.21. Measured dispersions in the 5σ CO levels of CO/Ag(111)

in a herringbone structure. However, there are geometries where one could expect a strong splitting for one σ-orbital and a very small splitting for the other σ-orbital. Consider a situation where the orbitals are arranged at right angles [2.72]. Assume one orbital to be bonding, the other one to be antibonding with respect to the two nuclei as is the case for the 5σ and 4σ orbitals, respectively. If we construct two-dimensional wave functions based on both orbitals, then at the Brillouin zone center the splitting of the bands is small or even zero for the 4σ ($2\sigma_u$) orbital, and large for the 5σ ($3\sigma_g$), orbital. It is quite possible that some observations in the literature [2.63] are due to such effects.

A particularly interesting observations has been made for herringbone systems if the temperature is increased. These physisorbed overlayers are known to undergo order-disorder transitions [2.70]. The CO molecules are then no longer locked into a herringbone structure but rotate freely on their site. This destroys the nonsymmorphic structure and, concomitantly, the splitting of the σ-levels disappears. CO/Ag(111) is a system where ARUPS can be used to study phase transitions in quasi-two-dimensional systems [2.62, 63].

The next step in our discussion is to consider the slightly more complicated situation for the weakly chemisorbed system CO/Cu(111) [2.61] with a heat of adsorption of 47 kJ mol^{-1} [2.67]. With respect to the CO/Ag(111) system the features in the spectrum shift and the intensities of the lines are altered considerably. Three lines are still found, but their assignment is, as we shall see further below, quite different from the one for the condensed molecular solid.

The assignment of the spectrum of the CO/Cu(111) system [2.61], given in Fig. 2.20, indicates that the considerations presented so far are not complete and sufficient to explain all experimental findings. It has been shown theoretically

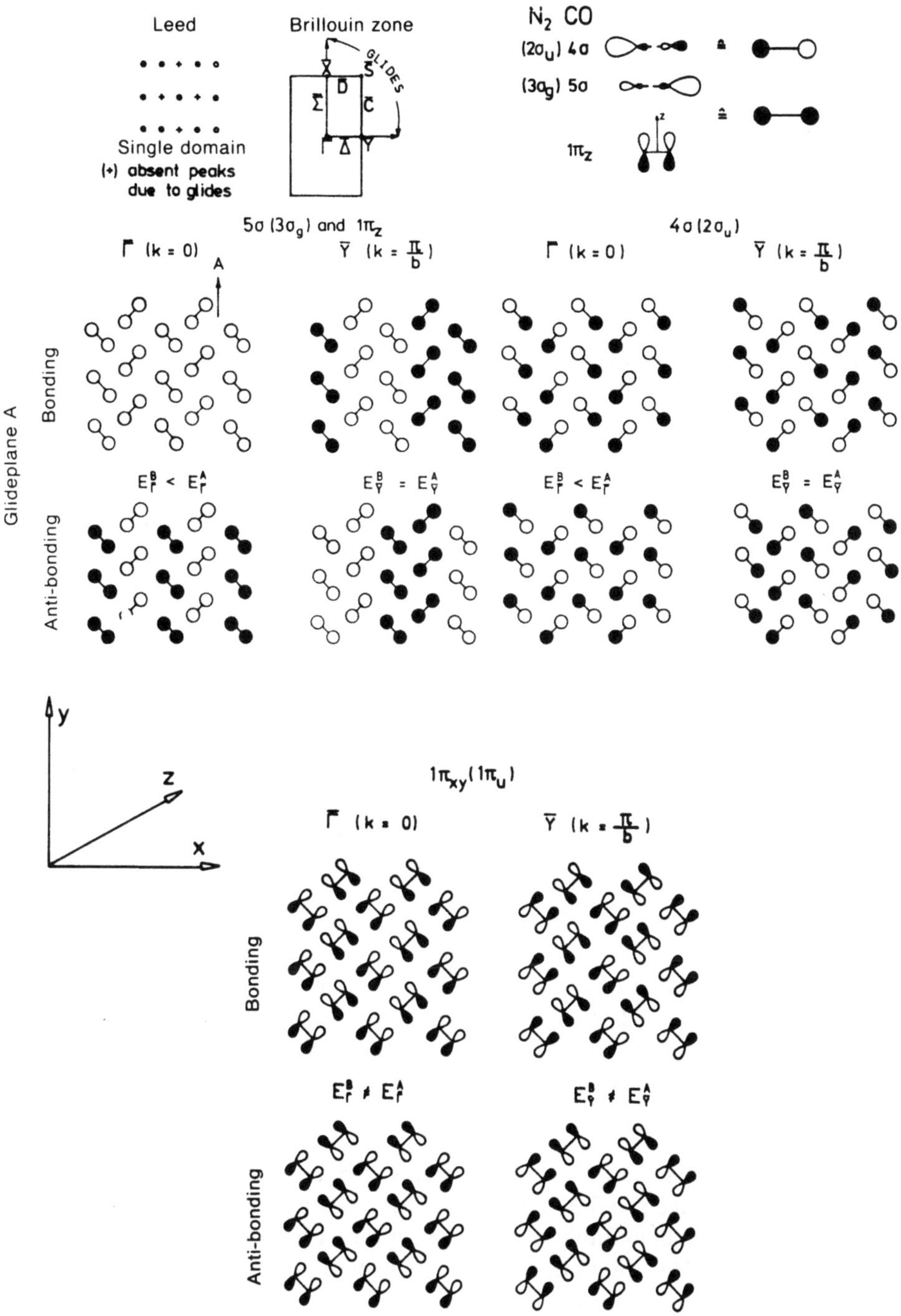

Fig. 2.22. Schematic representations of σ and π wave functions at two k-points for a herringbone structure

that for weakly chemisorbed systems so called 'shake up' excitations accompanying the 'normal' electron emission have to be considered [2.73–79]. These shake-up-excitations are manifestations of the fact that the ionization process is a rather complicated many-electron process [2.25, 80]. They can be assigned to electron excitations, in addition to electron emission. Their intensity is determined by the second matrix element in (2.2) whose magnitude is governed by the projection of the wave function of the shake-up state ${}^{N-1}\Psi_{e,E}$ onto the 'frozen' ion state $a_k\varphi_i$. Shake-up intensities are rather low for chemisorbed and for physisorbed systems but reach the maximum for intermediate metal-molecular coupling, i.e., weak chemisorption [2.73–80]. Again, ARUPS can be employed to support the assignment, as given in Fig. 2.20 for the CO/Cu(111) system [2.61]. If the most intense CO features were due to 1π-emission, as might be suspected by comparing the spectra of the CO/Cu(111) system with CO/Ag(111), a resonance behaviour for this particular peak would not be allowed. *Mariani* et al. [2.81] demonstrated that both bands at higher binding energy are due to states of σ-symmetry by investigating the shape resonance discussed above. The 4σ ion state as well as the accompanying shake up transition exhibited parallel resonance behaviour as expected according to the assignment in Fig. 2.20. A spectrum rather similar to the one of the CO/Cu system, but with even slightly more intense satellite structure, has been found for the system CO/Au [2.82]. In the latter system the adsorption energy is between the one for CO/Ag and CO/Cu which leads us to expect more intense satellites, and corroborates the ideas presented here.

Figure 2.23 exhibits the dispersions for the system $(\sqrt{7}\times\sqrt{7})$CO/Cu(111), for which Fig. 2.20 showed an electron distribution curve [2.35]. In this case the CO molecules are oriented perpendicular to the surface as in the case of the strongly chemisorbed systems. While the integrated $5\sigma/1\pi$-dispersion is compatible with the other CO overlayer systems, the 4σ-dispersion is considerably smaller than expected for the given intermolecular separation. The observed value is represented in Fig. 2.9 by the dashed circle. The shake-up which is – as noted above – associated with the 4σ-ionization shows almost no dispersion, but only a slight variation in relative intensity with respect to the 4σ-ionization. There are sum-rules [2.83, 84] relating intensity $A(\omega,k)$ and ionization energy (ω) of the peaks in the observed spectral function with the quasi-particle energy $(\varepsilon_k{}^{\mathrm{HF}})$. These sum-rules are of the type

$$\varepsilon_k{}^{\mathrm{HF}} = \int_{-\infty}^{+\infty} wA(w,k)dw. \tag{2.6}$$

We can apply this sum rule to the observed data and regain a dispersion indicated by the open circles in Fig. 2.23. This renormalized 4σ bandwidths can now be compared favourably with the values measured for the strongly chemisorbed systems. This shows that it is the ionization process that introduces the deviations in the observed bandwidths and not a different intermolecular interaction potential in this case.

With this example we finish the section on CO adsorbates and move on to chemically more complex systems.

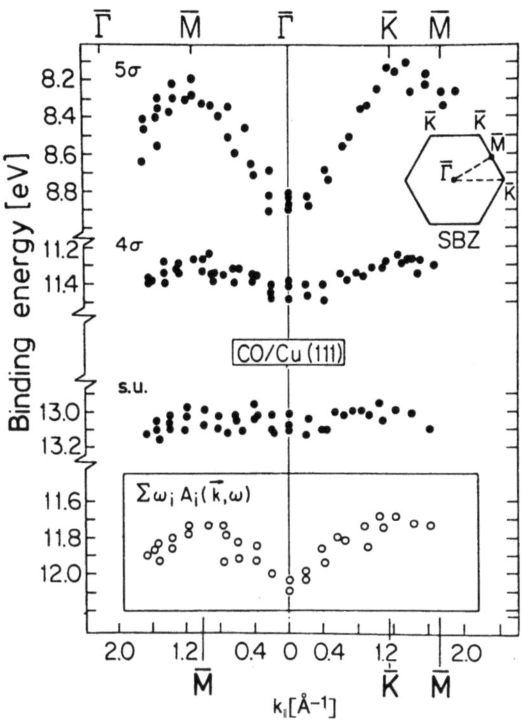

Fig. 2.23. Measured dispersion of the CO induced levels of the system CO/Cu(111). The renormalized dispersion of the 4σ-levels is plotted as open circles

2.3 Ethylene Adsorbates [2.85]

Rather complete studies of the band structure of more complex adsorbate systems, such as chemisorbed hydrocarbons are relatively rare.

Very recently, *Steinrück* and coworkers [2.86,87] published an interesting study of ethylene adsorption on Ni(110) where the aspect of local and global symmetry has been discussed. We want to summarize their results in the following.

Saturation coverage of Ni(110) by C_2H_4 at $T = 120$ K results in a $c(2 \times 4)$ LEED pattern. This pattern represents a very dense layer containing two molecules per unit cell (i.e., $\theta = 0.5$) as revealed through coverage determination with Temperature Programmed Desorption (TPD) [2.86].

The primary task, though, is, as in the case of the CO adsorbates, the determination of the local symmetry of the molecule at the surface. This is done for a dilute ethylene layer ($\theta \sim \theta_{sat}/2$) where lateral interactions are not important. We shall see that the local geometry prevails in the dense layer. There are indications from High-Resolution Electron Energy Loss Spectroscopy (HREELS) that the ethylene species is di-σ-bonded as opposed to π-bonded with the C–C axis parallel to the surface and that the site symmetry is actually lower than C_{2v} [2.88]. Theoretical results, however, indicate that the adsorbate bonding is quite similar in both the di-σ-bonded and the π-bonded cases, i.e., π-donation from C_2H_4 to the substrate is stronger as compared to π^*-back-donation [2.86] from the substrate towards the C_2H_4 molecule.

2.3.1 Symmetry Consideration and Ethylene-Substrate Interaction

Angle-resolved photoemission spectra of the dilute layer are shown in Fig. 2.24 [2.86]. The photon energy of 30 eV was chosen to optimize the intensity of the adsorbate-induced features. For comparison the angle-integrated photoelectron spectrum of gas-phase ethylene [2.89] including assignment is plotted. The spectra have been aligned at the $3a_g$ and $1b_{3u}$ levels (gas-phase nomenclature) as is frequently done in the literature [2.86]. The comparison reveals that the adsorbate spectrum is characteristic of a molecularly adsorbed species. It also shows that the $1b_{2u}$ ion-state is shifted by 1.1 eV to higher binding energy. This is by far the largest differential shift observed in this system. The shape of the ethylene orbitals in the gas phase is plotted in Fig. 2.25 [2.90]. It shows that the

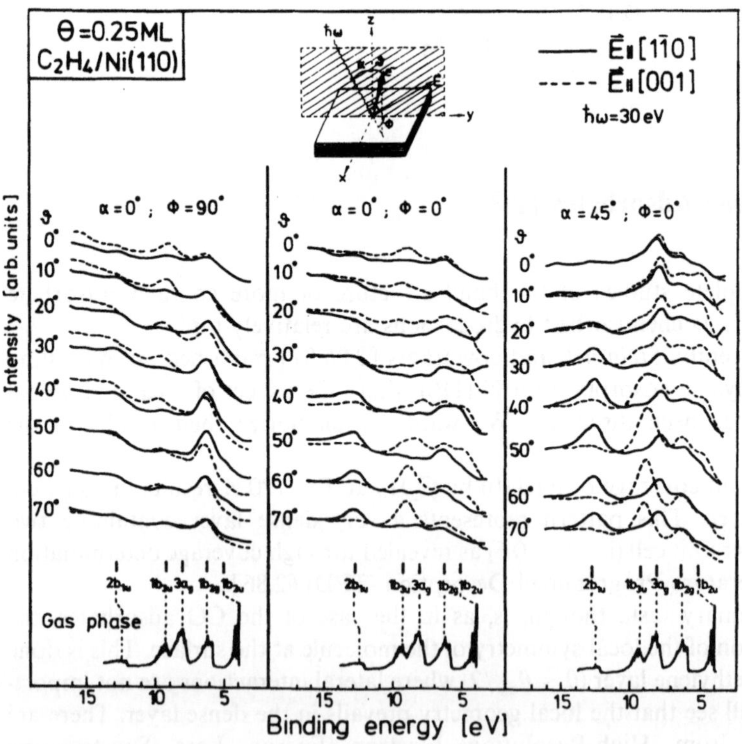

Fig. 2.24. Angle resolved electron distribution curves for a dilute C_2H_4 layer on Ni(110) in comparison to the gas phase [2.89]

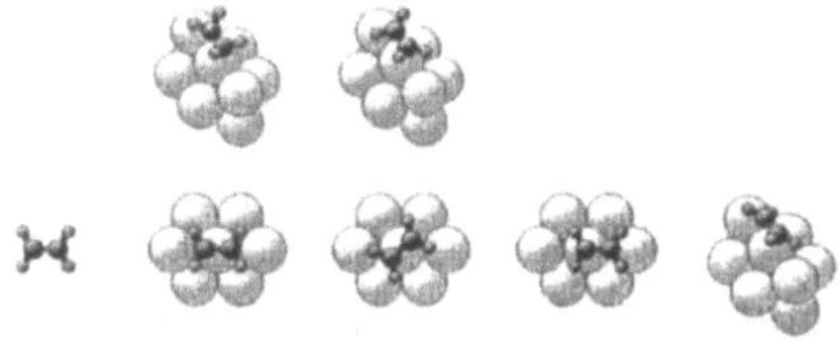

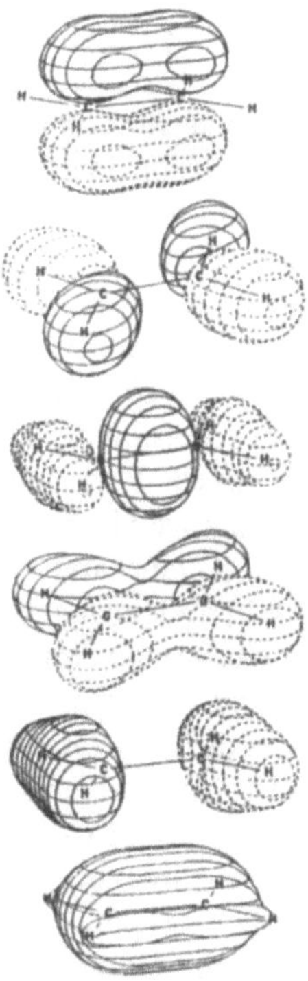

symmetry / orbitals	D_{2h}	C_{2v}	C_2	$C_s(xz)$	$C_s(yz)$
	$1b_{2u}$	a_1	a	a'	a'
	$1b_{2g}$	a_2	a	a''	a''
	$3a_g$	a_1	a	a'	a'
	$1b_{3u}$	b_2	b	a''	a'
	$2b_{1u}$	b_1	b	a'	a''
	$2a_g$	a_1	a	a'	a'

Fig. 2.25. Correlation table of ethylene orbital symmetries in the gas phase with those in varying site symmetries

$1b_{2u}$ orbital is characteristic for a π-bond. The large shift indicates that the main interaction of the molecule with the surface takes place through this orbital.

The site symmetry and the orientation of the adsorbed ethylene molecules in the dilute layer can be derived from the polarization, polar angle and azimuthal dependences of the observed ionization bands, which are displayed in Fig. 2.24 and are discussed below. Figure 2.25 shows how the irreducible representations of the molecular orbitals in the isolated molecule correlate with the corresponding irreducible representations when ethylene is bound to the surface in four different point group symmetries of the hypothetical sites. The real space geometries corresponding to the site symmetries are exhibited in Fig. 2.25. Due to the twofold symmetry of the Ni(110) surface there are two possibilities for a C_s site, as indicated.

There is an important experimental detail in the experiments by *Steinrück* and co-workers, see [2.86, 87], that must be mentioned at this point. The electron analyzer allows one to measure angular distributions of electrons by means of multichannel detection techniques including normal emission also at light incidence normal to the surface [2.91]. At these particular experimental conditions two bands ($3a_g$ at 8.0 eV and $1b_{2u}$ at 5.0 eV) show no normal emission intensity for normal incidence, but rather strong normal emission for incidence at 45°, with detection in the plane of incidence. It is clear from Fig. 2.25 that this experimental observation is strong evidence for a basically flat lying geometry. Both orbitals $3a_g$ and $1b_{2u}$ belong to the totally symmetric representation in any considered point group and are only allowed in normal emission for a non-vanishing z component of the electric field vector. Note that both orbitals show rather weak emission for normal incidence if the electrons are detected in a plane perpendicular to the incidence plane. This detection geometry is called "*forbidden*" *geometry* for diatomic adsorbates as discussed above. The azimuthal orientation can be extracted from the intensity variations in the $2b_{1u}$ (12.7 eV) and $1b_{3u}$ (9.2 eV) emission (Fig. 2.24). These states transform according to the b_1 and b_2 representations if C_{2v} point group symmetry is assumed. They exhibit opposite intensity variations if the plane of light incidence is changed from [001] to [1$\bar{1}$0]. The b_1 state shows strong emission in the [001] plane and weak emission in the [1$\bar{1}$0] plane, while the b_2 is weak along [001] and strong along [1$\bar{1}$0], respectively. Figure 2.25 depicts that the b_2 orbital has a node along the C–C while the b_1 orbitals has a node perpendicular to the C–C-axis. Consequently, if the electric field vector points along the C–C axis the orbital with the node perpendicular to the C–C axis (b_1) should lead to normal emission intensity, and if the electric field vector is oriented perpendicular to the C–C axis the b_2 orbital should yield electron ejection in the normal direction. From Fig. 2.24 it is therefore quite obvious that the C–C axis of the C_2H_4 molecule is to a first approximation oriented along the [1$\bar{1}$0] azimuth. However, the intensities do not completely go to zero in the appropriate directions indicating that the alignment is not perfect. In other words, the symmetry of the adsorption complex is lower than C_{2v}. This is also in line with intensity variations observed for the $1b_{2g}$ (6.2 eV) ion state which belongs to the a_2 representation in point group C_{2v}. In fact, a detailed analysis shows that the observed intensity of the a_2 ion state in normal emission can only be explained by either a slight tilting of the C—C axis, a lower symmetry adsorption site or a twisting of the molecule. *Steinrück* and the co-workers [2.86, 87] excluded the former due to the vanishing normal emission intensity of the $3a_g(a_1)$ ion state and therefore favour a lower adsorption site, i.e., slightly displaced from a high symmetry adsorption site or twisting of the molecule (Fig. 2.25).

Rösch and coworkers have analyzed the bonding via cluster calculations [2.86, 87] on the basis of the local density functional approach. Their analysis reveals that the molecule-substrate interaction follows the standard description of olefin bonding towards transition metal atoms: π-donation to the metal and backdonation from the metal ion to the π^*-orbital of C_2H_4. However, the calculations do not allow discrimination between the two main adsorption modes, π or di-σ.

2.3.2 Ordered Ethylene Overlayers

After having discussed the local site geometry as determined via ARUPS on a dilute C_2H_4 layer we come back to the study of the layer at saturation coverage. *Steinrück* and co-workers [2.86, 87] demonstrated that for the strongly laterally interacting C_2H_4 layer the local site is similar to the dilute layer so that we do not have to discuss the spectra in detail in this case.

From the variation of the binding energy with the emission angle, the two-dimensional band structure of the saturated ethylene layer can be evaluated. The two-dimensional band structure has been determined along the two high-symmetry directions of the substrate, i.e. $[1\bar{1}0]$ and $[001]$. Note that these directions are not necessarily mirror planes of the adsorbate Brillouin zone. The LEED Brillouin zone is exhibited in Fig. 2.26. The structure of the layer derived from thi: pattern is shown in the same figure. The molecules are put in the proposed asymmetric adsorption sites and are aligned along the $[1\bar{1}0]$ direction with 5 Å separation. Molecules in neighbouring rows are shifted with respect to each other by 1.25 Å. This results in translational equivalence at every fourth row, i.e., a $c(2 \times 4)$ structure. The ARUPS measurements combined with molecular dynamics simulations reveal that the active intermolecular C_2H_4 potentials lead to a band structure which is not directly compatible with the Brillouin zone derived from the LEED pattern [2.86, 87]. The plot of the data as $k_{\|}$ vs. E_B is exhibited in Fig. 2.27. Some high symmetry points in the extended Brillouin zone scheme deduced from the LEED pattern are indicated at the top of Fig. 2.27, and it is quite obvious that the observed band dispersions do not follow the true Brillouin-zone symmetry. The observed band structure may be explained if the interaction of the molecule with the substrate is neglected for all but the π $(1b_{2u})$ ion state and the molecules are displaced with respect to each other in neighbouring rows as indicated in the middle panel of Fig. 2.28. In other words, the periodic potential leading to the fourfold repeat unit is not important. The molecular dynamics simulations energetically favour this interpretation. It is known from other hydrocarbons that the intermolecular interaction is dominated by the repulsive hydrogen-hydrogen interactions which are minimized in the geometry shown in the center panel of Fig. 2.28 with respect to the other two limiting cases (left and right panels in Fig. 2.28). Without consideration of the substrate the adsorbate unit cell only contains a single molecule. The unit cell in reciprocal space is depicted in Fig. 2.26. The experimentally chosen $[001]$ and $[1\bar{1}0]$ directions of the substrate are included in the figure. They do not represent high-symmetry directions in the adsorbate mesh, but rather the paths form $\bar{\Gamma}$ to $\bar{Y}$ along $[1\bar{1}0]$. The calculated band structure along these paths

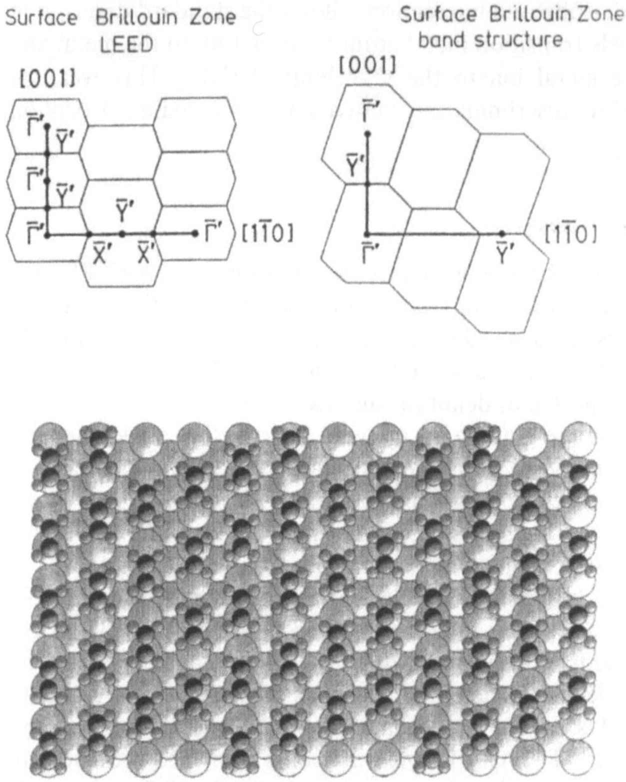

Fig. 2.26. Brillouin zones of the $c(4 \times 2)$ C_2H_4/Ni(111) system as observed with LEED (on the left) and as determined via band structure measurements (on the right). A real-space model is included

for the energetically favoured arrangement is depicted in Fig. 2.28 (center panel) and may be directly compared with the calculated band structures for other geometries. If we now compare the calculated band structure with the measured data, as done in Fig. 2.27, we find very good agreement for four out of five ion states. The only noticable exception is the b_{2u} ion state which couples the ethylene molecule to the substrate. It is obvious that none of the other considered geometries (left and right panels) would yield a band structure that agrees with experiment anywhere near to this one. The observed dispersion can be explained on the basis of nearest-neighbour interactions. Let us consider for simplicity the [001] direction. The highest binding energy ion state has b_{1u} symmetry. The molecular wave function can be taken from Fig. 2.25. For a two-dimensional array of molecules (Fig. 2.29a) the relative phases lead to a mainly nonbonding wave function at $\bar{\Gamma}$. If for every other row of molecules the phases change at $\bar{Y}$ the nonbonding character is not changed. Consequently, there is no dispersion of this orbital along the [001] direction. The two-dimensional wave function derived from the $1b_{3u}$ orbital (Fig. 2.29b) is strongly antibonding at $\bar{\Gamma}$.

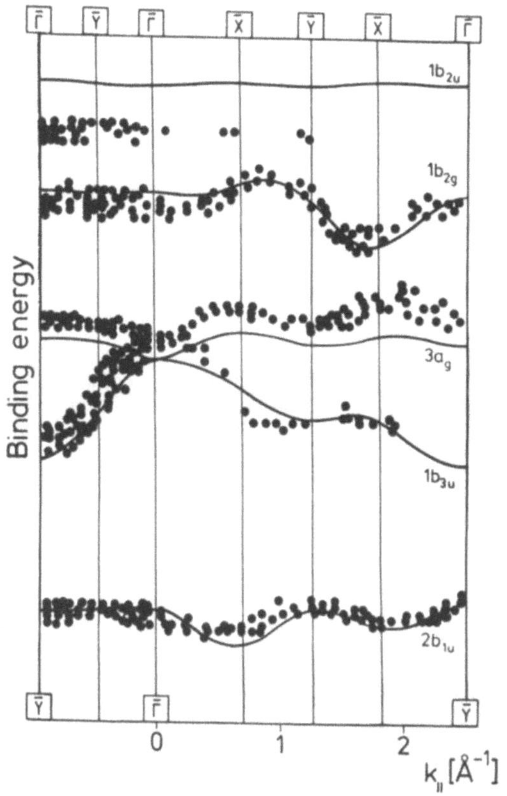

Fig. 2.27. Measured band structure for the system $c(4\times 2)$ C_2H_4/Ni(111). Points of high symmetry according to the LEED Brillouin zone (top) and the band structure Brillouin zone (bottom) are indicated

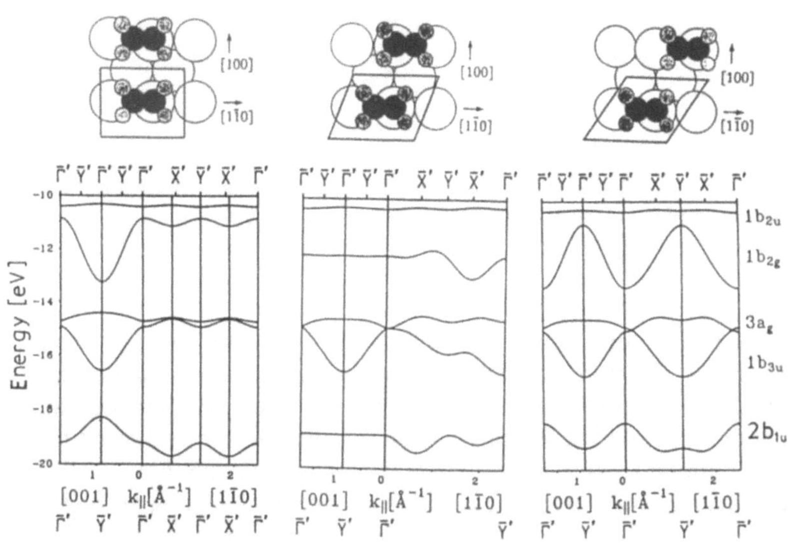

Fig. 2.28. Calculated band structures for three different lateral arrangements of C_2H_4 molecules

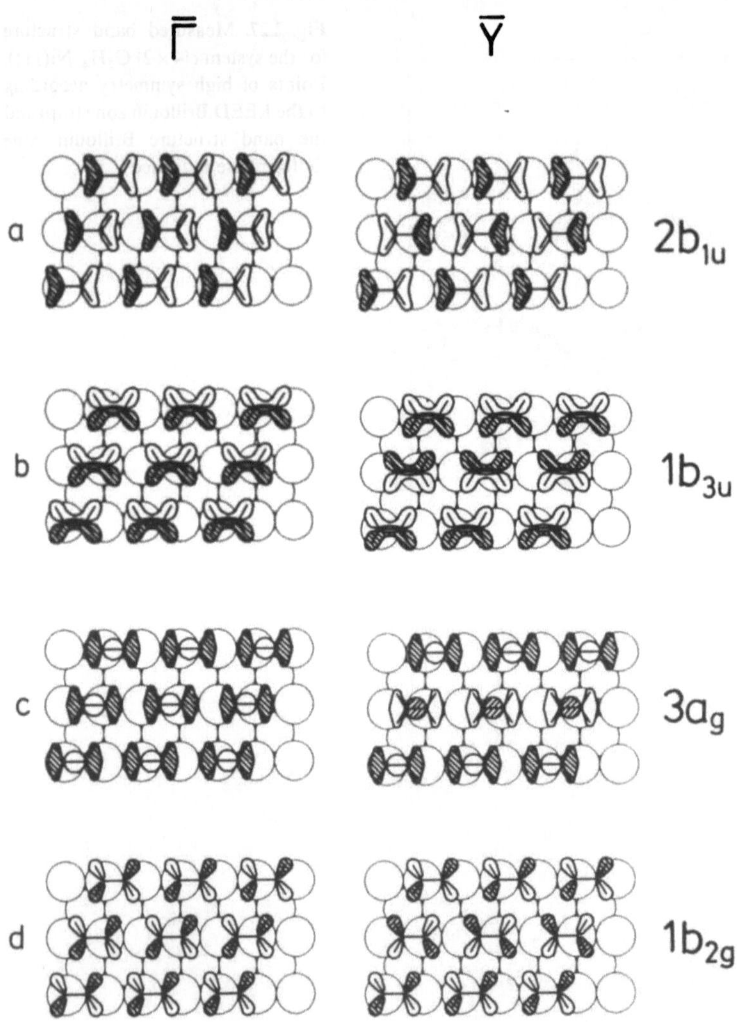

Fig. 2.29a–d. Schematic two-dimensional wave functions for the ethylene overlayer for the center (left) and the border (right) of the Brillouin zone **a)** $2b_{1u}$, **b)** $1b_{3u}$, **c)** $3a_g$, **d)** $1b_{2g}$

Upon going to the zone boundary along [001] the wave function becomes strongly bonding, therefore leading to strong downward dispersion. In the case of the $3a_g$ derived wave function (Fig. 2.29c), it is essentially nonbonding or weakly bonding at $\bar{\Gamma}$. At $\bar{Y}$ it remains basically nonbonding or weakly antibonding. The $1b_{2g}$ orbital leads to a two-dimensional wave function (Fig. 2.29d). The overall character is nonbonding at $\bar{\Gamma}$ and it remains nonbonding at $\bar{Y}$. We do not expect any strong dispersion for this orbital. The b_{2u} orbital couples to the substrate and thus we cannot expect correspondence to the simple band-structure calculation on an unsupported layer. Similar arguments hold for the other directions but they are slightly more involved because the [$1\bar{1}0$] direction is not a high symmetry direction within the adsorbate structure.

Summarizing this section it has been demonstrated that a close packed ethylene overlayer exhibits a pronounced band structure which is completely determined by nearest neighbour interactions. The overall structure documented in the band structure is determined via ARUPS. Calculations on a free unsupported ethylene layer completely account for the observations. It should be noted that independent of the packing of the molecules, C_2H_4 occupies, almost certainly, a nonsymmetric surface site and interacts with the surface through its π-electrons. The band derived from the π-electron level does not follow the calculated band structure. It behaves like a localized state created via interaction with the substrate. The interaction with the substrate is documented predominantly in a more pronounced differential binding energy shift of this ion state.

2.4 Benzene Adsorbates

Even more complex systems which nevertheless have been extensively studied during recent years are the various benzene adsorbates. Out of the many investigations performed so far (see, e.g., Table 2.3 [2.92–113] for a selection of structures) only rather few contain detailed band structure studies.

Table 2.3. Ordered benzene overlayers on metal substrates

Substrate	Structure	θ [ML]	References
Ni(111)	$(\sqrt{7}\times\sqrt{7})R19.1°$	0.143	2.92, 113
Ni(100)	$c(4\times4)$	0.125	2.93
Ni(110)	$c(4\times2)$	0.250	2.94, 95
Ru(0001)	$(\sqrt{7}\times\sqrt{7})R19.1°$	0.143	2.96
	$(2\sqrt{3}\times2\sqrt{3})R30°$	0.083	2.96
Rh(111)	$(2\sqrt{3}\times\sqrt{3})$rect	0.166	2.97–100
	$(\sqrt{19}\times\sqrt{19})23.4°$	0.159	2.101, 102
	$(\sqrt{7}\times\sqrt{7})R19.1°$	0.143	2.103
	$(2\sqrt{3}\times2\sqrt{3})R30°$	0.083	2.104
Pd(110)	$c(4\times2)$	0.250	2.105
Pd(100)	$c(4\times4)$	0.125	2.106
Ag(111)	(3×3)	0.111	2.107
Os(0001)	$(\sqrt{7}\times\sqrt{7})R19.1°$	0.143	2.108
	$(\sqrt{21}\times\sqrt{21})R10.9°$	0.19	2.109
Ir(111)	(3×3)	0.111	2.110, 111
Pt(100)	$c(4\times4)$	0.125	2.112

We shall review in the following data gained on adsorbates on hexagonal faces and choose as an example the Ni(111) [2.92, 113] and Os(0001) [2.108] surfaces. To further discuss the determination of additional azimuthal ordering, we turn then to the analysis of a surface with twofold symmetry, i.e. Ni(110) [2.94].

a) Symmetry Considerations on Molecule-Substrate Interaction in a Dilute Disordered Layer of C_6H_6/Ni(111) [2.112]

For Ni(111) *Huber* et al. [2.113] have determined the structure for dilute layers of C_6H_6 from angle-resolved photoemission. Figure 2.30 exhibits the comparison of the angle-integrated spectrum for a disordered C_6H_6 layer at $\theta = 0.1$ ML ($\theta_{sat} = 0.143$ ML) with the gas-phase spectrum [2.114]. The assignment of the observed spectral features in the adsorbate to the bands of the gas-phase photoelectron spectrum has been discussed in detail by many researchers [2.92–113]. The shifts observed between gas and adsorbate phase strongly suggest that the interaction between molecules and substrate basically involves the π electrons of benzene as is well known from transition-metal benzene complexes, i.e. $(C_6H_6)_2Cr$ [2.115]. Of course, the symmetry of the molecule, which is D_{6h} in the gas phase is reduced upon interaction with the substrate at least to C_{6v}. If there is moderate interaction between molecule and substrate the threefold symmetry of the substrate is likely to reduce the sixfold molecular

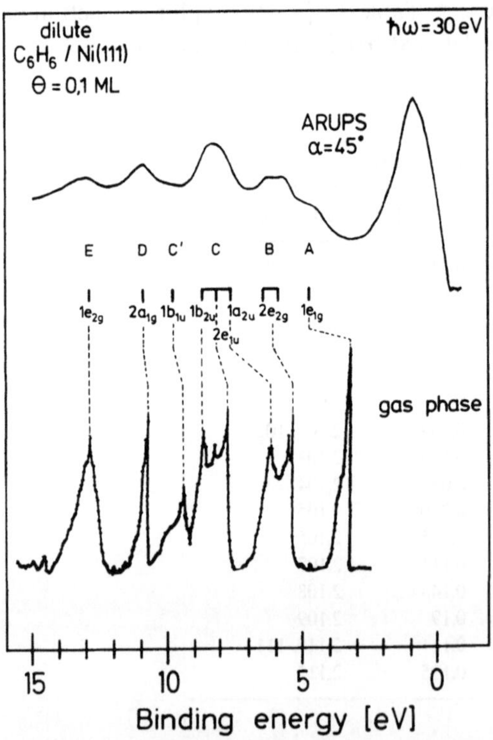

Fig. 2.30. Angle integrated photoelectron spectrum for a dilute C_6H_6/Ni(111) system in comparison with the gas phase [2.114]

symmetry to a threefold C_{3v} symmetry. The symmetry planes may either correspond to the σ_v or the σ_d symmetry planes of the sixfold symmetry. On a surface with twofold symmetry the point group could be further reduced to C_{2v}. Figure 2.31 contains a correlation table for the molecular orbital symmetries of C_6H_6 according to the various point groups. The + and − signs correspond to allowed and forbidden normal emission of the electrons for the two limiting light polarization directions, i.e. in the ring plane (x, y) and perpendicular to the ring plane (z). Due to the broad peaks in the adsorbate spectrum some ion states are superimposed. The bands in the experimental spectrum are therefore labelled in alphabetical order. Figure 2.31 can be used to establish the symmetry of the adsorption site by evaluating the experimental data (Fig. 2.32). The spectra show negligible emission from the $2a_{1g}$ orbital for $\alpha = 0$, i.e. normal incidence and normal emission, but strong emission for $\alpha = 45°$, i.e. a non vanishing z-component of the light. Together with the fact that the b_{1u} state suggests no normal emission for both light incidence angles this proves that the ring plane is oriented parallel to the surface. From the observation that the $1b_{1u}$ state shows negligible emission at all angles whereas the $1b_{2u}$ orbital is rather

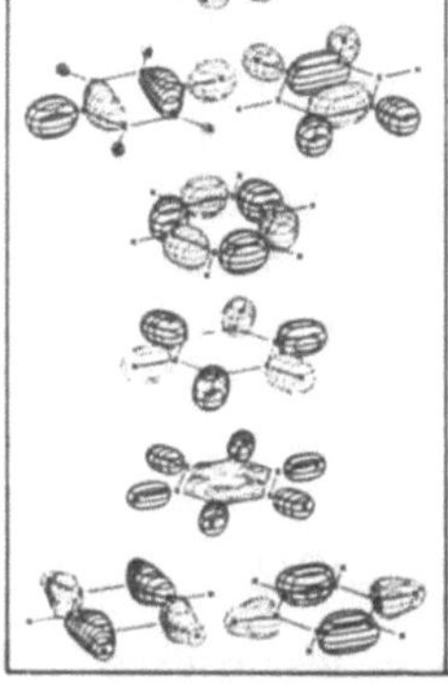

orbitals	symmetry												
	D_{6h}	C_{6v}			$C_{3v,\sigma v}$			$C_{3v,\sigma d}$			C_{2v}		
			xy	z		xy	z		xy	z		xy	z
	$1e_{1g}(\pi)$	e_1	+	−	e	+	−	e	+	−	b_1+b_2	+	−
	$2e_{2g}(\sigma)$	e_2	−	−	e	+	−	e	+	−	a_1+a_2	−	+
	$1a_{2u}(\pi)$	a_1	−	+	a_1	−	+	a_1	−	+	a_1	−	+
	$2e_{1u}(\sigma)$	e_1	+	−	e	+	−	e	+	−	b_1+b_2	+	−
	$1b_{2u}(\sigma)$	b_2	−	−	a_2	−	−	a_1	−	+	b_2	+	−
	$1b_{1u}(\sigma)$	b_1	−	−	a_1	−	+	a_2	−	−	b_1	+	−
	$2a_{1g}(\sigma)$	a_1	−	+	a_1	−	+	a_1	−	+	a_1	−	+
	$1e_{2g}(\sigma)$	e_2	−	−	e	+	−	e	+	−	a_1+a_2	−	+

Fig. 2.31. Correlation table of the orbital symmetries of C_6H_6 in the gas phase (D_{6h}) with phases in varying site symmetries. The + and − signs denote whether the orbitals can be exited with light polarized along the xy and z directions

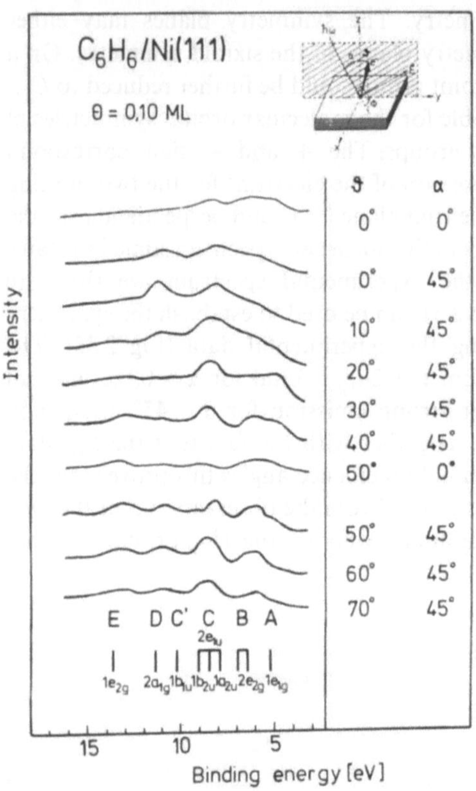

Fig. 2.32. Angle resolved electron distributions curves for a dilute C_6H_6 structures on Ni(111) surfaces a) two possible adsorption sites of isolated C_6H_6 molecules b) left panel $(\sqrt{7} \times \sqrt{7})R19.1°$ structure model; right panel $(\sqrt{21} \times \sqrt{21})R10.9°$ structure model

intense we conclude that the detector is placed in the σ_d mirror plane of the molecule, indicating that this plane acts as a mirror plane for the molecular symmetry. This is then compatible with a $C_{3v}(\sigma_d)$ symmetry of the adsorbate complex. Since the spectra have been collected along the $[1\bar{1}0]$ direction of the substrate *Huber* et al. [2.113] concluded that the benzene molecules are azimuthally oriented with the carbon-hydrogen bonds pointing along the $[2\bar{1}1]$ directions of the substrate as shown in Fig. 2.33a. The alternative local structure with $C_{3v}(\sigma_v)$ symmetry favoured by a recent cluster calculation, shown also in Fig. 2.33a is less likely but cannot completely be ruled out because measurements were only performed in one azimuth. Thus the structure marked $C_{3v}(\sigma_d)$ depicted in Fig. 2.33a is assumed to represent the interaction of an isolated benzene molecule with a Ni(111) surface.

b) Ordered Layers: C_6H_6/Ni(111) [2.113] and C_6H_6/Os(0001) [2.108]

C_6H_6 on Ni(111) as well as on Os(0001), which are examples of close packed surfaces of a fcc and a hcp metal, form ordered overlayers of the type $(\sqrt{7} \times \sqrt{7})R19.1°$ and sometimes, i.e. on Os(0001), even more close packed layers of $(\sqrt{21} \times \sqrt{21})R10.9°$ structure. Models of these structures are shown in

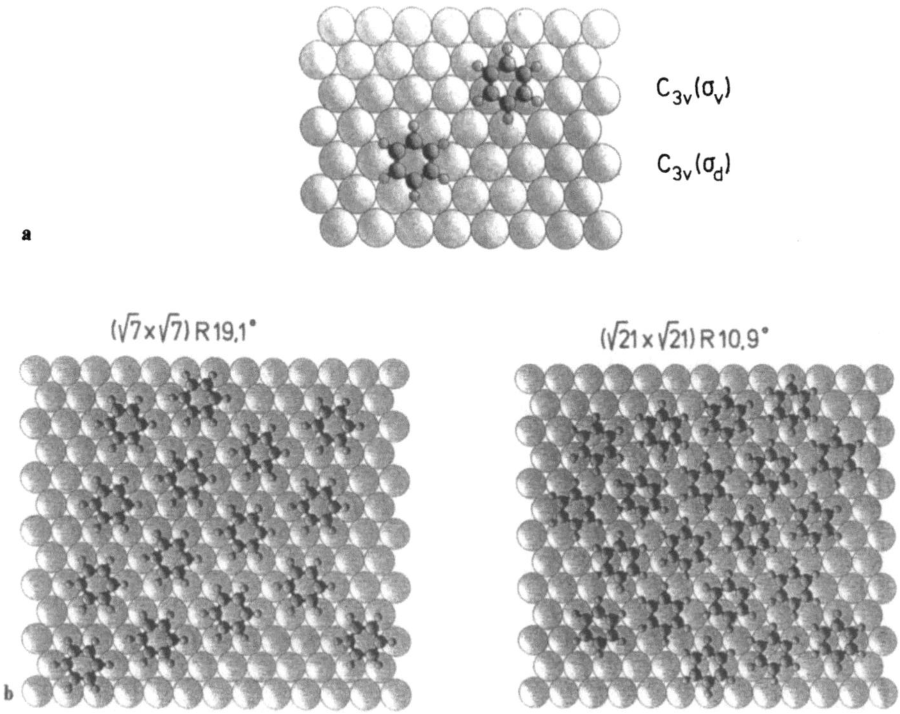

Fig. 2.33a, b. Schematic representation of the some C_6H_6 structures on (111) surfaces **a**) two possible adsorption sites of isolated C_6H_6 molecules **b**) left panel $(\sqrt{7}\times\sqrt{7})R19.1°$ structure model; right panel $(\sqrt{21}\times\sqrt{21})R10.9°$ structure model

Fig. 2.33b. The angle resolved spectra of a saturated $(\sqrt{7}\times\sqrt{7})R19.1°$ C_6H_6 layer on Ni(111) exhibit characteristic differences if compared with those of the dilute layer [2.113]. The spectra (not shown) indicate strong emission from the $1b_{2u}$ ion state except at particular angles. *Huber* et al. [2.113] took this as indications that the detection plane is no longer the σ_d mirror plane of adsorbed C_6H_6. An explanation would be an azimuthal rotation of the molecule due to strong lateral interactions. Figure 2.33b shows an anticipated model of the benzene layer keeping the adsorption site determined for the dilute layer, but allowing for a rotation by 30°.

Experimental evidence for strong lateral interaction is gained from the observation of dispersion of some of the ion states. Intermolecular interaction between planar hydrocarbons is mainly mediated through the hydrogen-hydrogen interaction. Therefore one expects those ion states to show the strongest effect that have large wave function coefficients at the hydrogen positions. Figure 2.31 reveals that there are several candidates. However, for a clear experimental identification of the dispersion an energetically isolated level should be used in order to avoid problems with overlapping levels. We consider the system $(\sqrt{7}\times\sqrt{7})R19.1°$ C_6H_6 on Os(0001) [2.108]. Figure 2.34 exhibits

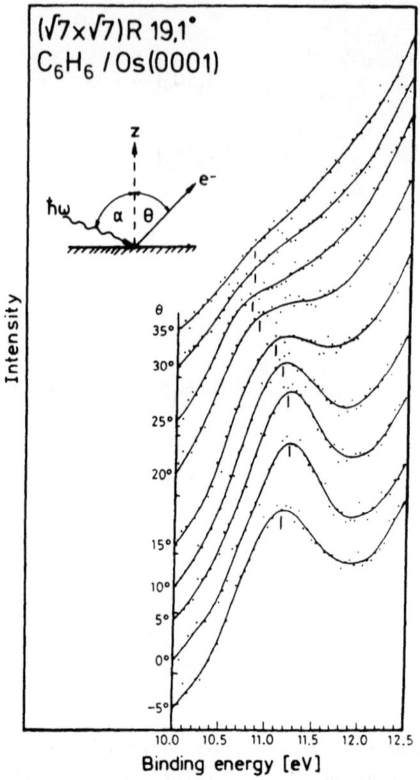

Fig. 2.34. Electron distribution curves for $(\sqrt{7} \times \sqrt{7})R19.1°$ C_6H_6/Os(0001) in the range of $2a_{1g}$ ion state as a function of the emission angle θ

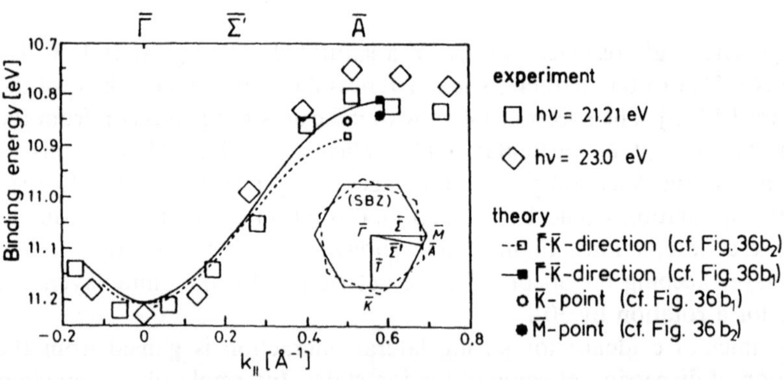

Fig. 2.35. Dispersion relation for the $2a_{1g}$ ion state of the system $(\sqrt{7} \times \sqrt{7})R19.1°$ C_6H_6/Os(0001)

photoelectron spectra of the $2a_{1g}$ level as a function of angle, and in Fig. 2.35 the dispersions are plotted for two photon energies along the direction indicated in the inset. The two possible domains of the overlayer are indicated. There exists a common direction $\bar{\Gamma} \rightarrow \bar{A}$ in both domains along which the experiments have been performed. Tight-binding calculations of an unsupported benzene layer have been performed in which the relative arrangement of the molecules within the layer has been varied in order to simulate the probable structures shown in Fig. 2.36a. Two extreme cases have been chosen, i.e. a parallel arrangement (Fig. 2.36b$_1$) and an interlocked arrangement (Fig. 2.36b$_2$). Figure 2.36a obviously represents an intermediate case where the repulsion between interacting molecules is minimized. The calculated dispersion curves [2.108] are included in Fig. 2.35. They correspond to the two azimuthal directions, i.e. $\bar{\Gamma} \to \bar{K}$ and $\bar{\Gamma} \to \bar{M}$ applying intermolecular interaction parameters derived from the geometries shown in Fig. 2.36b$_1$ and b_2. The energy position of the band at the $\bar{M}$ point is always higher than at the $\bar{K}$ point, independent of the absolute value of the interaction parameters. The former value can be estimated almost quantitatively via a simple procedure: we consider a parallel arrangement of C_6H_6 moieties (Fig. 2.36b$_1$). On both molecules a $2a_1$-orbital is placed. The overlap between the two molecules will be dominated by the overlap of the adjacent hydrogen-atomic orbitals. The reason is that the overlap depends exponentially on the distance. A calculation of the interaction parameter β yields 0.045 eV. Since the band constructed from the $2a_{1g}$-levels is of σ-type, we expect a band dispersion from high binding energy at $\bar{\Gamma}$ to lower binding energy at the zone boundary. In the simplest version of the tight-binding theory the band dispersion of a σ-type band in $\bar{\Gamma}$-$\bar{M}$ and $\bar{\Gamma}$-$\bar{K}$ direction are given by 9β and 8β, respectively. This yields 0.4 eV and 0.36 eV if we use the above mentioned β value. Those values are very close to the dispersion plotted in Fig. 2.35 as a result of the full calculation. A similar estimate can be made for the other geometry in Fig. 2.36b$_2$. This yields 0.36 eV and 0.32 eV, respectively. In conclusion we realize that the observed dispersion can be explained on the basis of simple considerations. Very similar results have been gained for the system $(\sqrt{7} \times \sqrt{7})R19.1°$ C_6H_6/Ni(111) and $(\sqrt{19} \times \sqrt{19})R23.4°$ C_6H_6/*Rh*(111). The observed dispersions are basically the same as for the Os(0001) surface layer and have been interpreted in a similar way by *Huber* et al. [2.113] and *Neuber* [2.102].

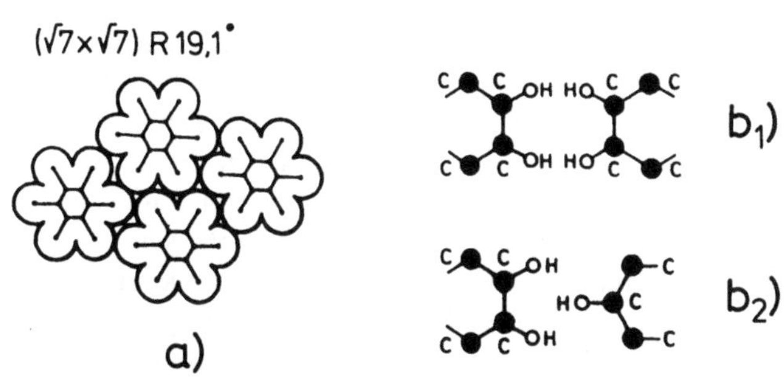

Fig. 2.36a, b. Arrangement of the C_6H_6 molecules in the $(\sqrt{7} \times \sqrt{7})R19.1°$ structure **a**) schematic representation of molecules in van der Waals contact **b**) 1: eclipsed arrangement of two C_6H_6 molecules 2: staggered arrangement of two C_6H_6 molecules

Graen and *Neumann* [2.116] have also determined the dispersion for a more densely packed layer with a $(\sqrt{21} \times \sqrt{21})R10.9°$ structure (Fig. 2.33b). Here the bandwidth increases due to the shorter nearest-neighbour distance. With such data one may now establish a bandwidth vs. distance plot quite similar to the one presented for CO adsorbates [2.102, 116].

c) Symmetry Considerations on Molecule-Substrate Interaction in a Dilute Disordered Layer of C_6H_6 of Ni[2.94]

A different situation is found for the adsorption of C_6H_6 on Ni(110). An ordered approximate $c(4 \times 2)$ structure forms for saturation coverage which shows pronounced intermolecular interactions. Before we discuss the band structure, the interaction of a C_6H_6 molecule with the surface of twofold symmetry has to be discussed. A priori, one might expect that the molecules are no longer adsorbed in a flat geometry due to the troughs of the surface, different from other close packed surfaces. In fact, such a tilt has been proposed for C_6H_6 on Pd(110) [2.117] while on the other surfaces such as Pt(110) (1×2) [2.118], Cu(110) [2.119] and Ag(110) [2.120, 121] flat lying geometries have been suggested. *Huber* et al. [2.94] have recently presented a very detailed study on the adsorption geometry of C_6H_6 on Ni(110). The dilute layer corresponds to a coverage of ~ 0.1 monolayer which is half of the saturation coverage. The layer is disordered. A series of angle resolved spectra is shown in Fig. 2.37. A parallel orientation of the molecule is concluded from the fact that the $2a_{1g}$ state is only observed for $\alpha = 45°$ and normal emission but at no emission angle in a plane perpendicular to the plane of light incidence. This means that the mirror planes of the benzene molecule must coincide with the [001] and $[1\bar{1}0]$ directions of the substrate which is a strong indication of an azimuthally oriented benzene molecule. The $1a_{2u}$-ion state behaves similarly and corroborates the above conclusion. Obviously, the adsorbate complex seems to exhibit C_{2v} symmetry. Consequently, the doubly degenerate *e*-levels should split. By comparing the normal emission spectra for normal incidences obtained at the two different azimuths there are indications (e.g., at 7.8 eV and 8.1 eV) for a split $2e_{1u}$ level. A characteristic angular intensity is exhibited by the $1b_{1u}$-ion state. Its intensity is very weak for the polarization of the light along the $[1\bar{1}0]$ azimuth but very strong for the light polarization parallel to [001]. Taking into account that for $\alpha = 45°$ the emission of the $1b_{1u}$ ion state is large along [001] and strongly attenuated along $[1\bar{1}0]$ corroborates the orientation of the molecule with its corners along the [001] direction. The behaviour of the $1b_{2u}$ ion state is in full agreement with this conclusion. It is therefore very likely that the symmetry of the adsorbate site is C_{2v}. We should note at this point that not all levels can be simply analyzed according to the above rules. For example, the $2e_{2g}$ level shows a splitting which is due to a Jahn-Teller effect [2.122–125], also known from the free molecule. Summarizing so far we realize that the plane of the molecule is parallel to the surface and the corners are oriented along the [001] direction. The mirror planes of the molecule coincide with the $[1\bar{1}0]$ and [001] directions

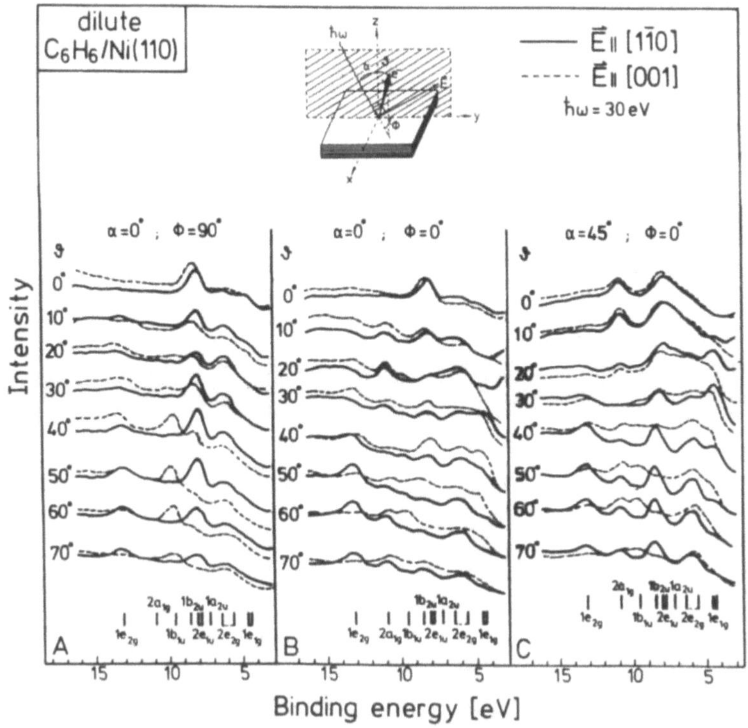

Fig. 2.37. Angle resolved electron distribution curves for a dilute C_6H_6/Ni(110) system

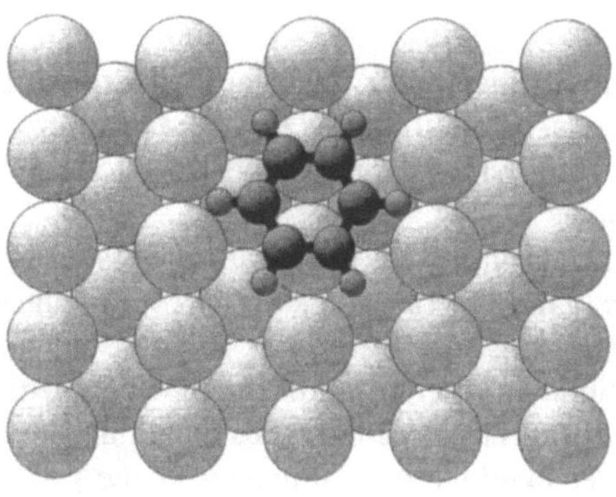

Fig. 2.38. Schematic representation of a single C_6H_6 molecule interacting with a Ni(110) surface

of the substrate and thus the symmetry of the adsorption complex is C_{2v}. This is depicted in Fig. 2.38.

d) Ordered Overlayer: C_6H_6/Ni(110) [2.94]

The saturated layer exhibits a coverage of $\theta \sim 0.25$ ML and a $c(4 \times 2)$ LEED pattern. The angle-resolved photoelectron spectra are displayed in Fig. 2.39. The

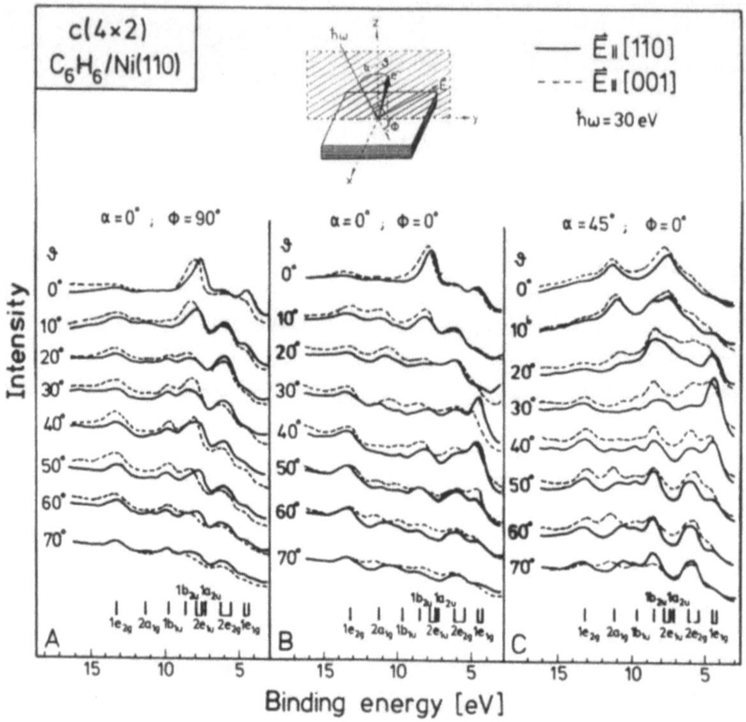

Fig. 2.39. Angle resolved electron distribution curves for a $c(4 \times 2)$ C_6H_6/Ni(110) system

various features in the spectra exhibit distinct polar angle and polarization dependences, but the differences are not as pronounced as for the dilute layer. Some ion states, in particular the $2a_{1g}$ orbital, show pronounced dispersions and we shall come back to this further below.

To determine the orientation of the C_6H_6 molecule in the saturated layer we resort again to the discussion of the $2a_{1g}$ and the $1a_{2u}$ ion states. From the fact that these states show significant normal emission intensity at $\alpha = 45°$, but no intensity at normal incidence, we conclude an orientation of the molecule with its molecular plane parallel to the Ni(110) surface. In contrast to the dilute layer the $2a_{1g}$ ion state of the saturated layer exhibits weak intensity in the [001] plane when the plane of light incidence is the [1$\bar{1}$0] plane. This is an indication for an azimuthal rotation of the benzene molecule with respect to the geometry in the dilute layer. Again the analysis of the $1b_{1u}$ and $1b_{2u}$ ion states provides further indications towards a structural determination. The $1b_{1u}$, as well as the $1b_{2u}$ ion state show emission intensities at all angles pointing towards an azimuthal rotation of the adsorbed molecules. A more detailed analysis of the spectra [2.94] finally leads to the conclusion that the symmetry of the adsorption complex is C_1, and that this symmetry is achieved by a molecule oriented parallel to the surface, azimuthally rotated with respect to the [1$\bar{1}$0] direction and at an adsorption site that is slightly asymmetric.

As mentioned above, the $2a_{1g}$ ion state exhibits significant dispersions. The band structure has been determined along the two high symmetry directions of the substrate Brillouin zone and is plotted in Fig. 2.40. Data have been obtained using various photon energies. The fact that these data coincide shows that the dispersion is caused by an electronic state with periodicity in two dimensions. The magnitude of the dispersion is 0.8 eV and thus considerably larger compared with, e.g. the $(\sqrt{7} \times \sqrt{7})R19.1°$ structure on the close packed surfaces. We shall come back to this point later. Real-space models for the approximate $c(4 \times 2)$ layer are depicted in Fig. 2.41. Figure 2.41a shows the situation if the molecules were not rotated with respect to the geometry in the dilute layer. Obviously, the van der Waals areas overlap strongly indicative of strong steric

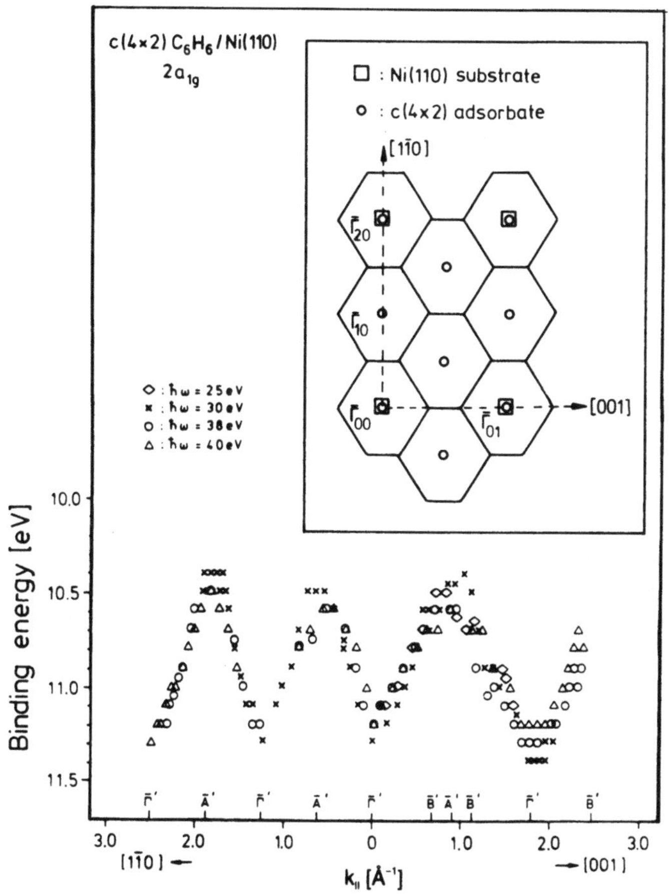

Fig. 2.40. Measured dispersion for a $c(4 \times 2)$ C_6H_6/Ni(110) system along two direction in k space as indicated in the inset

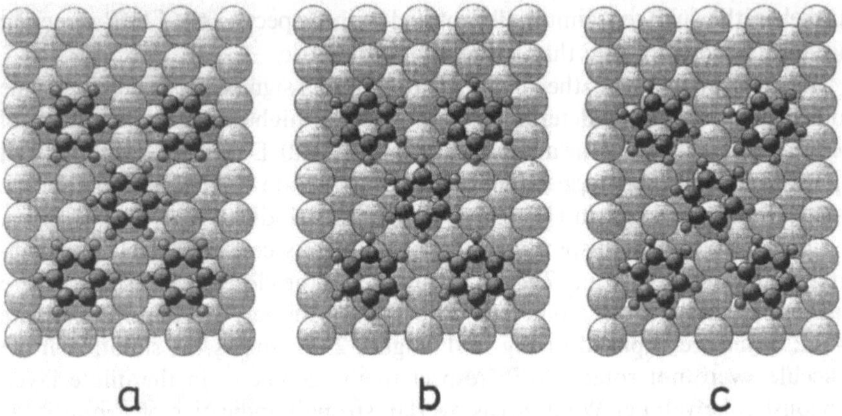

Fig. 2.41a–c. Schematic representations of the relative arrangements in a $c(4\times2)$ C_6H_6/Ni(110) layer **a**) assuming the same site as in the dilute layer **b**) rotation by 30° with respect to the orientation in the dilute layer **c**) arrangement intermediate between a) and b)

repulsion. While the geometry in Fig. 2.41b is sterically favourable, the molecules wants to assume the local geometry as in Fig. 2.41a. The arrangement in Fig. 2.41c is therefore a compromize with reduced steric repulsion and a geometry reasonably close to the local geometry. Such an arrangement has also been favored via a force field calculation by *Fox* and *Rösch* [2.126].

As mentioned above, the dispersion in the present system is by a factor of two larger compared to C_6H_6 on Ni(110) [2.112] and Os(0001) [2.108]. This must result from the significantly closer packing of the molecules. For C_6H_6 on Ni(110) the coverage of $\theta = 0.25$ ML correspond to a unit cell area of 35.2 Å^2 with a nearest neighbour distance of 6.11 Å. This should be compared with a unit cell area of 37.6 Å^2 for the $(\sqrt{7}\times\sqrt{7})R19.1°$ structure and a coverage of $\theta = 0.143$ ML for Ni(111). The 6–7% denser packing in the layer obviously has a strong nonlinear response. This is due to the fact that the intermolecular overlaps scale exponentially with distance and, in addition, the local arrangement is such that the repulsive interaction between the hydrogen atoms is rather large. This effect has been discussed above for the $(\sqrt{7}\times\sqrt{7})R19.1°$ structure. Clearly, the interaction energy of a single C_6H_6 molecule with the metal surface must be rather strong to tolerate the considerable intermolecular repulsions in the layer which in turn leads to the pronounced band structure effects in this system.

2.5 Co-Adsorbates

Some studies [2.16, 42, 44, 45, 127] on co-adsorbate systems have been published very recently. In the following we briefly review what is known about band dispersions in ordered co-adsorbates.

The first molecular co-adsorbate systems that have been studied with ARUPS with respect to level dispersions were ordered K/CO overlayers [2.16, 42, 45, 127]. The bandwidth found for the 4σ-level of K/CO/Ru(0001) is included in Fig. 2.9 as the filled triangle. Unfortunately, the structure model for the co-adsorbate is not unique [2.124]. The CO–CO distance used in the present case is based on the assumption that a hexagonal CO overlayer is co-adsorbed with a hexagonal K overlayer, which leads to the observed (3×3) overlayer structure [2.127]. There are other structures possible, which would give shorter CO–CO distances, but the result would always lead to a relatively large bandwidth, as compared with pure CO overlayers. Obviously, the co-adsorption of K cause the 4σ wave function to change considerably, in the sense that the CO–CO interaction is mediated via the co-adsorbed potassium.

An ordered CO/O co-adsorbate on Pd(111) has been studied using ARUPS [2.42]. Figure 2.42 presents a schematic picture of the structure proposed to explain the observed (2×1) LEED pattern. Early angle-integrated photoemission results [2.127] were basically reproduced. The co-adsorption of oxygen shifts the positions of the 4σ and $5\sigma/1\pi$ bands to higher binding energies. This has been taken as evidence for a strong oxygen CO-interaction. The 4σ dispersion is, however, basically caused by the strong lateral interaction along the CO rows. The ARUPS study shows that the 4σ dispersion, as presented in Fig. 2.9, is in line with those of pure CO adsorbates. Therefore, if there is any distortion of the wave function then it is smaller than in the case of K/CO adsorbates. Further comparison with other pure CO adsorbates on Pd(111) revealed that the observed chemical shift of the CO peaks can be explained exclusively via CO–CO interaction. Therefore, the reason for the high tendency of the CO + O/Pd(111) system to form CO_2 well below room temperature [2.128] must be due to CO–CO and O–O repulsive interactions rather than strong attractive CO–O interactions within the adsorbate.

Fig. 2.42. Schematic representation of the structure of the (2×1)CO + O/Pd(111) co-adsorption system

2.6 Synopsis

Intermolecular interaction in adsorbates on solid surfaces can be probed via angle-resolved photoemission through the determination of band structures. Theoretical modelling allows one to extract specific parameters from the experimental observations. Some aspects are:

a) direct intermolecular interaction scales logarithmically with intermolecular distance.
b) the short-range intermolecular interaction determines the periodicity of the observed band structure.

These aspects are discussed on the basis of several examples of increasing complexity starting with CO and going via C_2H_4 to C_6H_6.

While our knowledge on pure adsorbates is growing there is very little known on co-adsorbates. We anticipate that this is a field of research of some importance in the near future.

Acknowledgements. We would like to thank many colleagues for their input over a long time. In particular, we thank Manfred Neumann for a long-standing collaboration.

Financial support from various agencies, the 'Deutsche Forschungsgemeinschaft', the 'Bundesministerium für Forschung und Technologie', the 'Ministerium für Wissenschaft und Forschung des Landes Nordrhein-Westfalen', and the 'Fonds der chemischen Industrie' is gratefully acknowledged.

We would like to thank Jürgen Noffke for providing us with the results of a band-structure calculation and Marcus Donath for the calculation of the projected band structure of Ni(110). Heiko Hamann is gratefully acknowledged for a critical reading of the manuscript.

References

2.1 I. Langmuir: J. Am. Chem. Soc. **40**, 1361 (1918)
2.2 G.A. Somorjai, M.A. Van Hove: Adsorbed monolayers on solid surfaces. *Structure and Bonding*, Vol. 38, (Springer, Berlin, Heidelberg 1979)
2.3 J.C. Tracy, P.W. Palmberg: J. Chem. Phys. **51**, 4852 (1969)
2.4 J. Koutecky: Trans. Faraday Soc. **54**, 1038 (1958)
2.5 a) T.B. Grimley: Adv. Catal. **12**, 1 (1960)
b) T.B. Grimley: Ber. Bunsenges. Phys. Chem. **75**, 1003 (1971)
2.6 T.L. Einstein, J.R. Schrieffer: Phys. Rev. **B7**, 3629 (1973)
2.7 J.C. Tracy: J. Chem. Phys. **56**, 2736 (1972)
2.8 B.N.J. Persson: Surf. Sci. **258**, 451 (1991)
2.9 J.M. Szeftel, S. Lehwald, H. Ibach, T.S. Rahman, J.E. Black, D.L. Mills: Phys. Rev. Lett. **51**, 268 (1983)
J.P. Toennies: Experimental determination of surface phonons by helium atoms and electronic energy loss spectroscopy, in I, *Surface Phonons*, ed. by Wikress, G.W. de Wette, Springer Ser. Surf. Sci., Vol. 21 (Springer, Berlin, Heidelberg 1991) Chap. 5
2.10 J. Fink: "*transmission electron energy loss spectroscopy*", in *Unoccupied Electronic States*, ed. by J.C. Fuggle, J.E. Inglesfield, Topics Appl. Phys., Vol. 69 (Springer, Berlin, Heidelberg 1992), Chap. 7 and references therein

2.11 S.D. Kevan (ed.) *Angle Resolved Photoemission: Theory and Current Applications*, (Elsevier, Amsterdam 1992) and references therein
2.12 K. Horn, M. Scheffler, A.M. Bradshaw: Phys. Rev. Lett. **41**, 822 (1978)
2.13 H. Jones: *The Theory of Brillouin Zones and Electronic States in Crystals* (North-Holland, Amsterdam 1975)
2.14 H.-J. Freund, M. Neumann: Appl. Phys. **A47**, 3 (1988)
2.15 F. Greuter, D. Heskett, E.W. Plummer, H.-J. Freund: Phys. Rev. **B27**, 7117 (1983)
2.16 D. Heskett, E.W. Plummer, R.A. de Paola, W. Eberhardt, F.M. Hoffmann: Surf. Sci. **164**, 490 (1985)
2.17 N.D. Shinn: J. Vac. Sci. Technol. **A4**, 1351 (1986)
N.D. Shinn, T.E. Madey: Phys. Rev. Lett. **53**, 2481 (1984)
N.D. Shinn, T.E. Madey: J. Chem. Phys. **83**, 5928 (1985)
2.18 R.S. Saiki, G.S. Herman, M. Yamada, J. Osterwalder, C.S. Fadley: Phys. Rev. Lett. **63**, 283 (1989)
D.W. Moon, S.L. Bernasek, D.J. Dwyer: Surf. Sci. **163**, 215 (1985)
2.19 F. Zaera, E. Kollin, J.L. Gland: Chem. Phys. Lett. **121**, 464 (1985)
2.20 E.W. Plummer, W. Eberhardt: Adv. Chem. Phys. **49**, 533 (1982)
2.21 G. Odörfer: CO_2-Adsorption auf Palladium (111). Diplomarbeit, Universität Erlangen-Nürnberg (1987)
2.22 C.L. Allyn, T. Gustafsson, E.W. Plummer: Chem. Phys. Lett. **47**, 127 (1977)
2.23 J. Hermanson: Solid State Commun. **22**, 9 (1977)
2.24 K. Jacobi, M. Scheffler, K. Kambe, F. Forstmann: Sol. State Commun. **22**, 17 (1977)
M. Scheffler, K. Kambe, F. Forstmann, Solid State Commun. **25**, 93 (1978)
2.25 L.S. Cederbaum, W. Domcke: Adv. Chem. Phys. **36**, 205 (1977)
2.26 G. Borstel, M. Neumann, M. Wöhlecke: Phys. Rev. **B23**, 3121 (1981)
2.27 R.J. Smith, J. Anderson, G.J. Lapeyre: Phys. Rev. Lett. **37**, 1081 (1976)
2.28 G.J. Lapeyre, J. Anderson, R.J. Smith: Surf. Sci. **89**, 304 (1979)
2.29 G. Blyholder: J. Phys. Chem. **68**, 2772 (1964)
2.30 G. Blyholder: J. Vac. Sci. Technol. **11**, 865 (1974).
2.31 J.W. Davenport: Theory of photoemission from molecules in the gas phase and solid surfaces. PhD Thesis, University of Pennsylvania (1976), and Phys. Rev. Lett. **36**, 945 (1976)
2.32 E.W. Plummer, T. Gustafsson, W. Gudat, D.E. Eastman: Phys. Rev. **A15**, 2339 (1977)
2.33 T. Gustafsson: Surf. Sci. **94**, 593 (1980)
2.34 I.P. Batra, K. Hermann, A.M. Bradshaw, K. Horn: Phys. Rev. **B20**, 801 (1979)
2.35 H.-J. Freund, W. Eberhardt, D. Heskett, E.W. Plummer: Phys. Rev. Lett. **50**, 768 (1983)
2.36 K. Horn, A. M. Bradshaw, K. Jacobi: Surf. Sci. **72**, 719 (1978)
2.37 H. Kuhlenbeck, M. Neumann, H.-J. Freund: Surf. Sci. **173**, 194 (1986)
2.38 R. Miranda, K. Wandelt, D. Rieger, R.D. Schnell: Surf. Sci. **139**, 430 (1984)
2.39 K. Horn, A.M. Bradshaw, K. Hermann, I.P. Batra: Solid State Commun. **31**, 257 (1979)
2.40 P. Hofmann, J. Gossler, A. Zartner, M. Glanz, D. Menzel: Surf. Sci. **161**, 303 (1985)
2.41 D. Rieger, R.D. Schnell, W. Steinmann: Surf. Sci. **143**, 157 (1984)
P. Hofmann, S.R. Bare, D.A. King: Surf. Sci. **117**, 245 (1982)
2.42 G. Odörfer, E.W. Plummer, H.-J. Freund, H. Kuhlenbeck, M. Neumann: Surf. Sci. **198**, 331 (1988)
2.43 C.W. Seabury, E.S. Jensen, T.N. Rhodin: Solid State Commun. **37**, 383 (1981)
2.44 D. Heskett: Private commun. (1986)
2.45 C. Schneider, H.-P. Steinrück, T. Pache, P.A. Hemann, D.J. Coulman, E. Umbach, D. Menzel: Vacuum **41**, 730 (1990)
2.46 H. Kuhlenbeck, H.B. Saalfeld, U. Buskotte, M. Neumann, H.-J. Freund, E.W. Plummer: Phys. Rev. **B39**, 3475 (1989)
2.47 N. Memmel, G. Rangelov, E. Bertel, V. Dose, K. Kometer, N. Rösch: Phys. Rev. Lett. **63**, 1884 (1989)
2.48 G. Rangelov, N. Memmel, E. Bertel, V. Dose: Surf. Sci. **251/252**, 965 (1991)
2.49 D.A. Wesner, F.P. Coenen, H.P. Bonzel: Phys. Rev. Lett. **60**, 1045 (1988)

2.50 W. Riedl, D. Menzel: Surf. Sci. **163**, 39 (1985)
2.51 R.J. Behm, G. Ertl, V. Penka: Surf. Sci. **160**, 387 (1985)
2.52 S. Haq, J.G. Love, D.A. King: Surf. Sci. **275**, 170 (1992)
2.53 D.J. Hannamann, M.A. Passler: Surf. Sci. **203**, 449 (1988)
2.54 N. Pangher, J. Haase: Verhandlg. Dt. Phys. Ges. **17A**, SF37.6 (1993)
2.55 F. Hund: Z. Phys. **99**, (1936)
2.56 a) D.B. Litvin: Thin Solid Films **106**, 203 (1983)
b) D.B. Litvin: J. Phys. C **17**, L37 (1984)
2.57 H. Kuhlenbeck: Charakterisiermig des elektronischen Eigenschaften des Adsorbatsystems Ni(110)/CO(2 × 1)p2mg mittels Winkelaufgelöster Photoelektronenspektroskopie. Dissertation, Universität Osnabrück (1988)
2.58 B. Voigtländer, D. Bruchmann, S. Lehwald, H. Ibach: Surf. Sci. **225**, 1521 (1990)
2.59 M. Donath: Private commun. (1992)
2.60 J. Noffke: Private commun. (1992)
2.61 H.-J. Freund, W. Eberhardt, D. Heskett, E.W. Plummer: Ann. Isr. Phys. Soc. **6**, 185 (1983)
2.62 D. Schmeisser, F. Greuter, E.W. Plummer, H.-J. Freund: Phys. Rev. Lett. **54**, 2095 (1985)
2.63 W. Hansen, M. Bertolo, K. Jacobi: Surf. Sci. **253**, 1 (1991)
2.64 W. Eberhardt, H.-J. Freund: J. Chem. Phys. **78**, 700 (1983)
2.65 R.C. Weast: *CRC Handbook of Chemistry and Physics*, 54th ed., (The Chemical Rubber Company, Cleveland 1973)
2.66 G. McElhiney, H. Papp, J. Ritchard: Surf. Sci. **54**, 617 (1976)
2.67 J. Kessler, F. Thieme: Surf. Sci. **67**, 405 (1977)
2.68 K. Christmann, O. Schober, G. Ertl: J. Chem. Phys. **60**, 4719 (1974)
2.69 H. Conrad, G. Ertl, J. Koch, E.E. Latta: Surf. Sci. **43**, 462 (1974)
2.70 R.D. Diehl, S.C. Fain: Surf. Sci. **125**, 116 (1983)
2.71 M. Seel, H.-J. Freund: Solid State Commun. **55**, 895 (1985)
2.72 M. Gumbiowski, V. Staemmler: Private commun. (1991)
2.73 K. Hermann, P.S. Bagus, C.R. Brundle, D. Menzel: Phys. Rev. **B24**, 7025 (1981)
2.74 R.P. Messmer, S.H. Lamson: Chem. Phys. Lett. **65**, 465 (1979)
2.75 H.-J. Freund, E.W. Plummer: Phys. Rev. **B23**, 4859 (1981)
2.76 K. Schönhammer, O. Gunnarsson: Solid State Commun. **26**, 399 (1978)
2.77 D. Saddei, H.-J. Freund, G. Hohlneicher: Surf. Sci. **95**, 257 (1980)
2.78 E. Umbach: Surf. Sci. **117**, 482 (1982)
2.79 N. Mårtensson, A. Nilsson: J. Electron Spectrosc. Relat. Phenom. **52**, 1 (1990)
2.80 G. Wendin: *Structure and Bonding* Vol. 45, (Springer, Berlin, Heidelberg 1981)
2.81 C. Mariani, H.-U. Middelman, U. Iwan, K. Horn: Chem. Phys. Lett. **93**, 308 (1982)
2.82 H.P. Bonzel: Private commun. (1980)
A. Sandell: Autoionization studies of adsorbates, Thesis, Uppsala University (1993)
2.83 L. Hedin: Phys. Scr. **21**, 477 (1980)
2.84 B.I. Lundqvist: Phys. kondens. Materie **6**, 193 (1967)
B.I. Lundqvist: Phys. kondens. Materie **7**, 117 (1968)
B.I. Lundqvist: Phys. kondens. Materie **9**, 236 (1969)
2.85 N. Sheppart: Ann. Rev. Phys. Chem. 39, 589 (1988) and references therein
2.86 M. Weinelt, W. Huber, P. Zebisch, H.-P. Steinrück, M. Pabst, N. Rösch: Surf. Sci. **271**, 539 (1992)
2.87 M. Weinelt, W. Huber, P. Zebisch, H.-P. Steinrück B. Reichert, U. Birkenheuer, N. Rösch: Phys. Rev. **B46**, 1675 (1992)
2.88 J.A. Stroscio, S.R. Bare, W. Ho, Surf. Sci. **148**, 499 (1984)
2.89 K. Kimura, S. Katsumata, Y. Achiba, T. Yamakaki, S. Iwata: *Handbook of HeI Photoelectron Spectra of Fundamental Organic Molecules*, (Halsted New York 1981)
2.90 W.L. Jørgensen, L. Salem: *Orbitale Organischer Moleküle*, (Verlag Chemie, Weinheim 1974)
2.91 H.A. Engelhardt, W. Bäck, D. Menzel, H. Liebl: Rev. Sci. Instr. **52**, 835 (1981)
H.A. Engelhardt, A. Zartner, D. Menzel: Rev. Sci. Instr. **52**, 1161 (1981)
2.92 H.-P. Steinrück, W. Huber, T. Pache, D. Menzel: Surf. Sci. **218**, 293 (1989)

2.93 J.C. Bertolini, G. Dalmei-Imelik, J. Rousseau: Surf. Sci. **67**, 478 (1977)
2.94 W. Huber, M. Weinelt, P. Zebisch, H.-P. Steinrück: Surf. Sci. **253**, 72 (1991)
2.95 M.G. Ramsey, D. Steinmüller, F.P. Netzer, T. Schedel, A. Santaniello, D.R. Lloyd: Surf. Sci. **251/252**, 979 (1991)
2.96 P. Jakob, D. Menzel: Surf. Sci. **201**, 503 (1988)
2.97 E. Bertel, G. Rosina, F.P. Netzer: Surf. Sci. **172**, L515 (1986)
2.98 S. Chiang, R.J. Wilson, C.M. Mate, H. Ohtani: J. Microsc. **152**, 567 (1988)
2.99 K. Markert, K. Wandelt: Surf. Sci. **159**, 24 (1985)
2.100 a) C.M. Mate, B.E. Bent, G.A. Somorjai: J. Electron Spectrosc. Relat. Phenom. **39**, 205 (1986)
b) C.M. Mate, C.-T. Kao, G.A. Somorjai: Surf. Sci. **206**, 145 (1988)
2.101 M. Neuber, S Witzel, C. Zubrägel, H.H. Graen, M. Neumann: Surf. Sci. **251/252**, 911 (1991)
2.102 M. Neuber: Charakterisiering des reinen und des mit Kalium Versetzten Adsorptions-Systems Benzol/Rhodium (111), Dissertation, Universität Osnabrück (1992)
2.103 B.E. Koel, J.E. Crowell, C.M. Mate, G.A. Somorjai: J. Phys. Chem. **88**, 1988 (1984)
2.104 S. Uhlenbrock: Private commun. in [2.101]
2.105 F.P. Netzer: Vacuum **41**, 49 (1990)
2.106 P. Hofmann, K. Horn, A.M. Bradshaw: Surf. Sci. **105**, L260 (1981)
2.107 R. Dudde, K.H. Frank, E.E. Koch: Surf. Sci. **225**, 267 (1990)
2.108 H.H. Graen, M. Neuber, M. Neumann, G. Odörfer, H.-J. Freund: Europhys. Lett. **12**, 173 (1990)
2.109 M. Neuber: Private commun. (1993)
2.110 J.U. Mack, E. Bertel, F.P. Netzer: Surf. Sci. **159**, 265 (1985)
2.111 B.E. Nieuwenhuys, D.I. Hagen, G. Rovida, G.A. Somorjai: Surf. Sci. **59**, 155 (1976)
2.112 N.V. Richardsson, N.R. Palmer: Surf. Sci. **114**, L1 (1982)
2.113 W. Huber, P. Zebisch, T. Bornemann, H.-P. Steinrück: Surf. Sci. **258**, 16 (1991)
2.114 D.W. Turner, C. Baker, A.D. Baker, C.R. Brundle: *Molecular Photoelectron Spectroscopy*, (Wiley, London 1970)
2.115 F.A. Cotton, G. Wilkinson: *Advanced Inorganic Chemistry*, 4th ed., (Wiley, New york 1980)
2.116 H.H. Graen: Adsorption von Benzol auf Übergangsmetall oberflächen, Dissertation, Universität Osnabrück (1991)
H.H. Graen, M. Neumann: Private commun. (1991)
2.117 F.P. Netzer, G. Rangelov, G. Rosina, H.B. Saalfeld, M. Neumann, D.R. Lloyd: Phys. Rev. **B37**, 10399 (1988)
2.118 M. Surmann, S.R. Bare, P. Hofmann, D.A. King: Surf. Sci. **179**, 243 (1987)
2.119 M. Bader, J. Haase, K.-H. Frank, C. Ocal, A. Puschmann: J. Physique Coll. **C8**, 491 (1986)
2.120 S.R. Kelemen, T.E. Fischer: Surf. Sci. **102**, 45 (1981)
2.121 A.C. Liu, J. Stöhr, C.M. Friend, R.J. Madix: Surf. Sci. **235**, 107 (1990)
2.122 G. Herzberg: *Molecular Spectra and Molecular Structure, III. Electronic Spectra and Electronic Structure of Polyatomic Molecules*, (Van Nostrand Reinhold, New York 1966)
2.123 H. Köppel, L.S. Cederbaum, W. Domeck: J. Chem. Phys. **89**, 2023 (1988)
2.124 J. Eiding, W. Domcke, W. Huber, H.-P. Steinrück: Chem. Phys. Lett. **180**, 133 (1991)
2.125 M. Neumann, J.U. Mack, E. Bertel, F.P. Netzer: Surf. Sci. **155**, 629 (1985)
2.126 T. Fox, N. Rösch: Surf. Sci. **256**, 159 (1991)
2.127 J.J. Weimer, E. Umbach, D. Menzel: Surf. Sci. **159**, 83 (1985)
2.128 H. Conrad, G. Ertl, J. Küppers: Surf. Sci. **76**, 323 (1978)

3. High-Resolution Core-Level Photoelectron Spectroscopy of Surfaces and Adsorbates

N. Mårtensson and *A. Nilsson*

In the present chapter we will discuss some general aspects of core-level photoelectron spectroscopy for the investigation of surfaces and adsorbates. We will focus on a few aspects of this spectroscopy. In Sect. 3.2 some experimental comments will be given of a general nature and referring to the systems which have been used in most of the studies reviewed in this contribution. Sections 3.3–5 deal with core-level shifts, Section 3.3 gives a general introduction and discusses the role of initial and final state effects, Section 3.4 treats shifts of adsorbate lines, and Sect. 3.5 deals with surface core-level shifts for metals. Section 3.6–8 discuss core-level line shapes and core-level satellites. In Sect. 3.6 the most important contributions to the line shape are reviewed. Section 3.7 discusses vibrational broadening, and Sect. 3.8 then deals with shake-up satellites for adsorbates.

3.1 Background

Photoelectron spectroscopy can be used to investigate a variety of properties of surfaces, and adsorbed atoms and molecules [3.1–9]. Photoelectron spectroscopy provides direct information on the electronic states in a system. This chapter will focus on the investigation of core electrons, i.e. the electrons which are localized on a particular atomic site and which do not take part in chemical bonding.

There are a number of core-level spectroscopies which are used for the investigation of solids and surfaces. The most important of these are schematically reproduced in Fig. 3.1. The figure shows the valence and core levels for one of the atoms in a system. In core-level photoelectron spectroscopy (Fig. 3.1a), a photon (with energy $h\nu$) is absorbed by the system which emits a core electron. Its kinetic energy E_{kin} is given by

$$E_{\mathrm{kin}} = h\nu - E_{\mathrm{B}} - \phi, \tag{3.1}$$

where ϕ is the sample work function (this requires that the kinetic energy just outside the sample is considered. If E_{kin} denotes the kinetic energy in the spectrometer, ϕ is instead the work function of the spectrometer). E_{B} denotes the binding energy of the core electron relative to the Fermi level. In order to excite

Springer Series in Surface Sciences, Vol. 35
Applications of Synchrotron Radiation Ed.: W. Eberhardt

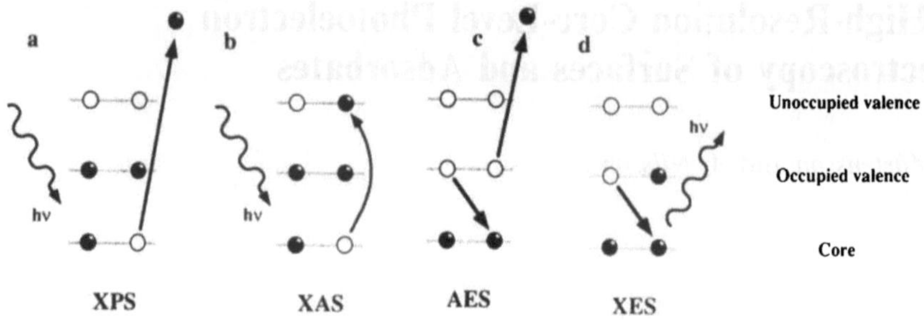

Fig. 3.1a-d. A schematic illustration of core level ionization, excitation and decay processes

a core electron, X-ray photons have to be used. The technique is referred to as XPS (X-ray Photoelectron Spectroscopy) or ESCA (Electron Spectroscopy for Chemical Analysis).

Figure 3.1b represents an X-ray absorption process where a core electron is excited to an initially empty valence orbital. The measurements are made by recording the absorption as a function of photon energy at and above the core-level threshold. This requires a tuneable photon source and is usually performed using synchrotron radiation. X-ray Absorption Spectroscopy (XAS) provides a method to study the empty valence-electron states of a system [3.10, 11]. When this method is used for surfaces it is often referred to as NEXAFS (Near Edge X-ray Absorption Fine Structure) [3.10].

Figures 3.1c and 3.1d depict the two types of processes by which a core ionized atom decays. For light elements and shallow core levels the Auger process dominates (Fig. 3.1c). The core hole is filled by an electron from one of the outer shells and the excess energy is taken up by an electron which is emitted from the system. For heavier elements and for deeper core levels the decay is instead dominated by X-ray emission processes (see Fig. 3.1d) in which the excess energy is emitted in the form of a photon. For all core levels both types of processes contribute. The added rates of these decay channels determine the life time of the core hole state and thereby the intrinsic width of the core electron line. The analysis of the emitted electrons or photons is the basis for Auger Electron Spectroscopy (AES) [3.12, 13] or X-ray Emission Spectroscopy (XES) [3.14], respectively. X-ray emission spectroscopy is traditionally a bulk technique but it has recently been shown that it can also be used to investigate the electronic structure of adsorbates [3.15]. The neutral core-excited XAS final states will decay by corresponding processes. The analogue of the Auger process is then denoted autoionization (Chap. 6).

The present chapter will almost entirely deal with photoelectron spectroscopy, although the relationship to the other spectroscopies will be mentioned in some cases. Each element in the periodic table has a unique pattern of core-level binding energies. This makes all core-level spectroscopies sensitive to what elements are present in a sample. A particularly useful property of XPS is that the detailed core-level positions depend on the chemical state of the probed atoms (chemical shifts). The chemical shifts can be used as fingerprints of different adsorbate species, such as for instance to monitor the dissociation of a molecular adsorbate. By analyzing the shifts within the framework of various theoretical models, it is possible to extract basic information on the electronic structure of the adsorbate complex as well as information of a thermochemical nature. This often requires rather elaborate theoretical considerations which, for instance, take into account the response of the complete system to the creation of the core hole (screening or relaxation). By combining photoelectron spectroscopy with other core-level spectroscopies, additional information can be obtained on matters like the dynamics of core-hole screening, vibrational interference, etc.

One of the important properties of core-level spectroscopies is that the core hole is localized at one atom. In this way all core-level spectroscopies probe local aspects of the sample. For instance, in the XAS and XES processes the obtained valence-electron information corresponds to a projection of the valence-electron structure on the core hole site. For an adsorbate, this implies that by recording spectra involving the adsorbate core levels, the adsorbate valence states are probed without overlapping contributions from the substrate. This distinguishes core-level spectroscopies from methods like valence band photoemission. In core-level photoemission the emitted electrons also undergo diffraction processes. The core ionized atom then acts as a point source of electrons for the scattering which makes the technique sensitive to the local structure. This is different from a regular electron scattering experiment like LEED (Low Energy Electron Diffraction) where the diffraction of an incoming electron wave is considered. XPD (X-ray Photoelectron Diffraction) is now an established structural technique [3.16]. It will be discussed further in Chap. 4.

Many of the basic properties of photoelectron spectroscopy can be visualized using the spectra in Fig. 3.2. This figure displays a series of Yb $4f$ photoelectron spectra from Yb layers of different thicknesses deposited on a Mo(110) substrate [3.17]. Yb grows in a layer-by-layer fashion and spectra are shown for 1, 2 and 3 MonoLayer (ML) films and for a thick epitaxial Yb layer. The $4f$ photoemission leads to a spin-orbit split $4f_{7/2}/4f_{5/2}$ doublet with an energy separation of 1.27 eV. The intensity ratio is 4/3, which is determined by the degeneracies of the two levels involved. Although the $4f$ level has a very low binding energy it can be treated as a core level. This is due to the small spatial extension of the $4f$ orbitals which implies that it does not take part in the bonding.

When one ML of Yb is deposited the spectrum shows one $4f$ doublet. For the 2 ML film two chemically shifted Yb $4f$ doublets are seen. The shift between these is due to the fact that the atoms in the two layers have different chemical surroundings; the top-most layer is an Yb surface layer with a layer of Yb atoms underneath, i.e., the atoms have the same distribution of nearest neighbours as a surface atom on a thick Yb sample. The atoms in the interface layer are surrounded by both Yb and Mo atoms.

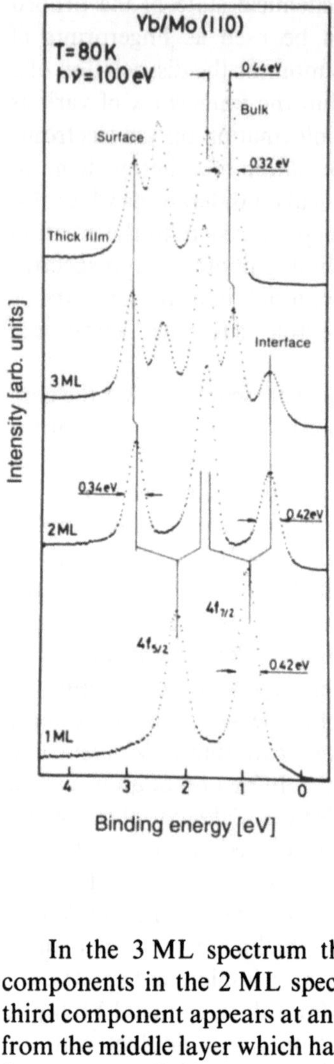

Fig. 3.2. Photoelectron spectra, showing the Yb 4*f*-region for different coverages of Yb on Mo(110). At a photon energy of 100 eV, the Yb 4*f* emission dominates totally over the contribution from the Mo substrate

In the 3 ML spectrum three shifted Yb components are seen; the two components in the 2 ML spectrum remain at nearly the same energies and a third component appears at an intermediate energy. The latter feature originates from the middle layer which has a complete nearest neighbour shell of Yb atoms. In a nearest neighbour sense the middle layer already represents a bulk Yb layer. Further depositions only result in intensity changes of these three components [3.18] and no new spectral features appear. Eventually the film gets so thick that the interface peaks are completely attenuated and the upper spectrum of Fig. 3.2, characteristic of a thick Yb film, is obtained.

This example immediately shows that the shifts are dominated by the nearest neighbours. This is seen from the fact that the surface and interface peaks have the same energies for the 2 and 3 ML films and from the fact that the bulk Yb binding energy is almost attained by three monolayers of Yb. This is a most useful property of the chemical shift which makes it possible to draw direct conclusions about the local coordination in a system by measuring core-level spectra. This approach has been described in detail for alloy systems [3.19] and will be further discussed below for adsorbates. Even if the nearest neighbours dominate the chemical shifts, there are some second-order effects. In the Yb spectra above, the Mo substrate is seen to shift also the second Yb layer by 0.05 eV (7% of the first-order shift). There are, however, cases where the electronic structure may be very sensitive to chemical modifications, which will lead to chemical shift effects that range over a much larger number of coordination shells.

Figure 3.2 also demonstrates the surface sensitivity of the technique. The interface peak is seen in all the situations except the thick film but its intensity is reduced by approximately a factor of two for each additional Yb layer. Hence, for the electron transport through one atomic layer of Yb, about 50% of the electrons undergo inelastic scattering and are lost in the spectrum. This property of the spectra is also seen directly in the 3 ML spectrum for which an intensity ratio of about 1:2:4 is observed for the three layers. For the thick sample almost 50% of the total intensity originates from the top-most atomic layer.

The information depth depends on the kinetic energy of the photoelectrons. The mean-free path as a function of electron energy has a minimum at about 50 eV. The exact position of the minimum, however, depends on the material.

When a core electron is emitted this causes a perturbation of the whole system. This leads to so-called shake-up and shake-off processes [3.20, 1]. Figure 3.3 shows the photoelectron spectrum related to the ionization of the Ne $1s$ level for the free atom [3.21, 22]. The one-electron removal of the $1s$ electron corresponds to the main peak at 870.3 eV [3.23]. However, there are a number of additional spectral features at higher binding energies and a continuum is seen which extends some 100 eV above the main line. The Ne atom has a $1s^2 2s^2 2p^6$ configuration. When a core electron is photoionized this may lead to the excitation of a $2p$ electron to a higher np shell. A number of such $2p$-np shake-up peaks are seen in the spectrum. At still higher excitation energies relative to the main line, $2s$-ns shake-up features are seen. There are also structures corresponding to the simultaneous excitation of two valence electrons. The continuous intensity ranging more than 100 eV above the main line is due to shake-off processes in which a second electron is also ionized. The shake-up/shake-off spectra can be used to probe the valence electron structure of a system. What makes this part of the spectrum interesting is that the processes correspond to valence electron excitations between the occupied and unoccupied levels in much the same way as in an optical absorption spectrum. However, since the excitations are caused by the creation of the core hole, the valence states are probed specifically on the core hole site. In Sect. 3.8 shake-up phenomena for adsorbates are discussed in more detail.

Figure 3.4 exhibits the C1s spectrum recorded for CH_4 in the gas phase [3.24]. The core ionization gives rise to several discrete features due to the excitation of vibrational motion [3.25]. The peak at lowest binding energy corresponds to the $v = 0$ final state. The peaks at higher binding energies are due to the excitation of one or more vibrational quanta. The splitting between the peaks is about 0.4 eV which is the energy of the vibrational quantum for core

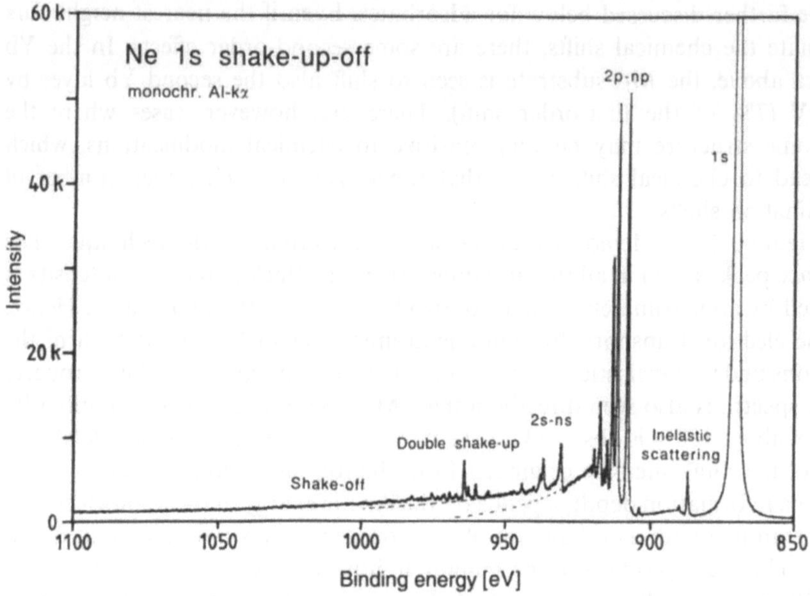

Fig. 3.3. The Ne1*s* shake-up spectrum. The shake-up spectrum consists of three groups of lines which correspond to different final state configurations. The lines at 35–55 eV shake-up energy originate from the $1s2s^22p^5np$ final state configurations. The $1s2s2p^6ns$ final state configurations appear at 55–80 eV. At very high shake-up energies, 80–130 eV, three hole two particle configurations (double shake-up lines) can be seen. At 46.7 eV above the main line the first shake-off continuum starts. The peaks denoted inelastic scattering are due to dipole excitations of other Ne atoms for the photoelectrons

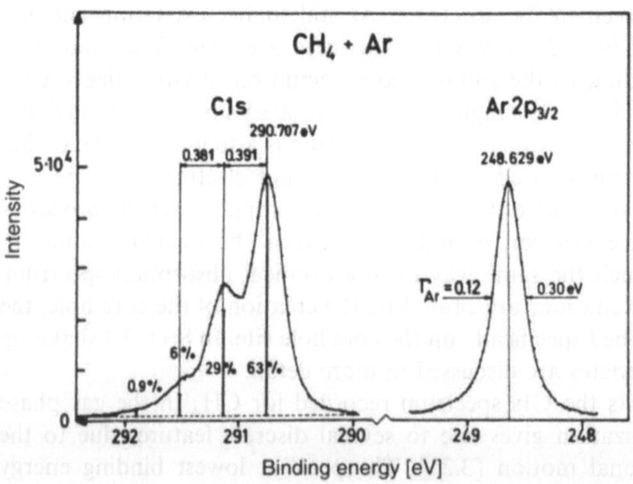

Fig. 3.4. Gas phase *C* 1*s* spectrum of methane with resolved vibrational levels. The bars in the figure are obtained by a numerical fit to the spectra

ionized CH_4. In an extended system a large number of vibrational modes may be excited and discrete features are no longer seen. Instead, a general broadening of the spectra is obtained. Vibrational excitations most often give the dominating contribution to the photoelectron line widths. This effect may also lead to temperature-dependent broadenings of the spectra [3.26, 27]. The role of vibrational broadening will be further discussed in Sect. 3.7.

3.2 Instrumentation

In this section we will briefly describe the instrumentation used for the high-resolution core-electron spectroscopy reviewed in the present chapter. The instruments have been constructed and built at the Department of Physics in Uppsala. The same type of equipment is now commercially available through Scienta Instrument AB in Uppsala.

The photon source generally used for the spectra in the present chapter is a monochromatized Al-K_α source. Some of the measurements have been performed with synchrotron radiation. In principle, there is no difference between a line source and synchrotron radiation for the measurement of core-level photoelectron spectra from adsorbates. In some cases it is important to have sufficiently high photon energy to ensure that one is close to the sudden limit (Sect. 3.8). For C 1s, N 1s, and O 1s excitations the laboratory source still produces core-level spectra of higher quality both regarding resolution and intensity. This is partly due to the fact that longer measuring times can be accepted in the home laboratory. This situation is now changed when the new third-generation synchrotron radiation sources come into operation. In the case of surface core-level shift measurements it is necessary to use tuneable synchrotron radiation in order to obtain the required surface sensitivity.

Figure 3.5 illustrates an outline of the 200 mm radius high-resolution electron energy analyzer [3.28]. The maximum resolution obtained for such an analyzer is 2.7 meV as determined in the gas phase for the Xe *5p* line using a helium lamp [3.28]. There have recently been large improvements in the energy resolution of gas-phase UV photoelectron spectra [3.29]. The analyzer in Fig. 3.5 has successfully been used at the MAX synchrotron radiation laboratory in Lund together with a soft X-ray grating monochromator [3.30]. Due to the high transmission even at high resolutions of the analyzer, spectra of very high quality have been obtained on a bending magnet beamline in spite of the rather low photon flux [3.30].

The high-resolution XPS system used for adsorbate core-level studies is depicted in Fig. 3.6 [3.31]. This is an ultra-high vacuum instrument similar to a gas-phase spectrometer described in detail in [3.32]. The X-ray source consists of a fine-focused high power electron gun, a water-cooled high-speed rotating anode and a wide-angle crystal monochromator mounted on a Rowland circle

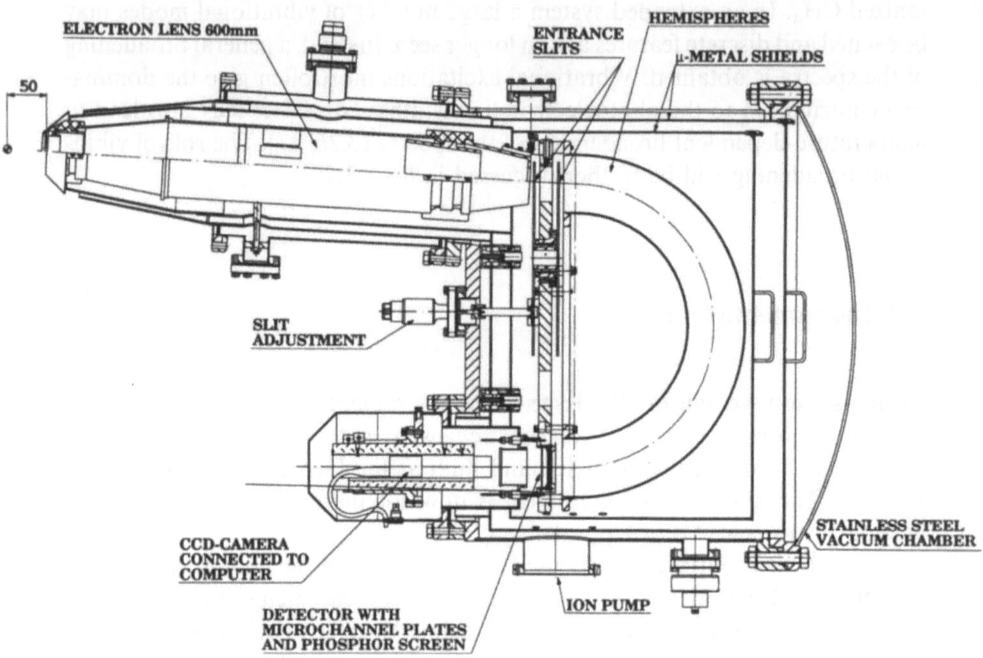

Fig. 3.5. Sectional view of the 200 mm hemispherical electron energy analyzer

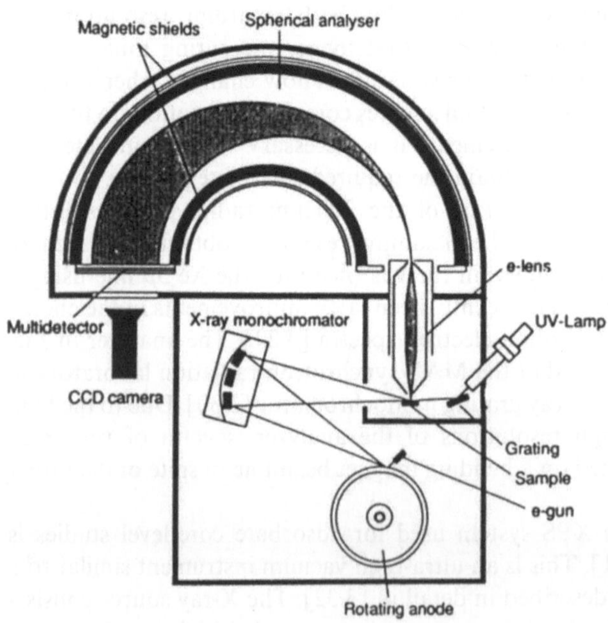

Fig. 3.6. Outline of the high resolution XPS system

(40 cm diameter) arrangement. The electron gun is a two-stage Pierce type gun which provides X-ray powers of 4 kW. The focused electron beam has a diameter of 1.5 mm. The disc of the rotating anode has a diameter of 30 cm and the outer perifery is water-cooled. The water flows in and out through the shaft of the anode. Rotational speeds of 5000 rpm were sufficient for the X-ray powers used. Both the water seal and the vacuum seal are contact-free. This is achieved by a series of narrow cylindrical slits with an engraved multithread system. The slits actively pump themselves and they also allow for efficient two-stage differential pumping. The X-ray monochromator consists of 25 spherically bent quartz single crystals, each of 36 mm diameter. The crystals are mounted in three rows on a large spherical concave surface formed in a large block of optical glass. The monochromator is kept at a constant temperature of 70°C.

The emitted photoelectrons enter through a four-element electrostatic lens into a large hemispherical electrostatic analyzer. These components are enclosed in a double μ-metal magnetic shield. The mean radius of the analyzer is 36 cm. The gap between the spherical electrodes is 15 cm which allows for a large area multidetection system. The detector consists of 2 channel plates in tandem and a phosphorous screen. The light pulses are registered by a CCD (Charge Coupled Device) camera and the readout system is interfaced to a computer.

An overall resolution of at least 0.25 eV can be obtained with the described system. However, the resolution is often reduced to 0.4 – 0.5 eV in order to increase the intensity from the weak adsorbate core levels.

3.3 Core-Level Shifts

The existence of chemical shifts is one of the most important aspects of core-level photoelectron spectroscopy. The chemical shifts can be used to distinguish between different chemical species in a sample. In many applications the shift scale is empirically established by studying different reference compounds. Often, however, the shifts have to be interpreted by a comparison to theory. The shifts can also be used to derive information on a more fundamental level about the electronic structure, the structure or the chemical stability of the system. This again requires that the shifts are treated within some theoretical framework.

This section will describe some general aspects of the theory of the chemical shift, especially relating to the question of initial and final states effects. The so-called equivalent core approximation is often useful in order to estimate the properties of the core ionized final state. This approximation will be described in the last part of this section.

3.3.1 Initial and Final State Effects

The ionization energy of a system can be expressed as the total energy difference between the ionized ($N-1$ electron) final state system and the initial (N

electron) system,

$$E_B = E_{tot}(N-1) - E_{tot}(N). \tag{3.2}$$

For the theoretical evaluation of an ionization energy various approaches can be used to derive the required total energies. When the total energies are calculated using the SCF (Self Consistent Field) method this is referred to as the ΔSCF approach. For the treatment of chemical shifts on this level, separate total energy calculations have to be performed for the initial and final states of the two systems which are to be compared.

There are also other ways to calculate the relevant total energy contributions to the terms in (3.2). Using a Born-Haber type of approach the ionization process is divided into a set of imaginary steps [3.33]. This is done in such a way that some of the energy terms are determined from independent experiments. In this way one can also isolate what type of energy terms are determining the shifts. Such decompositions will be discussed below for the case of adsorbate core levels and for the surface core-level shift in metals.

Using (3.2) we can express the chemical shift between two systems A and B as

$$\begin{aligned} \Delta E_B &= E_{tot,A}(N-1) - E_{tot,A}(N) - (E_{tot,B}(N-1) - E_{tot,B}(N)) \\ &= E_{tot,A}(N-1) - E_{tot,B}(N-1) - (E_{tot,A}(N) - E_{tot,B}(N)) \end{aligned} \tag{3.3}$$

The shift is thus determined by two total energy differences: one relating to the initial states and one relating to the final states. Depending on the system the chemical shifts may be dominated by either the initial or final state terms. When one compares the core ionization of two chemically inequivalent atoms in the same system the initial states are the same, and the binding energy shift is entirely due to the difference in total energy of the two final states. The chemical shift between the two chemically inequivalent nitrogen atoms for adsorbed N_2 is such a case [3.34, 35]. Similarly when one compares on top- and bridge-bonded CO on Ni(100) one knows that the two adsorption states have very similar total energies [3.36, 37]. In spite of this, large shifts are seen which again have to be attributed to the final-state total energy terms. In other situations, however, the shifts may be mainly determined by the initial-state terms.

Often chemical shifts are treated in a different way which leads to a completely different meaning of the terms initial- and final-state effects. Within the Hartee-Fock approximation the orbital energies have a direct interpretation in terms of ionization energies (Koopmans' theorem). This is obtained under the assumption that all the other orbitals are unaffected by the ionization. However, when a core electron is removed the remaining electrons respond to the suddenly created positive charge. This leads to a contraction towards the nucleus of all orbitals centered at the core-hole site. In a molecular or solid system there is, in addition, a flow of charge towards the core-hole site [3.38, 39]. In fact, in extended systems the ionization often leads to a final state in which the core hole is more or less completely screened by the valence electrons. The lowering of the core ionization energy due to these rearrangements is referred to as the relaxation energy. It is common to define the relaxation energy as the difference between the ionization energy calculated within Koopmans' approximation and the ionization energy calculated on the ΔSCF level. The ΔSCF level, however, is still only an approximation. In a molecular approach the difference between the experimental and the calculated SCF total energies is often referred to as the correlation energy. We will not discuss this term any further. Based on this separation of contributions it is common to separate chemical shifts into two parts, Koopmans' shifts and relaxation shifts (a third shift contribution should be due to differences in the correlation energy). These two parts are commonly referred to as initial- and final-state shifts, respectively. Note that this separation is completely different from the previous separation which was based on initial- and final-state total-energy terms. It is therefore important not to confuse these two approaches. It should also be noted that the initial-state shift defined as a shift on the Koopmans level is a theoretical construction.

There are several reasons why the latter approach has been quite common. Often it is simply assumed that the relaxation energy is the same for the ionization of different atoms in a system or that it is similar for core ionizations in similar systems. The experimental shifts are then compared directly to calculated Koopmans' shifts. This constancy of relaxation energies is quite often the case although it is most difficult to know when it really applies. Another reason for sticking to this type of approach originates from the earlier use of potential models for interpreting chemical shifts, i.e., the notion that the chemical shifts are due to the charge state of the ionized atom. The calculated orbital energies are sensitive to the charge state in much the same way as they are in these potential models. This approach therefore often leads to the interpretation of shifts in terms of atomic charges. The shifts are, in fact, often used as probes of charge states. However, this may lead to most erroneous conclusions. We will not discuss the relationship between these two approaches any further. In the following we will entirely discuss the shifts within a total energy framework.

3.3.2 Equivalent Core or $Z + 1$ Approximation

One way of getting total energy information relating to the core ionized states is to use the so-called *equivalent core approximation*. The radius of a core orbital is much smaller than that of a valence orbital. The core electrons are therefore located almost entirely inside the valence electrons. The effect on the valence electrons of a core ionization consequently will be nearly the same as if a unit charge were added to the nucleus. Many of the properties of a core ionized atom are therefore the same as the properties of the next element in the periodic table. This is the $Z + 1$ or equivalent core approximation. This approximation was used already long ago in the interpretation of optical spectra.

The accuracy of the $Z + 1$ approximation depends strongly on which properties of the core ionized system are considered. When calculating metallic screening energies it has been found that rather large corrections may have to be made [3.33]. All effects which are related to the open core shell are also missing. In those cases when multiplet effects are present, which involve the coupling to the core hole, the $Z + 1$ approximation gives information relating to multiplet averages. Usually when chemical shifts are considered the $Z + 1$ approximation is very accurate. In that case the total energy difference between the same core ionized atom in two different chemical surroundings is considered. In this way the largest errors in the $Z + 1$ approximation cancel out. The only terms which differ are the terms relating to the chemical interaction with the surrounding atoms and especially for this type of terms the $Z + 1$ approximation is quite accurate [3.33]. The validity for the equivalent core approximation for chemical effects is, for instance, well-known for the lanthanides. The properties of lanthanides are very similar over the whole series which means that even the $Z + 14$ approximation is quite accurate. The fact that the approximation is not perfect gives rise to the so-called *lanthanide contraction*.

For many systems information about the $Z + 1$ systems is available which can be used immediately for the interpretation of the shifts. In other cases it is possible to use general chemical know-how for the interpretation of the spectroscopic data. Furthermore, the accuracy of this approximation makes it possible to use the shifts to obtain total energy information on the Z and $Z + 1$ systems.

We will employ this approximation in several parts of the present chapter. We will use it for the interpretation of chemical shifts in Sects. 3.4 and 3.5. In the discussion of vibrational effects in Sect. 3.7 it will be utilized to obtain information about the final-state potential-energy surfaces. The $Z + 1$ approximation is also most helpful for the interpretation of shake-up spectra. The shake-up excitations correspond to electronic excitations in the core-hole state and this approximation can be used to obtain electronic structure information related to this state. In this case, however, one must consider the possibility that the core hole couples to the valence excitations.

3.4 Adsorbate Core-Level Shifts

In this section, we will show some examples of how adsorbate core-level shifts can be used to obtain information on the geometrical and electronic structure of small molecules adsorbed on surfaces. The shifts will be discussed in terms of differences in total energies between the initial and final states. In the first part of the section the two main screening mechanisms for adsorbates on metal surfaces are discussed. The remaining part is devoted to selected examples of the application of adsorbate shifts. These examples are the chemical shift between the two nitrogen atoms in N_2 which become inequivalent upon chemisorption, the shift between the different adsorption sites for CO, the site selectivity and molecular orientation sensitivity of the shifts for chemisorbed NO, and the layer dependent shifts for physisorbed O_2 on graphite.

3.4.1 Final State Effects

When an atom or molecule is adsorbed onto a surface, new final-state relaxation (screening) channels open up which are not present in the free atom or molecule. The surface induced relaxation can lower the final-state energy by several eV. There are two different types of substrate induced screening for an adsorbate: metallic screening and image potential screening.

Metallic screening involves a charge redistribution of the conduction electrons in the metallic system in order to achieve a locally neutral core hole site. The totally screened state can be looked upon as a state in which the screening electron has been taken from infinite distance. Since for a metallic system the appropriate reference level for the binding energy scale is the Fermi energy, we can consider the core ionization process as a core excitation to the Fermi level. The onset in X-ray Absorption Spectroscopy (XAS) corresponds to the creation of a final state where a core electron has been placed in the lowest unoccupied state, i.e. at the Fermi level. This means that the core ionized final state is more or less identical to the state obtained at the X-ray absorption threshold. There is only a negligible difference due to the fact that one electron is actually removed in the photoemission process which is not the case in XAS. This will change the Fermi level by an amount inversely proportional to the total number of valence electrons in the system. Due to this relationship the Fermi level position for the unoccupied valence states as probed by XAS can be derived from the corresponding photoemission binding energy.

Figure 3.7 indicates how the XPS and XAS spectra are related in two different cases of chemisorption. To the right, spectra are shown for a strongly chemisorbed atom, C on Ni(100). The adsorption energy is of the same order of magnitude as the cohesive energy of Ni metal. It is clearly seen how the XPS peak defines the threshold of the XAS spectrum. There is even a change in slope of the leading absorption edge at the peak position of the photoemission line. The same relationship holds also for molecular chemisorption. To the left, the C 1s spectra are shown for CO adsorbed on Ni(100). In this case the intensity is rather low at the X-ray absorption threshold since the main density of states in adsorbed CO is located well above the Fermi level. A similar relationship between XAS and XPS has been obtained in a number of weak chemisorption systems such as CO on Cu, Ag and Au surfaces, showing the existence of metallic screening for the main photoemission line. However, in these systems the main lines may have very low intensities, as discussed in Sect. 3.8.4.

The metallic screening of the adsorbate involves charge transfer from the substrate to the molecule. The degree of charge transfer does not have to correspond absolutely to one electron and the core ionized atom may in many cases be slightly ionic or polarized in the bond to the surface. Information on the electronic structure of the lowest core hole states can be obtained from the $Z + 1$

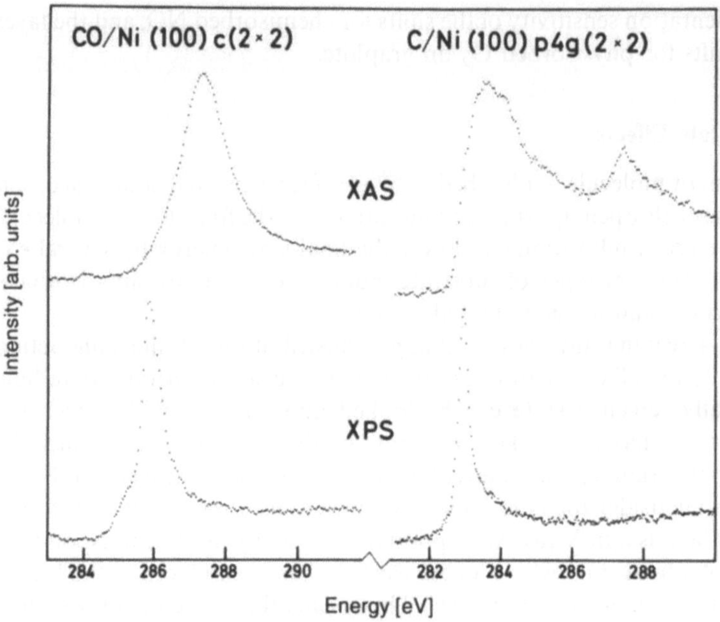

Fig. 3.7. C 1*s* *XA* and *XP* spectra for C/Ni (100)p_{4g} (2 × 2) and CO/Ni (100)c(2 × 2). The XA spectra are shown on a photon energy scale and the XP spectra on a binding energy scale referenced to the Fermi level. Note that the binding energy scale is reversed from the standard procedure

approximation. For adsorbed atomic carbon, the final state is replaced by an adsorbed nitrogen atom. The local electron population does not differ by exactly one unit between carbon and nitrogen since the two atoms have different electronegativities. When ionizing atomically adsorbed C, N and O, the final-state atoms will always have a larger electronegativity than the ground state atoms, leading to a screening by more than one electron. In the case of a chemisorbed rare-gas atom, the final state is similar to an adsorbed alkali atom which is ionic or extremely polarized, and we could anticipate that the charge transfer is much less than one electron.

For an adsorbed molecule such as CO, the metallic screening can be illustrated in terms of changes in the orbitals forming the bonds to the surface. Figure 3.8a shows the projected CO $2\pi^*$ derived density of states for CO adsorbed on Ni(100). In the free CO molecule the $2\pi^*$ orbital is empty, but upon adsorption it becomes slightly populated by the formation of 2π-metal hybrid states. This leads to a splitting into $2\pi^*$ derived states which are bonding and antibonding with respect to the surface (Fig. 3.8a). For CO on Ni the occupied bonding states have mainly Ni 3*d* character whereas the antibonding states have mainly CO $2\pi^*$ character. The core hole will modify the $2\pi^*$ derived states, as shown schematically in Fig. 3.8b for a C 1s hole. The $2\pi^*$ occupancy increases considerably due to the attractive core hole potential. This can be seen in

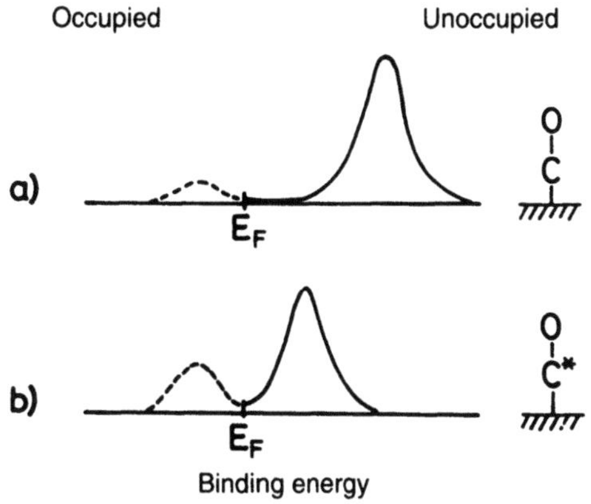

Fig. 3.8a, b. A schematic illustration of the CO derived $2\pi^*$ density of states for CO adsorbed on Ni (**a**) in the ground state and (**b**) in the presence of a core hole on the carbon atom

adsorbed NO which is the $Z + 1$ analogue for a C1s hole state. In NO, the $2\pi^*$ orbital is occupied by one electron already in the free molecule. The unoccupied $2\pi^*$ derived states are pulled down by 2 eV towards the Fermi level compared to CO. The $2\pi^*$ screening of the core hole in adsorbed CO corresponds to the increased weight of occupied $2\pi^*$ derived states in Fig. 3.8b. This increase can be viewed in two alternative ways; (i) some of the previously unoccupied $2\pi^*$ states are pulled down below the Fermi level due to the core hole potential and become populated, or (ii) the bonding d-π hybrid orbitals contract towards the adsorbate, thereby increasing their $2\pi^*$ character.

If the interaction between the adsorbate and substrate is weak as for a physisorbed molecule (small orbital overlap), the ionized state cannot be screened by electron transfer from the substrate. The relaxation occurs only by polarization of the substrate, which in the case of a metallic substrate is denoted image potential screening. The ionized atom or molecule is now left in a positively charged state and a charge redistribution in the metal causes a negative polarization charge to build up in the substrate.

Figure 3.9 compares the $1s$-$1\pi_g$ core excitation to the $1s$ core ionization spectrum for O_2 on graphite [3.40]. The $1\pi_g$ is a twofold degenerate orbital in the free O_2 molecule populated by two electrons. The two electrons have parallel spins giving rise to the paramagnetic state of the molecule. It is immediately seen that the XAS peak appears at lower energy than the photoemission peak. A metallically screened photoemission final state, neutralized by the transfer of an electron from the substrate to the $1\pi_g$ orbital, would have an energy corresponding to the X-ray absorption threshold. Hence, the XPS final state is not the lowest possible core hole state. It is a $1s^{-1}$ ionized state screened by the image potential. Further evidence of the lack of metallic screening in physisorbed O_2 is the fact that $1s$ line shows a doublet structure in the $1s$ photo emission spectrum. The molecular paramagnetic splitting of 1.1 eV arising from two final states with different spins ($^4\Sigma$ and $^2\Sigma$) [3.41], is preserved for the adsorbate.

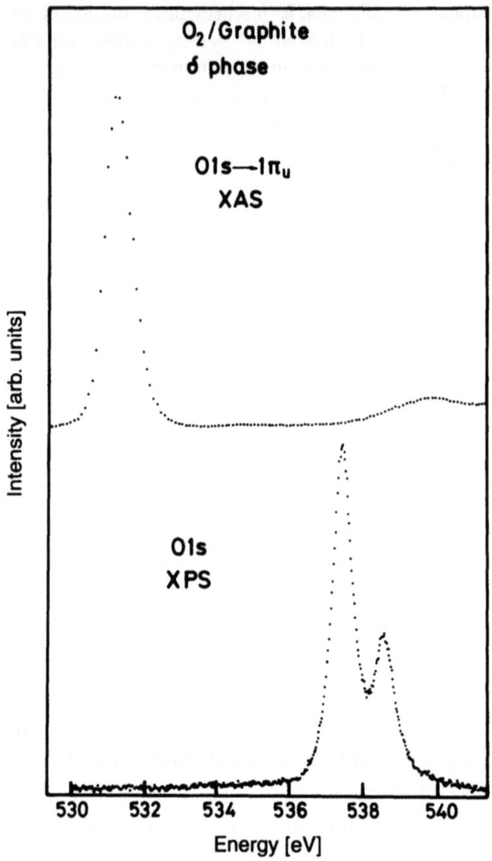

Fig. 3.9. O1*s* XA and XP spectra for O_2 physisorbed on graphite in the δ phase [3.40]. The XA spectra is shown on a photon energy scale and the XP spectra on a binding energy scale referenced to the substrate Fermi level. Note that the binding energy scale is reversed from the standard procedure. The spectra were recorded at 25 K

3.4.2 Chemical Shifts Between Inequivalent Atoms

The XP spectrum of molecular N_2 on Ni(100) in the $c(2 \times 2)$ structure is shown in Fig. 3.10 [3.34]. Due to the perpendicular adsorption geometry, the two nitrogen atoms are inequivalent and two chemically shifted N 1*s* peaks are seen at 399.4 and 400.7 eV binding energies. The component with the lowest binding energy corresponds to the outermost nitrogen atom [3.35]. This identification has been based on different observed angular variations of the photoemission intensities from the two nitrogen atoms due to photoelectron diffraction effects [3.34, 35]. The 1.3 eV shift can be understood in terms of the adsorption energies of the different final state species. Using the $Z+1$ approximation, the N 1*s* ionized final state can be replaced by an adsorbed NO molecule. Core ionization of the outer nitrogen atom leads to adsorbed NO with the nitrogen atom closest to the surface whereas the ionization of the inner nitrogen atom leads to NO with the oxygen end down. The stable configuration of adsorbed NO has the nitrogen atom downwards [3.42, 43]. Ionization of the outer nitrogen atom then

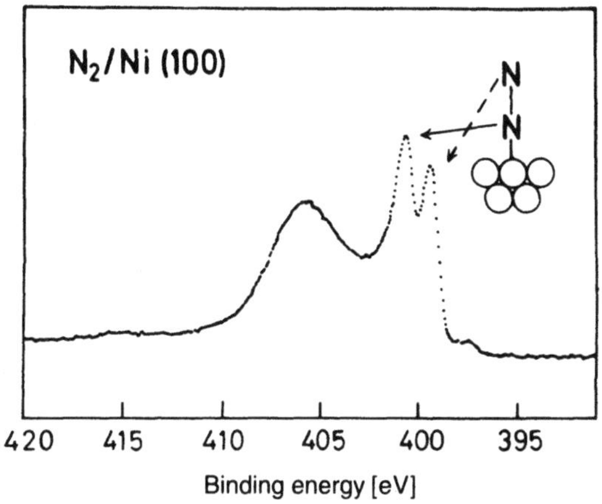

Fig. 3.10. The N 1*s* spectrum for N_2 adsorbed on Ni (100) in a $c(2\times2)$ structure. The spectra were recorded at 80 K. The two inequivalent nitrogen atoms are marked in the figure

leads to the most favourable final state, even though the core hole is further away from the surface. This demonstrates that the energetics of the final state screening in a chemisorbed molecule is not directly related to the distance of the core hole to the surface but that it depends on the overall electronic structure of the core-hole state. The difference in adsorption energies between the two NO orientations on Ni is equal to the binding-energy shift of 1.3 eV [3.35].

3.4.3 Adsorption Site-Dependent Shift in CO

We will now demonstrate how binding energy shifts can be used to determine the adsorption site of CO on Ni and Pt surfaces. Adsorption of CO on Ni(100) leads to the formation of a $c(2\times2)$ structure at $\theta = 0.5$ with all CO molecules in equivalent positions. In this structure the molecules adsorb in on-top sites with the molecular axis perpendicular to the surface and the carbon end downwards [3.44]. For higher CO coverages two new structures appear. This involves the population of other adsorption sites than the on-top sites. At saturation, $\theta = 0.67$, a $p(3\sqrt{2}\times\sqrt{2})R45°$ structure appears for which vibrational spectroscopy has proposed that only bridge adsorption sites are populated [3.44]. Between saturation and the $c(2\times2)$ structure a $c(5\sqrt{2}\times\sqrt{2})R45°$ structure develops with a mixture of on top and bridge sites [3.44].

Figure 3.11 shows C 1*s* and O 1*s* spectra of CO adsorbed on Ni(100) in the three different overlayer structures [3.46]. In the $c(2\times2)$ structure the C 1*s* and O 1*s* lines appear at 285.9 eV and 532.2 eV binding energy, respectively. For the $c(5\sqrt{2}\times\sqrt{2}\}R45°$ structure there are shifts towards lower binding energies and both lines become much broader than for the $c(2\times2)$ structure. Finally, in the $p(3\sqrt{2}\times\sqrt{2}\}R45°$ structure the line widths become similar to the $c(2\times2)$ spectra with total negative binding energy shifts of 0.5 eV for C1*s* and 0.9 eV for O1*s*. The increased line width for the intermediate structure is thus due to overlapping peaks from the two types of sites which are populated.

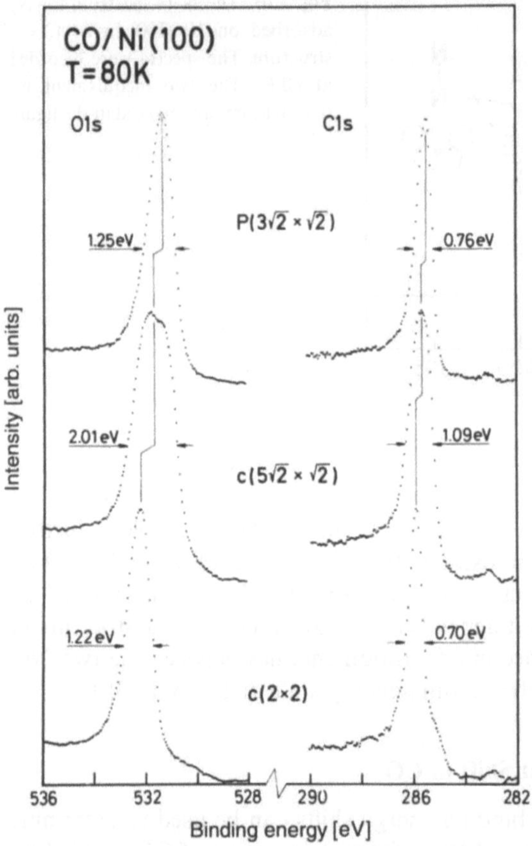

Fig. 3.11. C 1*s* and O 1*s* spectra for CO on Ni(100) in the $c(2\times 2)$, $c(5\sqrt{2}\times\sqrt{2})R45°$ and $p(3\sqrt{2}\times\sqrt{2})R45°$ structures. The spectra were recorded at 80 K. The half widths of the peaks are indicated in the figure

In the coadsorption system CO + H on Ni(100) a wider range of adsorption sites becomes accessible [3.45]. There are different phases of CO in which the molecules occupy on top, bridge and hollow adsorption sites as seen from vibrational spectroscopy [3.46]. The coadsorbed phases were prepared by first producing a $p(1\times 1)$H structure at 80 K and then adsorbing CO at the same temperature with subsequent annealings to different temperatures. At the bottom of Fig. 3.12 the spectra from the clean $c(2\times 2)$ overlayer of CO on Ni(100) are shown for comparison. Co-adsorption with hydrogen at 80 K also leads to a $c(2\times 2)$ structure with only on top population. The C1*s* and O1*s* peaks shift to higher binding energies by 0.4 eV and 0.7 eV, respectively, relative to the c(2 × 2) phase without hydrogen. It is interesting to note that the binding energy positions are similar to the $c(2\times 2)$CO phase on Cu(100) which also has on-top adsorption sites. Raising the substrate temperature to 170 K yields a $c(2\sqrt{2}\times\sqrt{2})R45°$ structure. In the spectra two peaks can now be seen with an energy separation of 1.2 eV for C 1*s* and 2.6 eV for O 1*s*. Vibrational spectroscopy shows that also hollow adsorption sites become populated in this phase [3.46]. The relative intensity between the two sites is 1 : 1. Increasing the temperature to

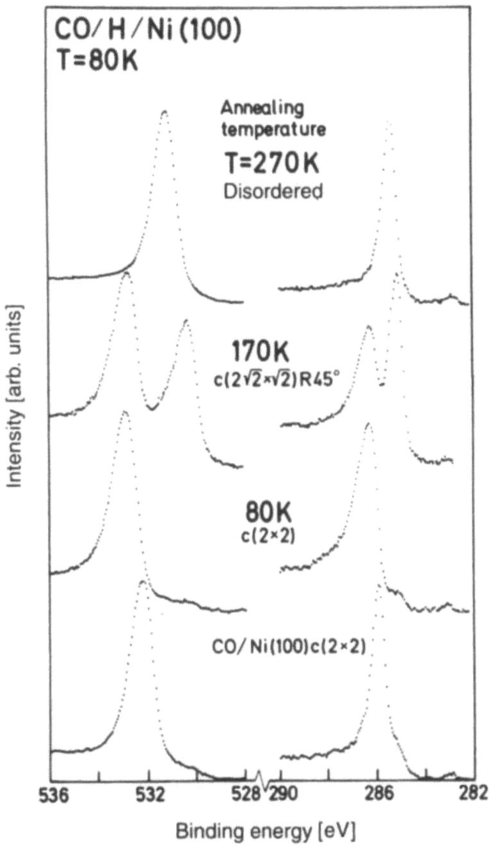

Fig. 3.12. XP spectra of the C 1*s* and O 1*s* levels from the CO–H coadsorption system on Ni (100). The spectra were recorded at 80 K after different annealing temperatures. The spectra from $c(2 \times 2)$-CO on pure Ni (100) are shown for comparison

270 K results in some desorption of CO as well as of H_2 (Σ-desorption), which leaves the remaining CO in bridge sites. The photoelectron spectra in Fig. 3.12 show peaks with intermediate binding energies. The binding energy is very close to the value obtained for the bridge populated $p(3\sqrt{2} \times \sqrt{2})R45°$ structure of the pure CO overlayer.

Similar results are also obtained for Co on Pt(111) [3.47]. Figure 3.13 exhibits the O 1*s* and C 1*s* spectra for three ordered structures of CO; (4×4), $c(4 \times 2)$ and $c(5 \times \sqrt{3})$ rect [3.48]. The corresponding Pt 4*f* spectra are shown in Fig. 3.20 together with structural models of the overlayer in Fig. 3.21 (Sect. 3.5.2). For the (4×4) phase, there is only one peak seen in each spectrum at 286.7 and 532.7 eV binding energy, respectively. The binding energies fall in the range typical of on-top sites. The $c(4 \times 2)$ and $c(5 \times \sqrt{3})$ rect structures involve both on-top and bridge sites and a second component is observed at 286.0 and 531.0 eV. The C1*s* binding energy is 0.6 eV larger than for the bridge site on Ni(100) whereas the O1*s* energy is exactly the same. In the $c(4 \times 2)$ phase ($\theta = 0.5$), equal amounts of sites are populated. For the $c(5 \times \sqrt{3})$rect overlayer ($\theta = 0.6$), 2/3 of the molecules occupy on-top sites.

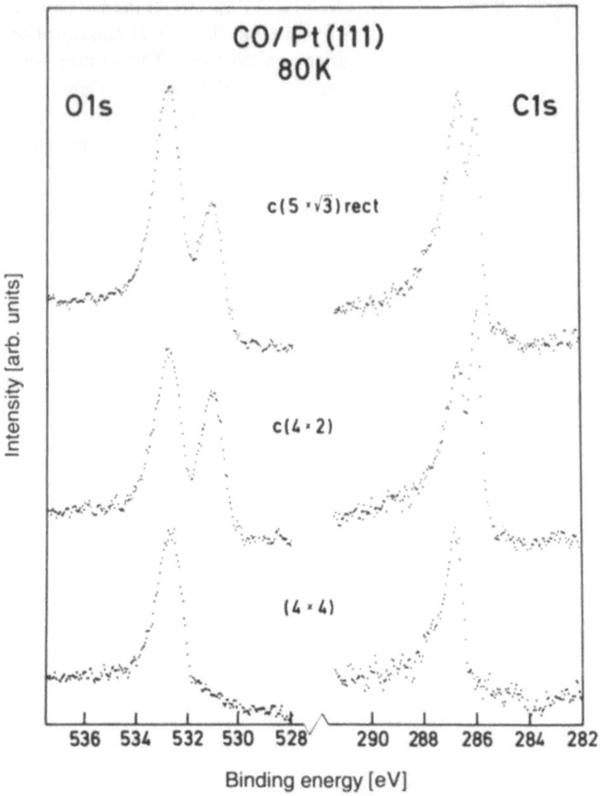

Fig. 3.13. C 1*s* and O 1*s* XP spectra from three ordered phases of CO on Pt(111); (4 × 4), *c*(4 × 2) and C($\sqrt{3}$ × 5) rect. The spectra were recorded at 80 K. Backgrounds from clean Pt have been subtracted in both core-level spectra

These examples clearly reveal that there are strong site-dependent binding-energy shifts for adsorbed CO. The shift is about twice as large for O1*s* as for C1*s*. The binding energy decreases with increasing coordination to the substrate atoms, i.e., in the order on-top > bridge > hollow. This is an empirical relationship which could be used in the study of CO adsorption on Ni, Pt and other surfaces. In vibrational spectroscopy a similar empirical rule has been used for many years based on the CO stretch vibration.

The shifts can be understood from total-energy considerations. In the initial state prior to ionization there are only small differences in adsorption energies (a few 100 meV) for the different sites. Hence, the major contribution to the shift must come from changes in the energy of the core ionized state. Within the $Z + 1$ approximation the C 1*s* ionization produces a NO-like final state and the O 1*s* ionization a CF-like final state. The carbon part of the CF molecule contains 'three unpaired' electrons available for bonding, while the 'fluorin' part of the molecule is completely saturated due to the large difference in electronegativity between the two atoms. For a free carbon atom with four unpaired electrons the adsorption site is such that the coordination to Ni is optimized. On Ni(100) the interaction is so strong that a reconstruction of the Ni lattice occurs with the carbon atoms in the hollow position. Also, CF can be anticipated to maximize the coordination in a similar way. From this it can be concluded that the O 1*s* final-state molecule has its equilibrium adsorption site in the hollow position, which is in agreement with the O 1*s* shift measurements. In order to move the CF molecule to the bridge and on-top sites the spectra show that an energy of 1 and 2 eV, respectively, is required. There is no large difference in electronegativity between the CO and NO molecules. NO is therefore expected to adsorb in a manner similar to CO, only with slightly larger adsorption energies due to the extra unpaired valence electron. Therefore, the variation in the final-state energy is smaller for C 1*s* than for O 1*s*. These chemical arguments have been supported by CI (configuration interaction) calculations of Ni clusters simulating the ionization for on-top and hollow adsorption sites of CO [3.36].

3.4.4 Molecular Orientation and Site-Dependent Shifts in NO

The adsorption of NO on Ni(100) is much more complex than the adsorption of CO. No ordered molecular overlayer structures are formed in this system. Figures 3.14 and 15 show how the N 1*s* and O 1*s* spectra develop for different doses of NO at 80 K [3.43]. Already at a dose of 0.3 L, which gives a coverage of 0.04, adsorbate spectra can be observed with a lot of details. At 0.8 L only one peak can be seen in each spectrum indicating a single adsorption site with N 1*s* and O 1*s* binding energies of 398.1 eV and 531.4 eV, respectively. For a dose of 1.2 L a second peak is seen in the N 1*s* spectrum at 1.8 eV higher binding energy. With increasing coverage the second N 1*s* peak increases in intensity and the first one vanishes completely at the saturated surface. In the O 1*s* spectra a second peak is observed at intermediate coverage with 1 eV lower binding energy. At saturation this O 1*s* peak decreases and leaves an asymmetry of the line towards lower energies. It can be noted that there is no clear connection between the double peak structures in the N 1*s* and O 1*s* spectra. The chemical shifts in the two spectra must therefore be of different origin.

The first adsorption state at low coverage has been found to be a lying-down or highly-tilted NO species from studies using X-ray photoelectron diffraction and UPS [3.49]. The second N 1*s* peak is related to a perpendicular NO adsorption state on the surface [3.49, 50]. The large N 1*s* binding energy shift of 1.8 eV between the standing-up and lying-down NO can again be understood using the $Z+1$ approximation, i.e. by replacing the final-state molecule by an adsorbed O_2 molecule. For molecularly adsorbed oxygen only the lying-down geometry has been found on different surfaces [3.51–53]. Therefore the lying-down final state must have the lowest total energy. This is in agreement with our results which give the lowest binding energy for the tilted

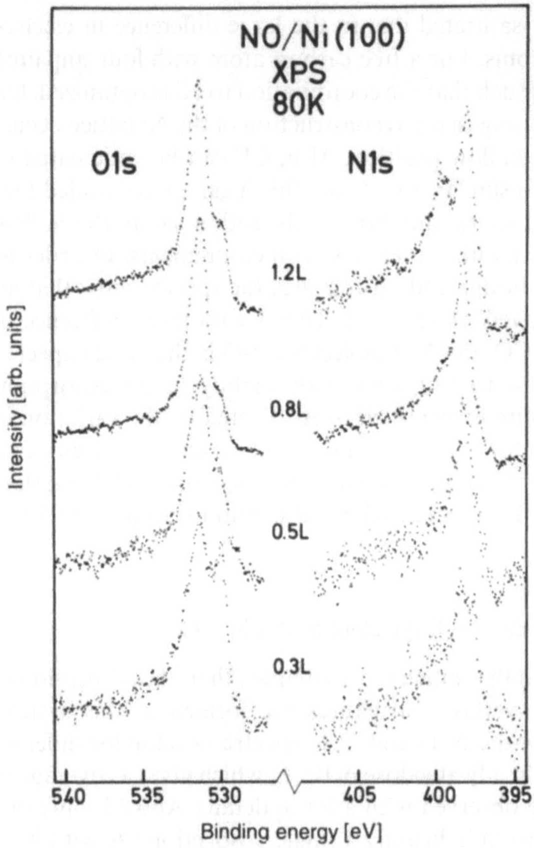

Fig. 3.14. O 1*s* and N 1*s* XPS spectra for low coverages of NO adsorbed on Ni(100) at 80 K. The different exposures are indicated in L (1L $= 1 \times 10^{-6}$ Torr s)

system. From the N 1*s* shifts we can estimate that an energy of nearly 2 eV is required to raise molecular oxygen into a perpendicular geometry.

In the O 1*s* spectra at intermediate coverage the second peak at lower binding energies grows when the perpendicular NO species becomes populated but then nearly vanishes at saturation coverage. From a comparison with vibrational spectra of this system there is an indication that the second peak is related to a hollow adsorption site and the first one to a bridge site [3.45]. For the perpendicular adsorbate the hollow site should then be first populated and with increasing coverage the molecules would be transferred to bridge sites. For perpendicularly adsorbed NO the adsorption site with lowest energy should correspond to the hollow site. This is also found using the $Z + 1$ approximation for the C 1*s* spectra of CO on Ni(100) where the final state NO-like molecule adsorbs with lowest energy in the hollow site (see Sect. 3.4.3). Applying the $Z + 1$ approximation to the O 1*s* final state in NO results in a NF-like final-state molecule. The observed shifts indicate that the site dependence of this molecule is similar to that of CF.

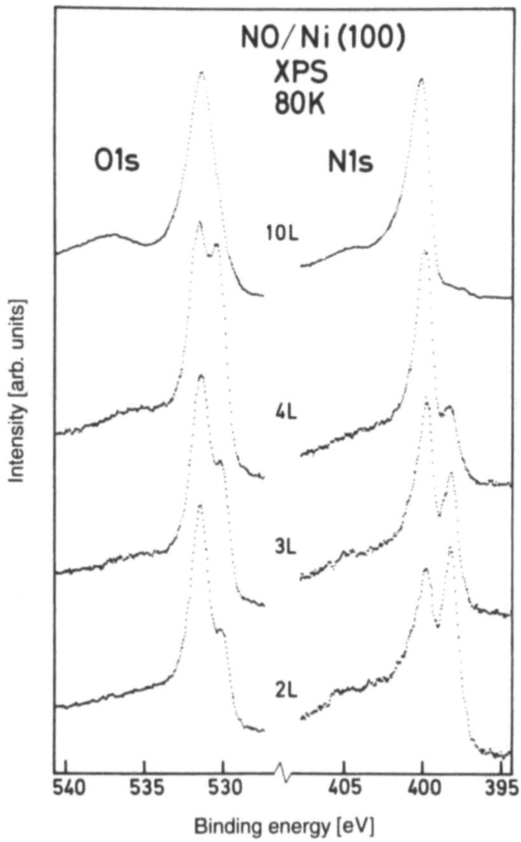

Fig. 3.15. O 1*s* and N 1*s* XPS spectra for high coverages of NO adsorbed on Ni(100) at 80 K. The different exposures are indicated in L

For adsorbed NO, core ionization of the different atoms provides complementary information. The N 1*s* energy is mainly sensitive to the molecular angle relative to the surface whereas the O 1*s* energy is more sensitive to the local adsorption site.

3.4.5 Physisorption-Induced Shifts in O_2

Next we will turn to a physisorption system, O_2 on graphite, where the ionized molecules are screened by an image potential. The phase diagram of O_2 on graphite is rather complex and involves many different phases. However, at 25 K only two MonoLayer (ML) phases are of interest. Starting at low coverage the δ phase is formed up to a coverage of 0.67. In this phase the molecules bond with the molecular axis nearly parallel to the surface [3.55, 56]. As the coverage is increased the molecules change their orientation. In the $\zeta 2$ phase, which completes the monolayer, the molecules are nearly standing up with a tilt angle of $37 \pm 10°$ from the surface normal [3.55, 56]. The molecules continue to grow in the perpendicular orientation when multilayers are formed. The angle of the molecular axis to the surface normal decreases with thickness [3.56].

Figure 3.16 depicts the O 1*s* spectra relative to the Fermi level from submonolayer coverages to roughly 10 ML of O_2 on graphite [3.57]. The paramagnetic splitting of the O 1*s* line was discussed in Sect. 3.4.1. The gas phase spectrum shown at the bottom of the figure has been shifted by 1.4 eV to lower binding energies in order to line up with the monolayer spectrum [3.41]. The lower binding energy for the physisorbate is due to the polarization of the surface (image potential screening) and the neighbouring molecules, as discussed in Sect. 3.4.1. The spectra from the submonolayer phases ($\theta_0 = 0.5$ and the δ-phase) are virtually identical showing similar local surroundings. Due to the change in adsorption geometry, the spectrum changes between the δ- and ζ2-phases. The more 'upright' adsorption geometry of the molecules in the ζ2-phase results in (ideally) two inequivalent oxygen atoms with respect to the surface. The atom closest to the surface will be more efficiently screened by the image potential [3.40]. By the same mechanism the librational motion (frustrated

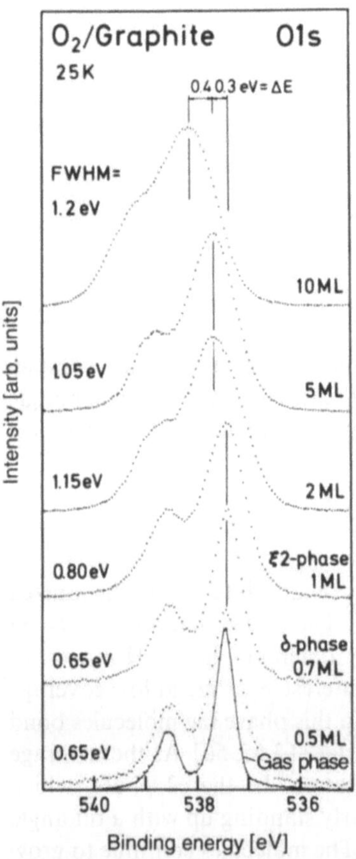

Fig. 3.16. The O 1*s* spectra for several coverages of O_2 physisorbed on graphite are shown. A shifted gas phase spectrum (from [3.41]) is inserted in the lower part of the figure. The number of different MonoLayers (ML) is indicated (right). The FWHM of a gaussian line profile of the quartet component is given (left), as well as the shifts in binding energy between different coverages (upper part)

rotations) of the molecules leads to a continuous range of binding energies which further broadens the spectrum [3.40].

At a coverage of 5 ML of O_2/graphite, a narrowing of the O 1*s* peak is observed. This is partly due to the fact that the intensity from the low-energy component (first monolayer) is suppressed by the covering layers. At the coverage corresponding to 10 ML, the growth mode is not known. The smeared out spectral features are due to inhomogenous sample charging, which is caused by the growth of crystalline oxygen. The shift in the second layer is 0.5 eV towards higher binding energy relative to the monolayer. This type of shifts has previously been interpreted in a simple model based on the distance between the ionized atom and the image plane [3.58–62]. However, it turns out that it is also important to consider the polarization screening of the surrounding molecules, both in the same layer and between adjacent layers [3.58]. Furthermore, it has been seen from calculations that the internal molecular screening depends on the external screening [3.63]. This indicates that it is not possible to separate the different screening channels, i.e. the internal screening, the image potential screening and the response from the surrounding molecule, into independent contributions only based on shift measurements. This requires studies of the satellite features related to the different screening contributions (Sect. 3.8.2).

3.5 Surface Core-Level Shifts

In Fig. 3.2 we saw for Yb that the core-level binding energy of a surface atom can be different from that of a bulk atom. The binding energy difference between the surface and the bulk is denoted Surface Core Level Shift (SCLS), ΔE_s.

$$\Delta E_s = E_B(\text{surface}) - E_B(\text{bulk}). \tag{3.4}$$

Using the SCLS the properties of the surface atoms can be probed [3.33, 64–68]. In this section we will at first briefly review how an expression for the SCLS can be derived which relates it to surface energies and surface segregation energies. The *Z*-dependence of the SCLS will then be discussed for the 5*d* transition elements and it is shown that the theory provides a straightforward explanation of the shift curve. Furthermore, the influence of adsorbates on the SCLS will be discussed. The different effects of physisorbed and chemisorbed molecules will be treated. Using spectra for CO/Pt(1 1 1) it will furthermore be demonstrated how the Pt 4*f* spectra can be used to probe what adsorption sites are populated.

3.5.1 Clean Metals

A most useful description of the surface core-level shift is obtained if one treats the surface and bulk ionization energies in terms of separate Born-Haber cycles

[3.33, 64]. This derivation is based on a total energy description of the initial and final states, as discussed in Sect. 3.3. The metallic screening of the final states is explicitly taken into account [3.33]. The Born-Haber cycles are used to decompose the total energy expressions into terms which can be obtained in independent ways. These terms can be divided into two categories: terms which represent the energies on the atomic level associated with the screening of the core hole site and terms which describe how the properties of the initial and final state atoms are renormalized in the solid state. When chemical shifts are considered, the first type of terms cancel out and we need only consider how the ionized atom bonds to the lattice before and after the ionization in the bulk and at the surface. In the initial state for a bulk ionization this defines a term which is equal to the cohesive energy of the metal. To derive the final-state energy we here use the $Z+1$ approximation. Note, however, that this approximation is not a necessary ingredient in the theory. The bonding energy of the final-state atom to the lattice can be expressed as the sum of two terms, the cohesive energy of the $Z+1$ element plus a term representing the solution energy of a $Z+1$ atom in the host system. Usually the latter term is rather small and we will, for the sake of clarity, neglect it in the present discussion. The core-level binding energy in the bulk of a metal can then be expressed as

$$E_B(\text{bulk}) = A + E_{\text{coh, bulk}}(Z) - E_{\text{coh, bulk}}(Z+1) \tag{3.5}$$

where A contains all terms in the Born-Haber cycle which do not depend on the detailed environment of the core ionized atom.

For the surface the same type of derivation can be made. The only terms which change, are the terms which describe the bonding, i.e. the cohesive energies. The bonding energy is reduced at the surface due to the reduced number of coordinating atoms. Instead of (3.5) we then obtain the equation

$$E_B(\text{surf}) = A + E_{\text{coh, surf}}(Z) - E_{\text{coh, surf}}(Z+1) \tag{3.6}$$

with the only difference that the bulk cohesive energies are replaced by surface cohesive energies. The surface cohesive energy can be expressed as $E_{\text{coh, surf}} = \alpha E_{\text{coh, bulk}}$ where α is a constant less than one. The cohesive energy for a surface atom is lower than for a bulk atom due to the reduced number of bonds at the surface. For a close-packed surface a typical value of α is 0.8. The difference between the bulk and surface cohesive energies is referred to as the surface energy $E_{\text{surf}} = E_{\text{coh, bulk}} - E_{\text{coh, surf}}$. The SCLS, ΔE_s, can thus be expressed as

$$\begin{aligned} \Delta E_s &\approx E_{\text{surf}}(Z+1) - E_{\text{surf}}(Z) \approx (1-\alpha)\,[E_{\text{coh}}(Z+1) - E_{\text{coh}}(Z)] \\ &\approx 0.2\,[E_{\text{coh}}(Z+1) - E_{\text{coh}}(Z)]. \end{aligned} \tag{3.7}$$

The value of α above will depend on the packing of the surface. The surface energies are larger for more open surfaces and hence one will also observe larger surface shifts.

Figure 3.17 shows the experimental surface shifts for the close packed surfaces of the 5d transition elements [3.69]. It is seen that there is a sign reversal

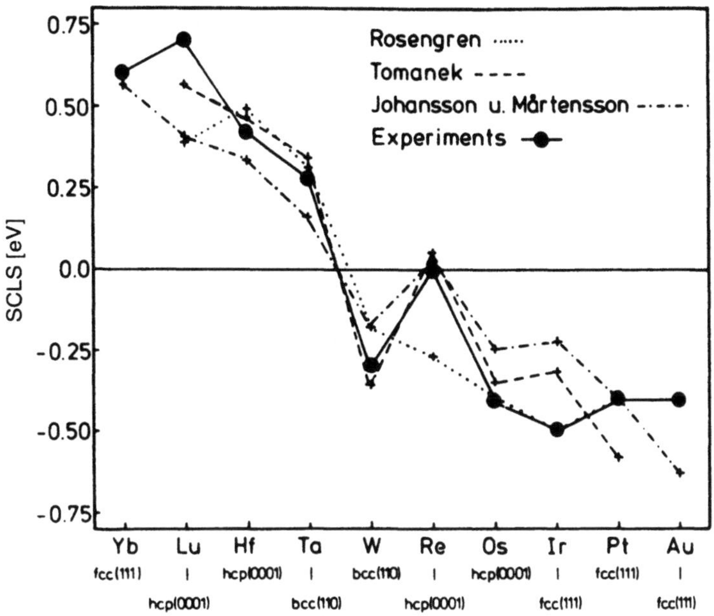

Fig. 3.17. Comparison of the experimentally determined SCLS of the 5*d* transition metals with the results of different theoretical approaches. Only the values of the close-packed surfaces are included

of the shift in the middle of the series. For the early elements the shift is positive while at the end of the series the shift is negative. The experiments are compared to three different evaluations of (3.7) [3.33, 70, 71]. As can be seen the theoretical curves reproduce the experimental results quite well. The details of this comparison are discussed in [3.69] where it is also shown that the real reason for the kink between W and Re is the fact that the Z-dependence of the surface energies is different for bcc, fcc and hcp metals.

From (3.7) one can directly see the reason for the general behaviour of the SCLS in Fig. 3.17. The cohesive energy for the 5*d* transition elements is dominated by the *d*-electron contribution. The 5*d* density of states can occupy a total of 10 electrons per atom. The character of the 5*d* states varies over the density of states. The states at the bottom of the band are bonding and the states at the top of the band are antibonding. For the first elements in the series only bonding states are occupied and the cohesive energy increases with atomic number. The cohesive energy has a maximum in the middle of the series and after this the antibonding states begin to be filled which reduces the cohesive energy as the *d*-band is further filled. With this functional form of cohesive energy it is directly seen from (3.7) that the shifts will be positive for the early elements and negative for the late elements.

Another interesting property of the SCLS is that it can be directly related to surface segregation energies. This can be seen from Fig. 3.18. The ionization of a bulk atom creates a final-state impurity in the bulk of the system. In terms of the

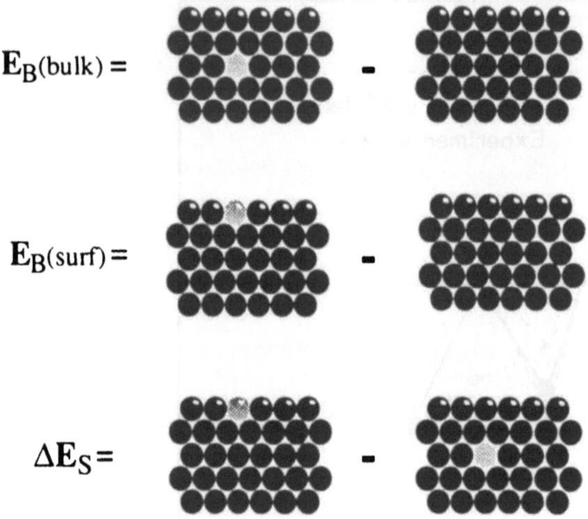

Fig. 3.18. Illustration of the correspondence between the SCLS, ΔES, and the surface segregation energy of the final state impurity. The final state can to a good approximation be represented by a $Z + 1$ impurity

bonding to the neighbours this atom behaves like a bulk $Z + 1$ impurity. In the corresponding way a core ionized surface atom behaves like a $Z + 1$ impurity at the surface. The two situations are sketched in the figure. It is then immediately seen that the energy difference between the two systems, i.e. the SCLS, corresponds to the surface segregation energy of the $Z + 1$ impurity. In an analogous way the shift between an atom in the bulk of a metal and an atom at a metallic interface, as was observed in Fig. 3.2, has a direct interpretation in terms of interface segregation energies [3.18]. This coupling to thermochemical data is particularly direct and powerful for metallic systems.

3.5.2 Adsorbate-Induced Shifts

In the presence of adsorbates the binding energies of the surface atoms change. The magnitude of the shifts is in some way connected to the interaction strength. For weak interaction as in physisorption the influence on the surface atoms will be very small. In the case of strong bonding, however, one can expect substantial shifts for the surface atoms although it is important to realize that a small shift does not exclude strong interaction. This can also be caused by compensating effects in the initial and final states.

Figure 3.19 shows the $4f_{7/2}$ spectrum for an Au(110) surface without any adsorbate and after the adsorption of N_2 and CO. The pure-metal spectrum shows a surface shifted component at lower binding energies. The adsorption of N_2 has no effect on the surface core-level shift. This indicates that N_2 physisorbs on the Au(1 1 0) surface which is consistent with other observations [3.72]. When instead CO is adsorbed the spectrum is seen to change drastically. This is a direct sign of chemisorption [3.73]. The original surface component decreases in intensity. Simultaneously, a new component appears at the high-binding-energy side of the main peak. The adsorption of CO affects the binding energies

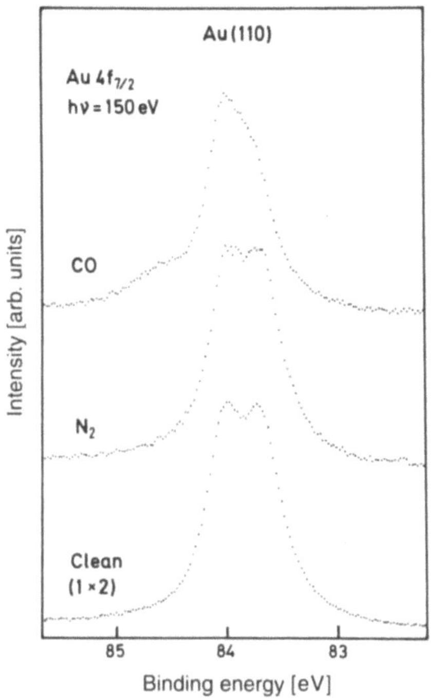

Fig. 3.19. $4f_{7/2}$ photoelectron spectra for Au(110) with and without adsorbates. The gold surface exhibits a 2×1 reconstruction. No ordered CO overlayers could be observed with LEED

of the surface atoms to the extent that the sign of the SCLS is even reversed. The fact that the whole surface layer does not shift, again demonstrates the very local character of the chemical shift.

In the theoretical treatment of SCLS outlined above the presence of an adsorbate enters by adding terms which represent the adsorption energies in the initial and final states, respectively. To a good approximation the adsorption energies can be associated with the adsorbate and with the atom or atoms to which it is directly coordinated. For on-top adsorption only one surface atom is coordinated to the adsorbate. In this case the SCLS will be modified by the adsorption-energy difference between the molecule adsorbed on the Z-metal substrate and the molecule adsorbed in the same geometry on the $Z+1$-metal substrate. If the adsorbate is more strongly bonded to the Z-metal this will introduce a positive contribution to the shift. This is the expected direction for CO on Au since CO will bond more strongly to gold than to the $Z+1$ element which is mercury. In fact, in the latter case we expect no chemisorption bond at all.

For other geometries of the adsorbate the treatment becomes more involved. For bridge adsorption each molecule is coordinated to two substrate atoms. The adsorption energy per coordinated substrate atom (or per bond) is therefore half the adsorption energy. In the final state one substrate atom is

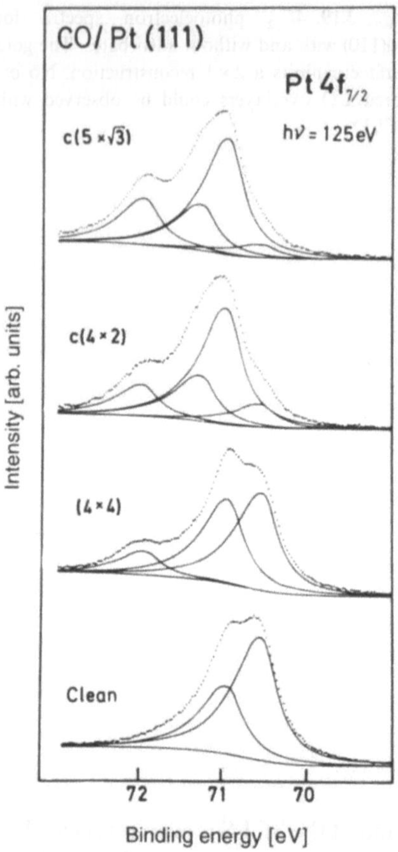

Fig. 3.20. $4f_{7/2}$ photoelectron spectra for Pt(111) for the clean metal and for three different CO adsorption phases. The overlayer structures are described in Fig. 3.21. The spectra were recorded at liquid nitrogen temperature

replaced by a $Z + 1$ atom. This defines a term which represents the energy of a bond between the $Z + 1$ atom and the bridge-bonded adsorbate. If we assume that this energy is not sensitive to the fact that the other substrate atom, to which the adsorbate coordinates, is still a Z atom and not a $Z + 1$ atom, then this energy will define an energy which is half the bridge-site adsorption energy on the $Z + 1$ substrate. With these assumptions the modification of the surface shift for bridge adsorption is, thus, half the difference between the bridge-site adsorption energy on the Z- and $Z + 1$-metals. If for a certain system the adsorption energies (in the initial and final states) are similar for the on-top and bridge adsorption sites (which is usually the case where both types of sites can be populated) this leads to an expected shift between a free surface atom and a bridge-bonded molecule being about half of the shift to an on-top bonded molecule.

In Sect. 3.4.3. the C 1*s* and O 1*s* spectra for CO on Pt(111) were discussed. These spectra could be used to separate on-top and bridge bonded species. In Fig. 3.20 the Pt $4f_{7/2}$ spectra are shown for the same phases as were discussed

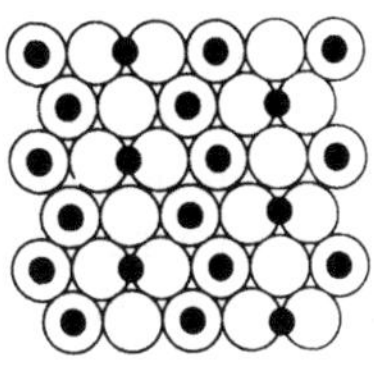

c(5 × √3)

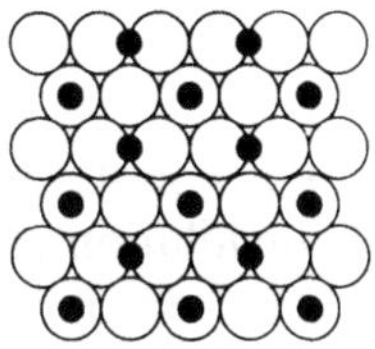

c(4 × 2)

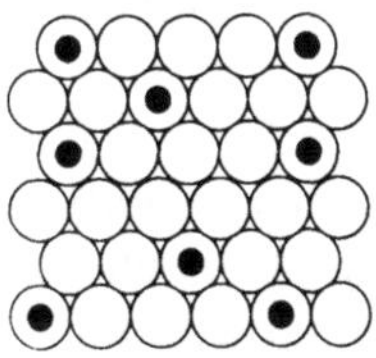

(4 × 4)

Fig. 3.21. Illustration of the investigated CO/Pt(111) adsorption phases. The three phases correspond to adsorbate coverages of $\Theta_{CO} = 0.19$ (4×4), $\Theta_{CO} = 0.5$ $(c(4 \times 2))$ and $\Theta_{CO} = 0.6$ $(5\sqrt{3} \times \sqrt{3})$ in substrate units. The fraction of on-top adsorbates for the three phases is 100%, 50% and 67%, respectively

above, i.e. the (4×4), $c(4 \times 2)$ and $c(5 \times \sqrt{3})$rect phases [3.74]. The proposed structures for these phases are depicted in Fig. 3.21 [3.75]. At the bottom of Fig. 3.20 the spectrum for the clean surface is shown. As can be seen there is a surface shift towards lower binding energies in accordance with the discussion above. It can also be noted that the surface sensitivity is such that the surface peak actually dominates.

In the (4×4) phase the intensity of the original surface peak is reduced. At the same time a new peak appears on the high-binding-energy side of the bulk peak. The adsorbate induced shift of this peak relative to the unperturbed peak is 1.41 eV. In the $c(4 \times 2)$ and $c(5 \times \sqrt{3})$rect phases the unperturbed surface peak decreases even further. In these cases there are two new surface components: the one interpreted as linearly coordinated CO and a new peak which has a binding energy almost half way between the on-top coordinated peak and the unperturbed surface peak (shifted 0.73 eV relative to the pure surface). This new peak corresponds to the Pt surface atoms which coordinate bridge bonded CO. In accordance with the discussion above this leads to an intermediate shift. Actually, the shift is reduced by a factor close to two. This might naively be anticipated.

The surface shifts are thus directly related to the local coordination of the adsorbates. It is also interesting to note that the surface peak corresponding to the uncoordinated Pt surface atoms is unaffected by the adsorption. This demonstrates the extremely local character of the shifts. It is clear from this example and the previous discussion of the shifts in the adsorbate levels that the core-level data provide most detailed information on the geometrical arrangement of the adsorbates. Furthermore, the local character of the probe makes it independent of long-range order of the adsorbate structures.

3.6 Core-Level Line Shapes

There are a number of effects which determine the shape of a core-level spectrum. Most often the observed spectrum can be viewed as a convolution of the fundamental spectral shapes due to these individual effects. Some contributions, however, may be interconnected in a more complicated way which makes it necessary to consider the effects in a unified manner. In this section we will assume that the effects are separable. We will briefly discuss the various contributions both in the language used for molecules and in a solid-state language.

The different line-shape contributions are sketched in Fig. 3.22 for a free molecule as well as for a solid [3.9]. These are due to the spectrometer resolution function, the finite life-time of the core-hole, inelastic scattering of the photoelectrons, electronic shake-up and shake-off processes and vibrational excitations.

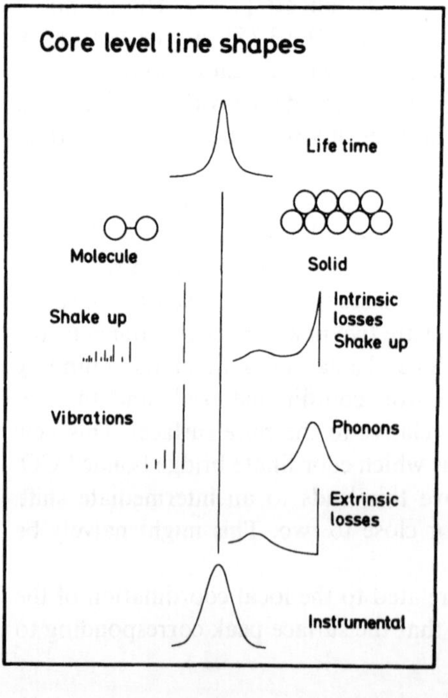

Fig. 3.22. Summary of different line shape contributions for a free molecule and for a solid

Each core level has an intrinsic width due to the finite lifetime of the hole state. The decay is due to radiative and non-radiative processes involving electrons in orbitals outside the core hole (Sect. 3.1). There is usually very little chemical influence on the decay rate and the intrinsic width is essentially an atomic property. The lifetime broadening has a Lorentzian shape. For the C 1*s*, N 1*s* and O 1*s* lines, which are mainly treated in the present contribution, the FWHM (Full Width at Half Maximum) of this contribution is only of the order of 0.1 eV [3.76] and can usually be neglected. For other core levels, however, the Lorentzian broadening may be much larger. There are also rare cases when the intrinsic line shape is more complicated than a simple Lorentzian function. This occurs for instance when the one electron state interacts strongly with other states of similar energy [3.77, 78].

As discussed in the preceding sections, core ionization leads to a modification of the other orbitals in the system. This perturbation gives rise to shake-up and shake-off satellites as shown for the Ne 1*s* spectrum in Sect. 3.1. Figure 3.22 schematically shows typical shake satellite spectra for a molecule and for a solid. In the molecule the shake-up part of the spectrum consists of a number of discrete transitions due to excitations from occupied to unoccupied orbitals like in Ne. In solids the electronic states form continuous bands. In metals there are filled states up to the Fermi level and empty states right above the Fermi level. The ionization will thus create excitations with energies all the way down to zero. This shake-up contribution gives rise to characteristic core-line asymmetries [3.79]. Other shake-up satellites are due to the creation of collective plasmon excitations or interband transitions and are seen as discrete features in the spectra. It could be noted that for metals the distinction between shake-up and shake-off becomes less meaningful because of the continuous character of the transitions for electrons excited to states below the vacuum level. The shake-up satellites are discussed in detail in Sect. 3.8.

For a molecule or a solid the removal of a core electron usually leads to the excitation of vibrational motion. This is due to the fact that the potential energy curves for the atomic motion are different before and after ionization. The effects of these processes for the spectral shape are indicated in Fig. 3.22 for a molecule and for a solid. In a molecule only a limited number of modes can be excited and the final vibrational excitations can be described in terms of discrete states. In the solid the atomic motion is instead treated in terms of phonons. The band character of the phonons creates a continuous broadening function, which is often considered to be of Gaussian shape. Note, however, that the phonon excitations may generate an asymmetric broadening as well. An identification of the different broadening mechanisms simply on the basis of the type of line profile (Lorentzian, Gaussian or asymmetric) may be dangerous. The vibrational broadening effects are further discussed in Sect. 3.7.

In solids there are satellites due to energy loss events experienced by the escaping photoelectron. These are denoted extrinsic losses. It is often difficult to distinguish between intrinsic (shake-up/shake-off) and extrinsic satellites since they may have very similar energies. Such a separation may, in fact, be rather superficial since it is based on the assumption that the photoemission process can be treated as a sequence of well separated events. In Fig. 3.22 a typical extrinsic loss spectrum is shown. For a more detailed discussion of extrinsic loss processes see [3.80].

For adsorbates the energy loss contribution can, in general, be neglected. The photo-ionization takes place in the outermost layer of the sample and the photoelectrons are emitted directly into the vacuum. The detection of inelastic loss satellites would require that electrons which are emitted into the solid or along the surface undergo large angle scattering processes. However, both the elastic and inelastic scattering are strongly peaked in the forward direction for typical XPS kinetic energies (Chap. 4). The probability that an electron from the adsorbate layer reaches the spectrometer after only one inelastic scattering event is therefore low (unless the detected electrons are emitted at grazing angles). This makes it reasonable to interpret the complete satellite spectrum in terms of shake-up processes.

The final contribution which is illustrated in Fig. 3.22 is the instrumental-resolution function which in many cases is the limiting factor for what broadenings can be detected. The resolution may vary considerably but is typically of the order of 0.4–1 eV in core-level photoelectron spectra, although a resolution down to about 0.1 eV can be achieved under certain conditions. The main broadening is usually due to the exciting radiation. Using monochromatized Al-K_α a width of the order of 0.2 eV can be obtained. Similar or smaller widths can also be achieved using synchrotron radiation, especially for shallow core levels which require relatively low photon energies. The situation for synchrotron radiation will, however, further improve also for higher photon energies in the near future with the new third-generation synchrotron-radiation sources [3.81].

In a photoelectron spectrum from a condensed system there is always a background corresponding to the ionization of other core and valence levels and Auger transitions (this contribution is not included in Fig. 3.22). The background is due to photoelectrons and Auger electrons which have undergone several inelastic loss processes during their way through the material. The exciting X-rays penetrate rather deep into the material and the individual loss processes usually involve relatively low energies. This means that the electrons which contribute to the energy-loss tails may originate from large depths and the total inelastic intensity in a spectrum is therefore quite significant. This may cause difficulties when investigating weak satellite spectra from adsorbate layers.

3.7 Vibrational Effects

For adsorbates the core-level line widths are often dominated by vibrational effects. In this section we will discuss how these effects can be described and quantitatively understood. After a general introduction the widths of the adsorbate 1s lines will be discussed for C, N and O on Ni(100). For these systems the vibrational broadenings can be modelled by employing the $Z + 1$ approximation and using data from surface structure and surface phonon measurements. The temperature dependence of these lines is also described. After this, the line profiles for adsorbed CO are discussed. In this case it is important to consider the intramolecular motion as well as the motion of the adsorbate relative to the surface. Finally, the importance of vibrational broadening for metal spectra is discussed in connection with the observation of temperature-dependent Yb 4*f* line widths.

3.7.1 The Franck-Condon Principle

If the core ionization leads to changes of the interatomic distances and/or the force constants, then vibrational motion is excited in the final state. The ionization is usually treated within the Born-Oppenheimer approximation, which implies that the nuclear motion can be decoupled from the electronic motion. This allows the total wave function to be factored into one nuclear and one electronic part which are solutions to separate equations. Figure 3.23 depicts the potential-energy curves for the nuclear motion for two different cases. The lower curve in each figure corresponds to the initial state and the upper curve to the final state of a particular electronic transition, such as a core ionization. The vibrational states are characterized by quantum numbers v and v' for the initial and final states, respectively. In Fig. 3.23 it is assumed that only

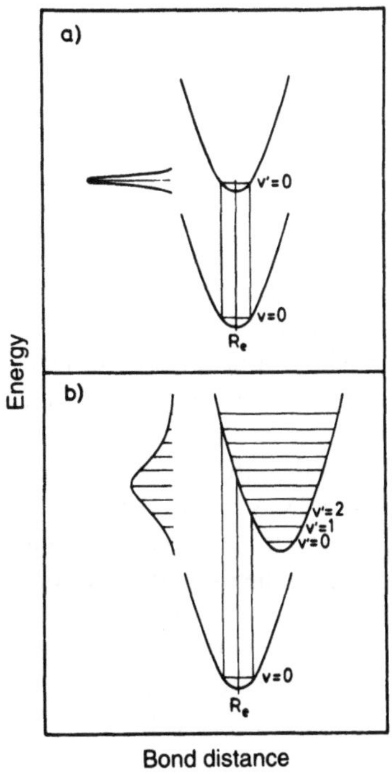

Fig. 3.23a, b. Initial and final state potential energy curves describing a photoionization in two different cases: (**a**) corresponds to a small and (**b**) to a large geometry change due to the ionization, giving rise to small and large line widths, respectively

the $v = 0$ vibrational state is populated. For the transition probabilities to the various vibrational states the Franck-Condon principle is generally applicable. The principle states that the relative transition probabilities for the same electronic transition are given by the square of the overlap between the initial and final-state nuclear-motion wave functions. The situation in Fig. 3.23a refers to a situation where the initial and final state potential energy curves are very similar. The two sets of wave functions are thus approximately the same. In this case only the $v = 0$ to $v' = 0$ overlap is non-zero. In Fig. 3.23b the potential energy curves are quite different and the $v = 0$ wave function is nonorthogonal to a whole series of final-state wave functions with $v' = 0, 1, 2$ etc. This leads to the excitation of a number of different vibrational final states. The C 1*s* ionization for CH_4, (Fig. 3.4), corresponds to the latter situation. The C 1*s* ionization leads to a shortening of the CH bond by 0.05 Å [3.25].

In the case of a valence photoionization, Fig. 3.23a corresponds to the ionization of a non-bonding orbital while the situation in Fig. 3.23b pertains to the ionization of a bonding or antibonding molecular orbital. When these arguments were transferred to the case of core ionization one was at first led to the conclusion that the situation in Fig. 3.23a applies. With common definitions of bonding and non-bonding orbitals, which are based on the overlap of the atomic orbitals involved, any core orbital is clearly non-bonding. However, this type of argument was later found to be misleading. The situation in Fig. 3.23b may well occur also for a core ionization. The removal of a core electron leads to substantial modifications of the electronic structure as discussed in Sect. 3.4, which may influence the equilibrium bond distances and interatomic force constants.

3.7.2 Vibrational Broadening in Adsorbates

When atoms and molecules are adsorbed on surfaces the vibrational modes which are present in the free molecule remain also in the adsorbate, usually with only slightly modified parameters. Furthermore, due to bonding to the substrate, new vibrational modes will appear. These are related to the translational and rotational degrees of freedom of the adsorbate. These additional modes may lead to large differences between the line profiles of free and adsorbed molecules [3.9].

Another consequence of the vibrational effects is that the line widths may depend on temperature. This is easily seen from Fig. 3.23. A higher temperature leads to thermal population of excited vibrational states. If the initial- and final-state potential energy curves are very similar (Fig. 3.23a) the excited vibrational states will ionize to the corresponding final states ($v = v_i \rightarrow v' = v_i$ transitions). The energies of these transitions are essentially the same as for the $v = 0 \rightarrow v' = 0$ transition and there is no additional broadening of the core-electron line. However, if the potential-energy curves are different (Fig. 3.23b) a higher temperature will lead to larger broadening.

The $c(2 \times 2)$ overlayers of C, N and O on Ni(100) are in many respects similar. They all occupy fourfold hollow sites and they form strong bonds to the substrate. They are interesting in connection with the discussion of vibrational effects since independent information on the initial-state as well as the final-state potential energy curves can be obtained. For the final states the $Z + 1$ approximation can be employed. In this way the C 1*s*, N 1*s* and O 1*s* final states behave very much like adsorbed nitrogen, oxygen and fluorine atoms, respectively. In particular, this implies that the potential-energy curves for the final states can be represented by the curves for the $Z + 1$ elements.

Figure 3.24 compares the 1*s* spectra for C, N and O recorded at 150 K [3.27]. The width FWHM is 0.67 eV for the C 1*s* line, 0.94 eV for the N 1*s* line and 1.42 eV for the O 1*s* line. Figure 3.25 shows the spectra for the $p(2 \times 2)$ and $c(2 \times 2)$ phases of oxygen for two different temperatures (150 K and 600 K). The O 1*s* peak is significantly broader for the $c(2 \times 2)$ phase (1.42 eV compared to 1.18 eV for the $p(2 \times 2)$ phase as measured at 150 K). Furthermore, the O 1*s* line width increases with temperature. Also, for nitrogen there is a temperature effect whereas the C 1*s* line width is essentially independent of temperature.

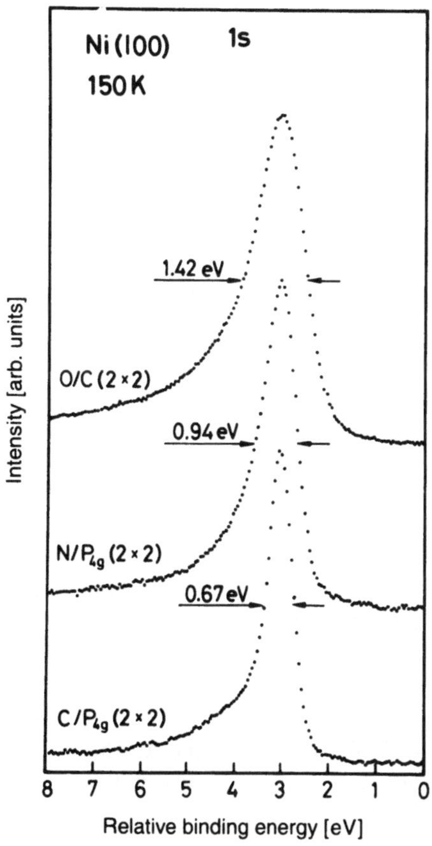

Fig. 3.24. C 1*s*, N 1*s* and O 1*s* core line profiles for the atomic $c(2 \times 2)$ adsorbate layers measured at 150 K. The adsorption of carbon and nitrogen leads to a *p4g* reconstruction of the Ni surface layer

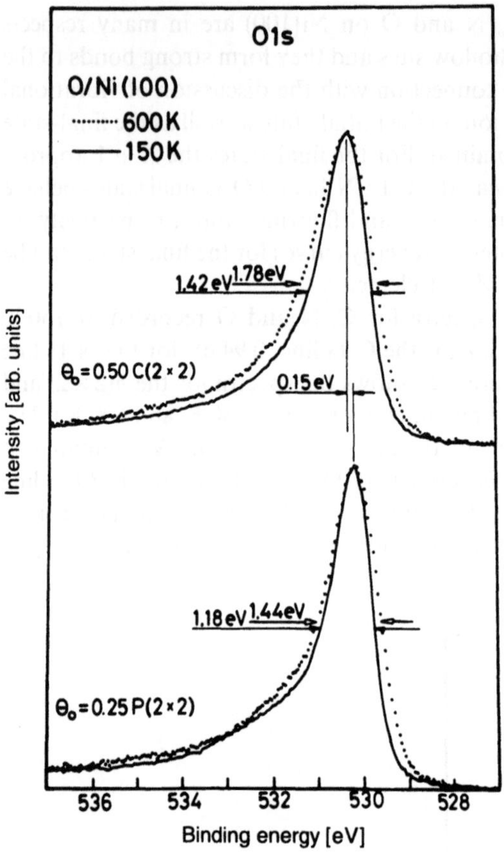

Fig. 3.25. O 1*s* core line profiles for the $c(2\times2)$ and $p(2\times2)$ phases of O/Ni(100) recorded at 150 K and 600 K

As seen from the $Z+1$ approximation and the fact that the whole series of atomic adsorbates occupy fourfold sites, the equilibrium geometry does not change parallel to the surface upon core ionization. Hence, for the parallel motion the situation in Fig. 3.23a applies and these modes will not contribute significantly to the broadening. The potential energy curves for the perpendicular motion of C, N and O in $c(2\times2)$ overlayers, as derived in [3.27], are displayed in Fig. 3.26. The curves were obtained within the harmonic approximation using bond distances from LEED and SEXAFS determinations [3.82–84] and force constants from electron-energy-loss data [3.85–87]. If the $Z+1$ approximation is used the curves in Fig. 3.26 are sufficient to estimate the C 1*s* and N 1*s* widths. The C and N potential energy curves are very similar which immediately explains the small C 1*s* broadening. For the N 1*s* ionization, however, there is a substantially larger shift in equilibrium bond distance between the initial and final (oxygen-like) states which leads to calculated vibrational broadenings of 0.55 and 0.84 eV for the N 1*s* line at 150 K and 600 K, respectively. Adding the other broadening contributions yields values in reasonable agreement with experiment [3.27].

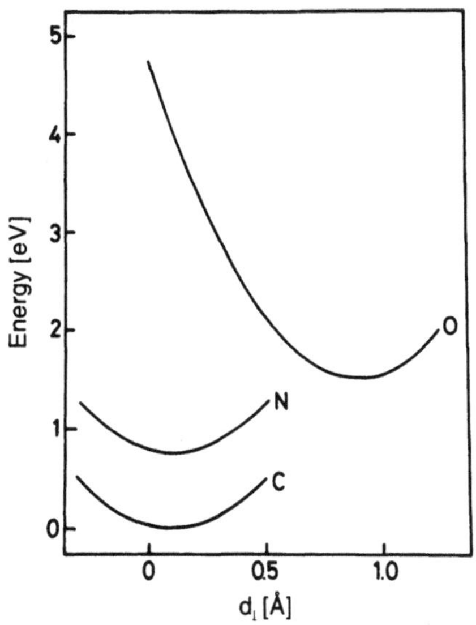

Fig. 3.26. Potential energy curves for the vertical motion of C, N and O in the $c(2 \times 2)$ adsorbate phases derived within the harmonic approximation (see text)

For the O 1*s* ionization, detailed knowledge of the (fluorine-like) final state is lacking. However, it was shown that the observed widths can be reproduced in a consistent way with the same type of model [3.27]. The calculations suggest that the larger widths and the more pronounced temperature effects for the $c(2 \times 2)$ phase are connected to a shallower potential energy minimum for this phase.

Next we will turn to the line widths of adsorbed CO. Chemisorption of CO leads to substantial broadenings of the core level spectra and it has been demonstrated that these broadenings are almost entirely due to vibrational effects [3.26, 36, 88]. Figure 3.27 depicts the O 1*s* and C 1*s* spectra for the CO/Ni(100) $c(2 \times 2)$ structure recorded at 80, 230, 300 and 340 K [3.26]. At 340 K the O 1*s* peak has a FWHM of 1.47 eV. The peak has a pronounced asymmetry towards lower binding energies with a relatively distinct cut-off of the low energy tail. As the temperature is lowered the peak becomes narrower (FWHM = 1.22 eV at 80 K) and more symmetric and it shifts slightly to larger binding energy. The full width at the base of the line, however, remains the same. Similar, but smaller effects are seen also in the C 1*s* spectrum. At 340 K the FWHM for the C 1*s* peak is 0.96 eV. Also in the C 1*s* spectrum a temperature variation is seen with a peak width of 0.71 eV at 80 K.

In order to see what effects determine these widths we first consider the free molecule. In the photoelectron spectrum of CO in the gas phase, which has been recorded under similar experimental conditions, widths of 0.71 eV and 0.56 eV are found in the C 1*s* and O 1*s* spectra, respectively [3.89]. The C 1*s* broadening is due to a shortening of the equilibrium bond length upon core ionization. The effect of an O 1*s* hole is instead to lengthen the bond. A difference in the case of the ionization for adsorbed CO is that the core hole is screened by charge

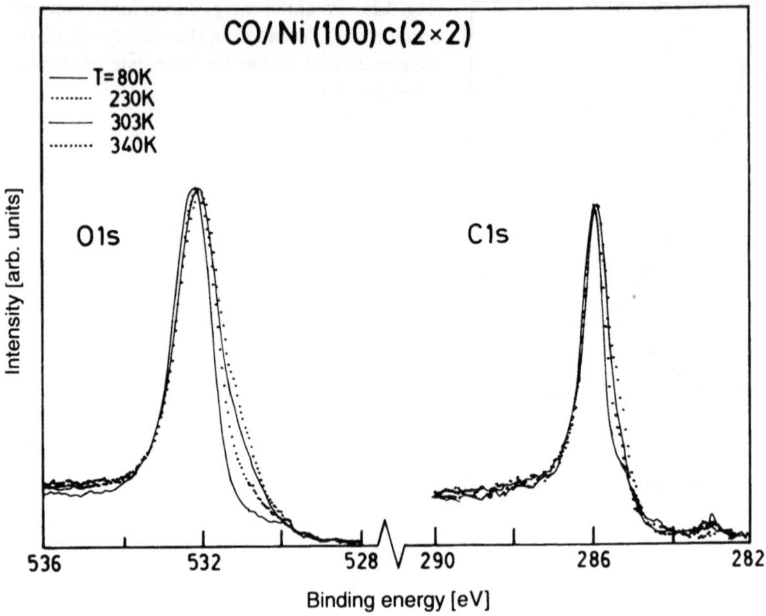

Fig. 3.27. O 1*s* and C 1*s* core line profiles for Co/Ni(100) *c*(2 × 2) recorded at four different temperatures. Both core lines become broader with increasing temperature

transfer from the substrate. To a first approximation the core ionized adsorbate is therefore similar to a core excited neutral CO molecule. The widths of the C 1*s*-2p and O 1*s*-2p transitions have been measured by electron-energy-loss spectroscopy for the free molecule [3.90] and optical absorption (Chap. 6). For the C 1*s*-2p excitation there is no vibrational broadening at all. This is due to the fact that the population of the antibonding 2*p* orbital leads to a bond lengthening which compensates the shortening introduced by the core hole. In the case of the O 1*s*-2*p* excitation the bond lengthening instead becomes even more pronounced, resulting in a width of 1.3 eV. Already, here, we see an explanation for the larger widths of the O 1*s* lines. This conclusion is supported by a comparison to the spectra for $Cr(CO)_6$. The metal-CO bond in this molecule is in many respects similar to the bond for an on top bonded adsorbate and the observed C 1*s* and O 1*s* widths are 0.82 eV and 1.23 eV, respectively [3.91]. Compared to the core-valence excitation spectra for CO it is seen that there is an additional broadening in the C 1*s* spectrum which we interpret as due to geometry changes in the Cr–C bond upon core ionization. The O 1*s* excitation is still dominated by the large bond length increase in the C–O bond.

When CO adsorbs on the Ni(100) surface, new vibrational modes appear which are related to the translational and rotational degrees of freedom of the free molecule. There are, in all, six vibrational modes for a diatomic adsorbate. One vibrational mode v_1 is related to the free molecule C–O stretch. There are three frustrated translation modes: one mode v_2 is connected to the translation normal to the surface (Ni–C stretching mode) and two modes are connected to translations parallel to the surface. The remaining two modes correspond to frustrated rotations. The last modes appear as degenerate pairs denoted ν_4 and ν_3, respectively, in the $c(2\times 2)$ structure.

The ν_1 and ν_2 vibrational energies have been determined to be 256.5 meV and 59.5 meV, respectively [3.92]. The ν_4 mode has been studied by inelastic helium atom scattering, giving a vibrational energy spacing of 3.5 meV [3.93]. From cluster calculations [3.94] and He atom scattering data for CO on Pt(111) [3.95] it can be concluded that the ν_3 mode has a higher energy.

Both the ν_3 and ν_4 modes could give large contributions to hot bands. It has been argued that the frustrated translation modes are of primary importance in the present case. A vibrational energy spacing of 3.5 meV corresponds to a very shallow potential energy surface. This is consistent with the fact that the mobility of CO is large even at 80 K and the general observation that the difference in total energy for different adsorption sites is small [3.96].

In Fig. 3.28 we have schematically drawn initial and final state potential energy curves which qualitatively explain the O 1s spectra. These are shown along lines passing through the top and bridge sites and along the top and hollow sites, respectively. The periodic functions correspond to independent CO molecules at the surface. For the motion of an individual CO-molecule in a more close-packed adsorbate layer one would also have to include the interaction with the neighbouring CO molecules. These differences are, however, not important for the present qualitative discussion.

The ground-state potential energy surface is drawn with a minimum in the top site, an intermediate energy in the bridge site and the highest energy in the

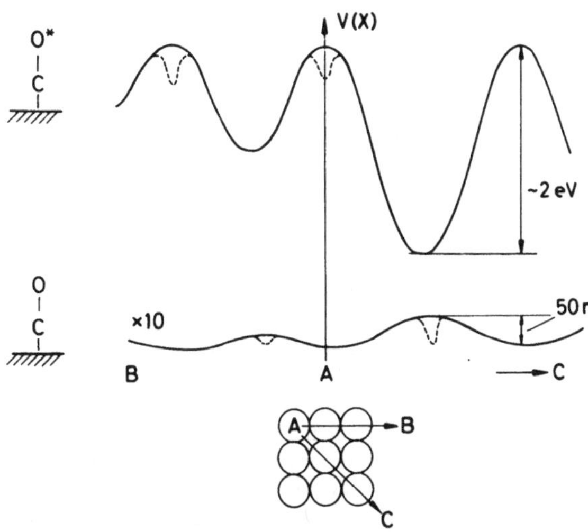

Fig. 3.28. Tentative initial (lower) and O 1s hole state (upper) potential energy curves for CO on Ni(100) along the [10] (arrow A to B) and [11] (arrow A to C) directions

hollow site. This potential-energy surface will lead to a temperature-dependent probability function for the location of the CO molecules at the Ni surface. At low temperatures the molecules will be confined to the region around the top sites. For increasing temperatures there will be an increasing population at larger distances from the top positions, i.e., around the bridge sites and at other intermediate energy sites. In a semi-classical model the spectral line shapes are generated by projecting this distribution onto the final-state potential energy surface [3.88]. The main spectral weight (i.e., the peak position at 532.2 eV) corresponds to transitions from the most probable locations for the CO molecules, i.e. around the top site. The tail towards lower binding energies shows that there is a range of lower final-state energies that can be reached. There are accessible final states with energies as low as 530 eV. The lower intensity of this part of the spectrum is due to the lower population of the corresponding sites in the initial state. From the existence of this low-energy tail it can be concluded that the equilibrium adsorption site (the adsorption site with the lowest energy) for the final state is not the top site. To explain the spectral shape and its

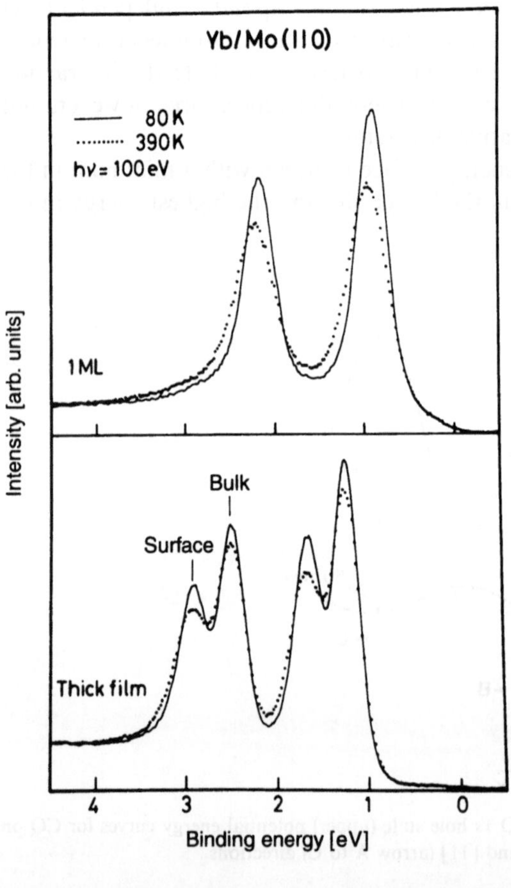

Fig. 3.29. Temperature dependence of the Yb 4*f* photoelectron spectra obtained from one monolayer of Yb on Mo(110) and from a thick epitaxial Yb film. The depositions were made at room temperature. The full-drawn spectra were recorded at 80 K and the dotted ones at 390 K. A photon energy of 100 eV was used

temperature dependence it was proposed that the O 1s ionized state has its lowest final-state energy in the hollow position, the highest energy in the top site and an intermediate energy in the bridge site and that there is a considerable difference in adsorption energies between these sites. This assumption is totally confirmed by the chemical-shift measurements reviewed in Sect. 3.4.

In Fig. 3.27 it is seen that the C 1s line is much narrower than the O 1s line. This difference can be explained directly within the present model. What is required is that the final-state potential energy curve is much shallower for the C 1s final state. As was seen in Sect. 3.4 this is indeed the case. From the discussion in that section it is also clear that in this case one may anticipate a larger influence from frustrated rotational motion. The $Z + 1$ analogue of C 1s ionized CO is NO and, as was discussed in Sect. 3.4, this molecule may, in some cases, adsorb in a tilted fashion. In that case the ionization represents the situation in Fig. 3.23b for the rotational coordinate and that rotational modes may give a larger contribution. To conclude the discussion on CO/Ni(100) we see that a completely consistent description of line shifts and line shapes is obtained.

Temperature effects are also seen in metal spectra. Figure 3.29 exhibits spectra recorded at 80 K and 390 K for 1 ML of Yb on Mo(110) and for a thick Yb film [3.17]. In the monolayer it can be seen that the line width increases with increasir g temperature from 0.42 to 0.57 eV. For the thick film the surface peak shows a similar increase in width from 0.36 eV at 80 K to 0.46 eV at 390 K while the bulk peak only reveals a small change from 0.32 eV to 0.36 eV. The larger width of the surface peak as well as the more pronounced temperature dependence indicates that there are larger vibrational amplitudes at the surface. This can be understood as being due to the increased possibility for movements perpendicular to the surface.

3.8 Shake-up Satellites

In this section we will describe the dynamic response of a system caused by the emission of a core electron. Due to this effect each core line is associated with a rich shake-up spectrum, as shown in the Ne 1s spectrum in Fig. 3.3. The satellites observed in atoms and molecules undergo large changes upon adsorption on surfaces. The intensity and spectral shape of the satellite features depend strongly on the interaction energy and the local character of the adsorption bond. It will be shown that the energy distribution of these shake-up satellites is determined to a large extent by the local electronic structure of the adsorbate-substrate complex in the presence of the core hole. The shake-up intensities can be used to probe certain properties of the ground state system.

After a general discussion of the shake-up process, a number of examples will be given for different types of adsorption ranging from physisorption to strong chemisorption for atomic as well as molecular adsorbates. We will describe how the image-potential screening and the polarization of the surrounding adsorbate atoms are manifested in the core-level spectra from weakly adsorbed Ar atoms. For strongly chemisorbed atoms, excitations related to the chemisorption bond can be identified. For oxygen on Ni, a localization of the valence states on the core ionized atom leads to satellite shapes which are dominated by the exchange splitting between the core and valence shells. The spectra from adsorbed molecules such as CO and N_2 give rise to intense shake-up satellites. In the case of CO adsorbed on Ag, the so-called *main line* is only a weak feature compared to the giant satellite lines. The satellites in molecular adsorbates on various substrates often have the character of excitations between bonding and antibonding states related to the chemisorption bond.

3.8.1 General Aspects

The creation of shake-up satellites is due to the fact that the removal of a photoelectron is much faster than the time required for the rearrangement of the valence electron charge distribution. For the theoretical treatment of the photoionization process the Hamiltonian of the system is considered to change abruptly upon ionization whereas the charge distribution due to the remaining electrons is continuous as a function of time. As a consequence the wave function of the remaining electrons does not correspond to an eigenstate of the final-state Hamiltonian. It has to be expressed as a linear combination of eigenstates of the new Hamiltonian. These eigenstates are observed as separate structures in the spectrum [3.97, 98].

The assumption of an infinitely short time for the removal of the photoelectron is referred to as the sudden approximation. The time for the photoemission event depends on the velocity of the escaping photoelectron and the sudden limit therefore corresponds to high excitation energies. Closer to the core ionization threshold the adiabatic limit is instead approached with, for instance, reduced satellite intensities. The sudden approximation is, in general, quite appropriate in the case of XPS and is the basis for most theoretical treatments of the photoionization process [3.99–101]. It is, however, difficult to formulate strict criteria for the applicability of the sudden approximation, and the energy range within which it is valid is still, to a certain extent, an open question.

Some important sum rules can be derived within the sudden approximation [3.99–101]. As mentioned above, one cannot observe directly the energy of removing an electron from the system while letting all the other electrons in the system remain in their original states. This hypothetical state would be observed if the final-state orbital wave functions were the same as in the initial state. This ionization energy is generally called the *Koopmans energy*. The deviation from this situation depends on how much the orbitals relax as a function of the removal of the core electron. However, it can be shown that the Koopmans energy corresponds to the center of gravity of the complete spectrum including the shake-up satellites. In general, the main line corresponds to the energetically lowest accessible eigenstate of the core ionized system. If the relaxation energy is large for the lowest state, the sum rule implies that there is considerable intensity for the higher excited shake-up and shake-off states.

Shake-up in atoms and molecules is usually described in terms of orbital excitations in the core-hole system. Only shake-up states which have the same total symmetry as the main core-hole state can appear in the spectra. In general, there are a few transitions which dominate the spectra. Usually it is found that the strongest shake-up transitions involve monopole excitations of the valence electrons, i.e., excitations from occupied orbitals (nl) to higher orbitals with the same angular momentum ($n'l$) quantum number.

In solids it is common to distinguish between a few types of shake-up processes depending on the character of the excitations. In metals the core-electron lines are always asymmetric (Fig. 3.22). This is due to shake-up excitations of low-energy electron-hole pairs. Since the density of states is finite at the Fermi level, the excitations have energies ranging all the way from zero. This type of process generates an asymmetric tailing of the core-electron lines towards the high-binding-energy side. Several theoretical treatments of the resulting line shapes have been performed. The most common parametrization which is due to *Doniach* and *Sunjic* [3.79] is based on an analysis of the asymptotic behaviour for long response times of an electron gas to the creation of the core-hole. This type of line shape is often found to describe the experimental peaks quite well and is often used when numerically fitting experimental spectra. There may, however, be other mechanisms as well (Sect. 3.7) which produce asymmetric broadenings which are then artificially taken care of by this model line profile.

Other shake-up excitations are due to plasmons and interband transitions. The energies of these transitions are often rather insensitive to the presence of the core hole and therefore coincide with the energies of the inelastic loss processes. For this reason it may often be difficult to distinguish between intrinsic (shake-up) and extrinsic loss processes for solids. Some of the interband transitions, however, may have a more localized character and are thereby to a larger extent influenced by the core hole. In Ni, for instance, there are shake-up satellites which correspond to different local $3d$ configurations in the presence of the core hole.

A major difference between core ionization of a free and an adsorbed molecule is connected to the additional screening in the latter case (Sect. 3.4.1). The charge redistribution connected with the screening from the substrate leads to additional shake-up features connected to the new screening channels. There are two different types of shake-up excitations that become important for an adsorbate. First, there are local excitations confined to the adsorbate-substrate complex. In this case it is important to explicitly take the effect of the core hole into account. Second, there are excitations in the substrate due to the long-range screening of the adsorbate.

3.8.2 Substrate Excitations for Adsorbed Argon

The image charge screening for weakly bonded adsorbates was described in Sect. 3.4.1. The ionization of an adsorbed rare-gas atom will be screened in this way. There are cases when the bonding of rare-gas atoms to the substrate could involve some chemical contribution. However, even if this is the case the charge transfer part of the screening will only involve very small charges and the image potential screening will dominate, as discussed in Sect. 3.4.1.

Figure 3.30 compares Ar 2*p* spectra recorded for the free atom [3.22] and three different situations of Ar adsorbed or graphite: a multilayer (14 ML) of Ar, 1 ML adsorbed on two spacer layers of Xe, and 1 ML adsorbed directly on the

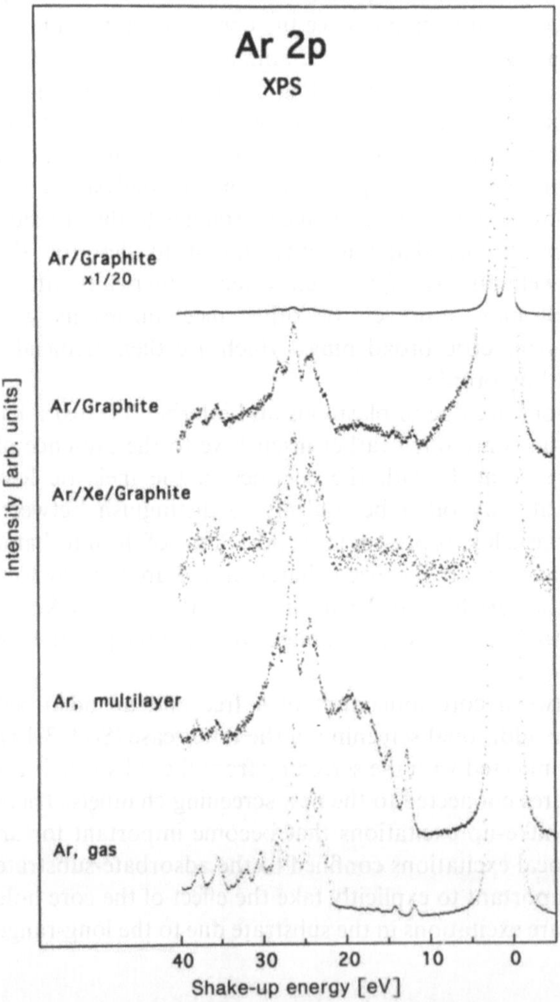

Fig. 3.30. XP spectra showing the Ar 2*p* shake-up region for Ar in the gas phase, 14 monolayers on graphite, 1 monolayer adsorbed on two layers of Xe condensed on graphite and 1 monolayer adsorbed on graphite. The spectra have been aligned to the position of the Ar $2p_{3/2}$ peak and normalized to the intensity of the 2*p* main peaks. The energy scale is relative to the $2p_{3/2}$ ionization energy

graphite surface [3.102]. The gas-phase shake-up spectrum is strongly dominated by monopole transitions [3.22]. The shake-up energy region 20–35 eV is due to excitations of $3p$ electrons into Rydberg levels, i.e. $3p$-np transitions. At higher shake-up energies, satellites due to $3s$-ns transitions appear. These excitations are caused by the relaxation of the atomic orbitals upon the creation of the core hole. The weak features appearing in the region 12–20 eV are attributed to inelastic scattering of the outgoing electrons through $3p$-ns dipole excitations.

In the multilayer situation, the atomic shake-up features at 20–40 eV are still present. The main lines maintain their Lorentzian shapes and there is a large band gap in the spectrum. The most pronounced difference compared to the gas-phase spectrum is that the inelastic scattering contribution above 12 eV has increased considerably. This is to be expected for a condensed Ar film where the outgoing electrons have to pass through many Ar layers. The Ar/Xe/graphite spectrum has a number of similarities with the multilayer spectrum. The main lines are Lorentzian-shaped. There is a large band-gap in the spectrum and the atomic shake-up features in the region 20–40 eV are essentially identical for the two situations. Note, also, that there is still some intensity in the energy region corresponding to the atomic dipole excitations.

The spectrum obtained when a monolayer of Ar is adsorbed directly on graphite resembles that of the Ar/Xe/graphite system. The atomic monopole shake-ups are identical and appear at the same shake-up energy as in the Ar/Xe/graphite spectrum. The atomic excitations in the energy region 12–20 eV are also present for Ar/graphite. However, it is immediately seen that extra intensity has emerged on the high-binding-energy side of the main lines, which is not present in the previous situations.

For a physisorbed system, there are two screening contributions which have to be considered, in addition to the atomic relaxation: screening via the formation of an image charge in the surface and screening due to the polarization of the surrounding atoms in the adsorbate layer (Sect. 3.4.5). If these excitations contribute to the screening of the core hole they have to be connected to characteristic excitations in the substrate as well as in the surrounding Ar atoms. The first process explains the shake-up intensity close to the main lines, whereas the latter one gives rise to the atomic excitations observed in the region 12–20 eV.

For a number of reasons it was concluded that the atomic dipole excitations are not dominated by inelastic processes. They are part of the shake-up spectrum [3.102]. For instance, the spectral features disappear at low coverages where Ar adsorbs in a disordered phase with well separated atoms indicating that these excitations are due to the polarization of the surrounding Ar atoms.

The shake-up processes involving substrate excitations can be identified by comparing with inelastic scattering data. For graphite, the electron-energy-loss spectrum reveals a distinct loss at 6 eV, which is interpreted as due to a plasmon exciton [3.103]. If the ionization of an Ar $2p$ electron is accompanied by this plasmon excitation, a 6 eV satellite is expected for each of the two spin-orbit split main lines, i.e., at shake-up energies of about 6 and 8 eV relative to the

$2p_{3/2}$ ionization energy. This is in agreement with the spectrum in Fig. 3.30. This satellite is therefore interpreted as due to the image charge screening.

It is now interesting to compare the image potential screening induced shake-up excitations for Ar adsorbed on other substrates. Figure 3.31 displays the spectra for 1 ML of Ar adsorbed on graphite, Ni(100) and Ag(110). The shake-up peaks related to the atomic relaxation are still present, but they are considerably broadened for adsorption on Ni and Ag. Such broad shake-up peaks are generally associated with chemisorbed species, as will be explained in Sect. 3.8.3. Ar on Ag(110) has recently been found to behave spectroscopically like a chemisorbed system [3.104]. The broadening is thus explained by a hybridization between the unoccupied atomic Ar levels and the unoccupied bands of the metal substrate. The same conclusion can be made for Ar/Ni(100).

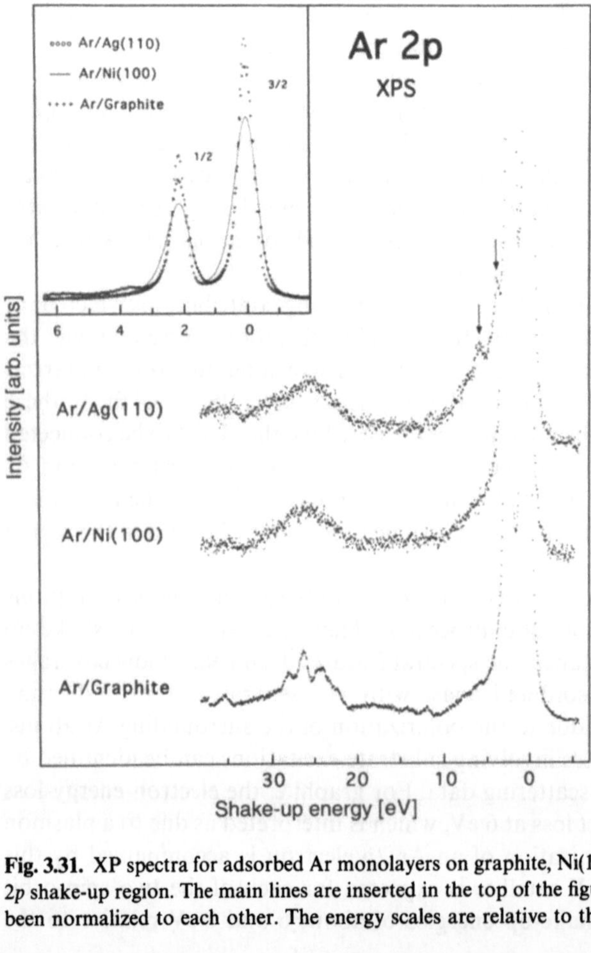

Fig. 3.31. XP spectra for adsorbed Ar monolayers on graphite, Ni(100) and Ag(110) showing the Ar $2p$ shake-up region. The main lines are inserted in the top of the figure. The areas of the peaks have been normalized to each other. The energy scales are relative to the $2p_{3/2}$ ionization energy

If we now turn to the low-energy shake-up energy region, two sharp peaks in the Ar/Ag-spectrum are observed at 3.7 and 5.8 eV, respectively. In the Ag 3*d* XP spectrum, sharp satellites appear at 3.8 eV below the main peaks corresponding to unresolved bulk (3.87 eV) and surface (3.63 eV) plasmon losses [3.105–107]. The two peaks in the Ar 2*p* spectrum are interpreted as due to Ag plasmon excitations accompanying the ionization of the two Ar 2*p* components. Since the polarization charge is set up at the surface, it is most likely that the shake-up process is determined by excitations of Ag surface plasmons.

A comparison between the Ar 2*p* main peaks for the three systems is shown in the inset of Fig. 3.31. The 2*p* peaks are much broader in the case of Ar/Ni, whereas the widths of the 2*p* peaks for Ar/Ag are similar to those of Ar/graphite. The broadening effect for Ar/Ni is explained in terms of substrate excitations which have a much larger contribution at low energies than for Ar/Ag [108]. In addition to these low-energy excitations, there are substrate excitations stretching down to about 10 eV for this system, as seen in Fig. 3.31.

The features due to substrate excitations and the excitation of surrounding Ar atoms have a total intensity of $7 \pm 4\%$. The part due to atomic relaxation amounts to $10 \pm 5\%$, which can be compared to the value for the free atom ($13 \pm 5\%$). It may be added that the decrease of the atomic contribution is significant and is an effect of the presence of the substrate. The shake-up intensities are related to the degree of modification of the valence orbitals upon ionization. Therefore, it seems that the presence of an image screening charge in the substrate leads to a smaller contraction of the orbitals for the adsorbate than for the free atom.

3.8.3 Local Excitations in Strongly Chemisorbed Atoms

For chemisorbed atoms the possibility of metallic screening will give rise to new shake-up processes connected to the charge transfer from the substrate to the core ionized adsorbate. This extra screening channel will change the valence electronic structure of the adsorbate significantly when the core hole is created. From the series of atomic adsorbates, C, N and O on Ni(100), we will demonstrate that the dominating shake-up processes in these types of systems are mainly localized to the adsorbate-substrate complex.

Figure. 3.32 shows the satellite spectra from the $c(2 \times 2)$ overlayers of C, N and O on Ni(100) [3.109]. Both the energy and intensity of the satellite peaks increase from C to N to O. The main shake-up peaks in the C1*s* and N1*s* spectra appear at 7.0 eV and 10 eV above the main lines, respectively. In the O 1*s* spectrum the satellite is split into at least three states with shake-up energies 10.6 eV, 13.2 eV and 20 eV with maximum intensity for the 13.2 eV state.

The atomic adsorbates form hybrid states between the 2*p* and metal 4*sp* states [3.110, 111]. The occupied 2*p*–4*sp* bonding states appear in the region 4–6 eV below the Fermi level and have mainly adsorbate 2*p* character. There is also some hybridization between 2*p* and metal 3*d* states giving rise to states which are located closer to the Fermi level [3.111]. It has been shown by XAS on

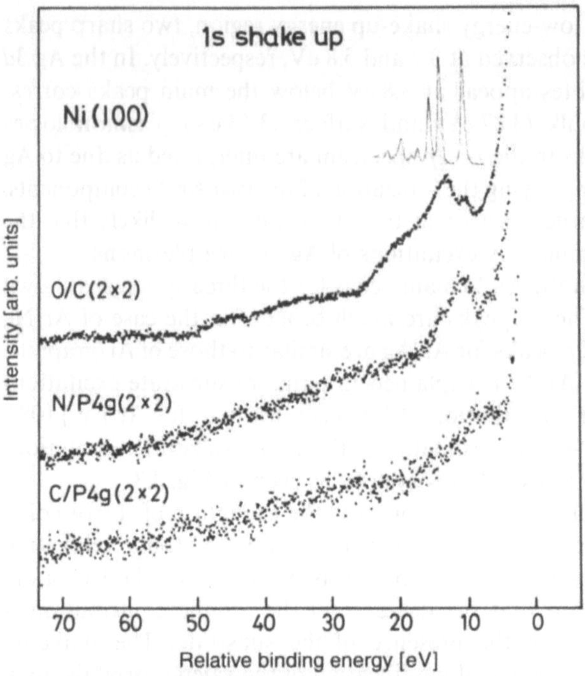

Fig. 3.32. 1*s* shake-up spectra of C, N and O adsorbed on Ni(100). The backgrounds from the clean surface have been subtracted. The spectra are aligned according to the main lines. The 2*p*-*np* region of the Ne 1*s* shake-up spectrum from [3.21] has been inserted for comparison with the oxygen shake-up spectrum. The Ne spectrum is shifted by 27 eV to allow the first shake-up peaks in the spectra to coincide

atomic adsorbates, that there exist states with 3*p* character on the adsorbate in the region 3–5 eV above the Fermi level [3.110]. The satellite peaks in Fig. 3.32 are related to shake-up excitations from the bonding 2*p*–4*sp* hybrid states to 3*p* states [3.109]. For the free atom this corresponds to monopole excitations. The energies of these transitions can be estimated using the $Z+1$ approximation. The shake-up excitations in the C 1*s*, N 1*s* and O 1*s* spectra are then viewed as valence electron excitations in adsorbed nitrogen, oxygen and fluorine, respectively. The valence excitation energy increases with atomic number for the present adsorbates which is also seen from a comparison with optical data for the free atoms [3.112].

In the O 1*s* spectrum there is some fine structure which is not seen in the spectra from the other adsorbates. The final state in the O 1*s* ionization is a fluorine-like species. The adsorption of fluorine on metals has a strong ionic character, leading to an F^- ion with a closed 2*p* shell [3.113]. This implies that the O 1*s* final state of the adsorbate is essentially isoelectronic to a free Ne atom. This makes it interesting to make a comparison with the shake-up spectrum of Ne (Fig. 3.3). A shifted part of the 2*p*–*np* shake-up region in Ne [3.21] is inserted in Fig. 3.32 above the O 1*s* spectrum. A striking similarity between the two spectra may be noted. The interpretations of the individual peaks in the shake-up spectra are similar. The first two peaks at 10.6 and 13.2 eV in the oxygen spectrum are assigned to 2*p*–3*p* excitations split into upper and lower states by exchange interaction with the 1*s* core hole. The shoulder of the 13.2 eV peak and the broad structure at around 20 eV are interpreted as excitations into higher members of the *np* lower and upper Rydberg series. There are also some tendencies of fine structure in the shake-off continuum related to 2*s*–*ns* excitations and double shake-up processes in Ne (see Fig. 3.3).

The 1*s*–2*p* exchange splitting in the O1*s* satellite spectrum is a direct signature of a localization tendency of the 2*p* shell. This may be compared with the well-known satellite in the Ni 2*p* spectrum where a similar localization of the 3*d* shell in presence of the core hole leads to multiplet splitting of the local 2*p*–3*d* configuration [3.114].

There are parts of the O 1*s* spectrum which are more directly related to the interaction with the substrate. In Fig. 3.33, the O 1*s* spectra from oxygen on Cu(100) and Ni(100) are compared [3.110]. There are clear differences both in line widths and line profiles between the two oxygen core levels. The peak from O/Ni is substantially broader than the O/Cu peak and also more asymmetric. For O/Cu, the line is almost symmetric. In this case a weak but distinct shake-up

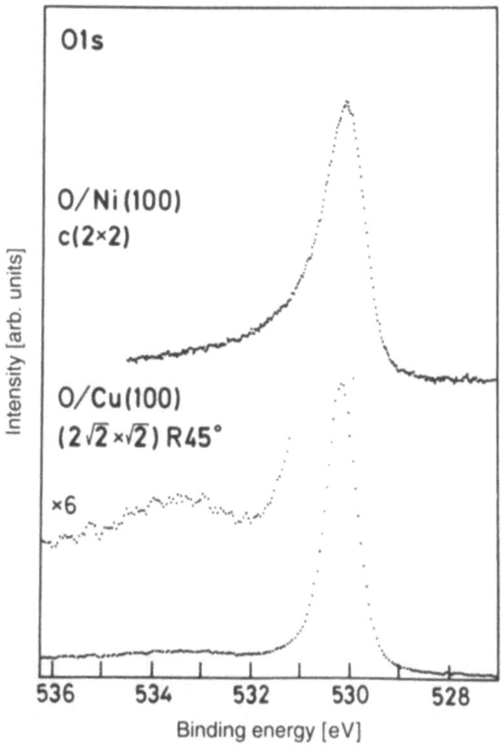

Fig. 3.33. XP spectra of the O 1*s* level in the $c(2\times2)$ and $p(2\sqrt{2}\times\sqrt{2})R45°$ phases from O on Ni(100) and Cu(100), respectively. The spectra were recorded at 80 K

structure may be observed about 3 eV from the main line. The 2*p*–3*d* hybrid states can explain these differences [3.110]. In Ni, these states are located in the vicinity of the Fermi level. This enables low energy shake-up excitations which give rise to the asymmetric line profile. For Cu these states are nearly filled and located 2 eV below the Fermi level resulting in a discrete shake-up feature.

3.8.4 Chemisorption-Induced Excitations in Adsorbed CO and N_2

The adsorption of CO and N_2 leads to spectacular shake-up spectra with intensities which are much higher than for the main lines. The interpretation of these giant satellites has been much discussed for more than 15 years [3.31, 34, 115–125]. The various models were recently reviewed [3.126]. Only the interpretation proposed by the present authors will be given below.

Figure 3.34 shows the C 1*s* and O 1*s* shake-up spectra for the CO/Ni(100) $c(2 \times 2)$ structure and for $Cr(CO)_6$ measured in the gas phase [3.31, 91]. The comparison with carbonyls is interesting since the local bonding between adsorbed CO and the substrate is similar to the bonding in the carbonyls. The features in the carbonyl spectra are sharper and the shake-up peaks are better resolved. However, there is a striking similarity between the positions of the main features for adsorbed CO and for the carbonyl. The same characteristic differences between the C 1*s* and O 1*s* spectra are seen for both systems. The

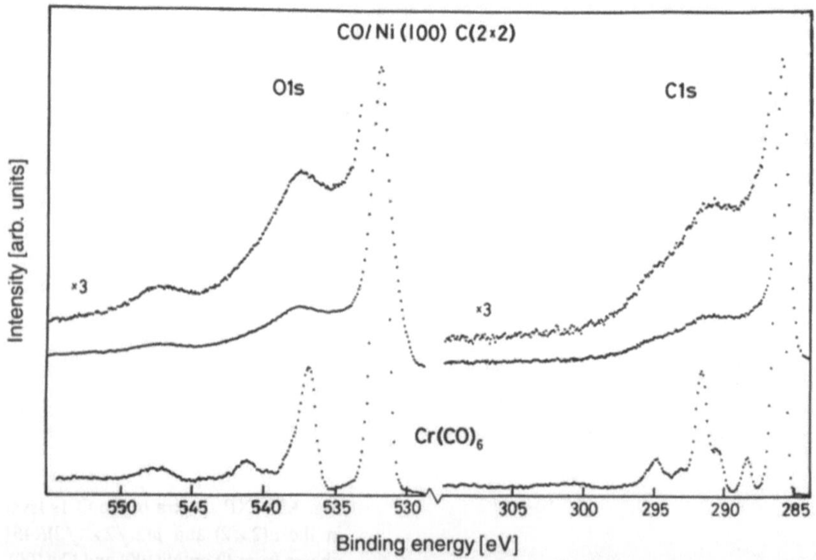

Fig. 3.34. XP spectrum of the C 1*s* and O 1*s* regions for CO/Ni(100) $c(2 \times 2)$ and $Cr(CO)_6$ in the gas phase. The clean Ni background spectrum has been subtracted. The energy scale of the $Cr(CO)_6$ spectra has been shifted by 7.3 eV (C1*s*) and 7.5 eV (O1*s*) in order to allow the main lines to coincide. The spectra were recorded at room temperature

similarity between the two sets of spectra demonstrates that the shake-up energies are dominated by the local electronic structure of the metal-CO system.

The C 1*s* and O 1*s* shake-up spectra are significantly different. Three shake-up structures are observed in both spectra, although the shake-up features appear at quite different energies. The difference between the C1*s* and O1*s* shake-up spectra demonstrates the necessity of taking the effects of the core hole explicitly into account when treating the shake-up transitions.

Before discussing the interpretation of the shake-up spectra, we will make a comparison with spectra from N_2 on Ni(100) [3.34]. The N_2 molecule is isoelectronic with CO and the spectra show some similarities in the gas phase. The adsorption of N_2 in a perpendicular geometry leads to two inequivalent nitrogen atoms, as discussed in Sect. 3.4.2. The shake-up spectra from the two different atoms have been separated using the forward scattering of the photoelectrons emitted from the innermost nitrogen atom [3.34]. The separated spectra from the two different nitrogen atoms are shown in Fig. 3.35. The shake-up spectra are significantly different, which further demonstrates the local character of the core ionization process.

It is interesting to compare the spectrum from the outer nitrogen atom with the C 1*s* shake-up spectrum from adsorbed CO. This can be seen by using the $Z + 1$ approximation of the core ionized final states. Both lead to a final state similar to an adsorbed NO molecule bonding with the nitrogen end towards the

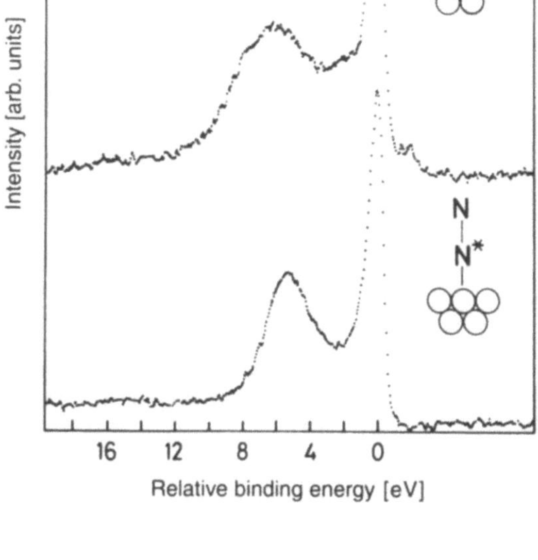

Fig. 3.35. Separated N 1*s* spectra for N_2 on Ni(100) in a $c(2 \times 2)$ structure. The spectra are aligned according to the main line positions. The separation was carried out by measuring the N 1*s* spectra at two different angles with respect to the surface normal, 5° and 35°, respectively, and making two subtractions. The intensity from the inner nitrogen atom is strongly enhanced for emission angles along the surface normal

surface. Figure 3.36 compares the C 1*s* spectrum in CO and the N 1*s* in N_2 from the outer nitrogen atom [3.34]. There is a clear similarity between the main shake-up features in the two spectra which show the same three-peak structure. Only the third state appears at slightly different energies. However, the intensities of the various shake-up states in the two spectra are quite different. The C 1*s* satellites are generally about a factor of two less intense.

The three shake-up states have different origins [3.34]. The first one is related to excitations between adsorbate-substrate bonding to antibonding states, the second peak involves higher and more extended adsorbate states derived from the molecular np Rydberg states, and the third is derived from an internal molecular excitation which is also seen in the free molecule.

In Sect. 3.4.1, the interaction of the CO $2\pi^*$ orbital with the Ni substrate was described [3.11, 31]. This interaction leads to splitting of the $2\pi^*$ mainfold into bonding ($2\pi_b$) and antibonding ($2\pi_a$) states with the Ni 3*d* states. Furthermore, these states are strongly perturbed by the presence of the core hole which is illustrated in Fig. 3.8. The core-hole and the charge transfer in the metallic screening process change the character of the $2\pi_b$ states. It can therefore be anticipated that the $2\pi_b$ states become strongly involved in the various shake-up excitations. The first peak at around 2 eV, Fig. 3.36, can be interpreted as transitions between the $2\pi_b$–$2\pi_a$ states in the final-state molecule, as depicted in Fig. 3.8b.

We can now describe the difference in the N 1*s* spectra for the two different nitrogen atoms in Fig. 3.35 [3.34]. As was shown in Sect. 3.4.2, the final state

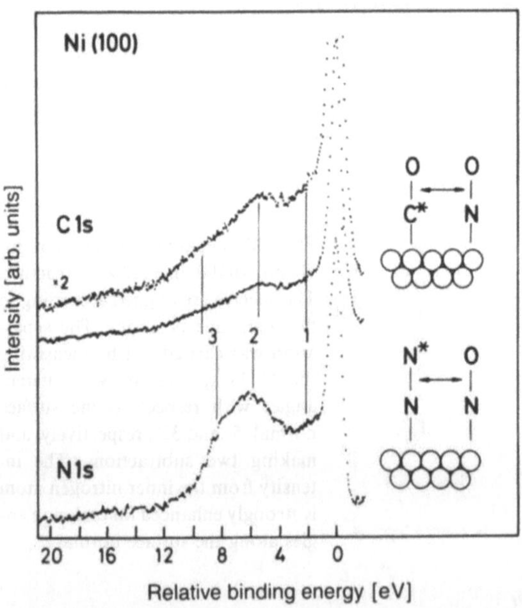

Fig. 3.36. C 1*s* spectrum from CO adsorbed on Ni(100) in a *c*(2 × 2) structure and N1*s* spectrum from the outer nitrogen atom of N_2 on Ni(100). The spectra have been recorded at 80 K. The spectra are aligned according to the main line positions

NO-like molecule interacts very weakly with the substrate in the orientation with the oxygen end down. The weak adsorbate-substrate interaction leads to negligible splitting of the $2\pi^*$ level, which is why no satellite associated with the $2\pi_b$–$2\pi_a$ transition is observed at 2 eV.

The strong satellites at around 5–6 eV which are seen in all spectra are assigned to transitions of the $2\pi_b$ screening states into 3*s*, 3*p* and 4*p* derived states [3.31]. These Rydberg derived states have recently been observed in XA spectra of adsorbed CO and N_2 on Ni [3.127]. The Rydberg orbitals observed in the free molecule survive upon chemisorption and form new hybridized adsorbate-substrate states. Similar valence-electron excitation energies have been observed in the analogue NO/Ni. These spectra show a broad electronic transition centered around 5.5 eV [3.128]. This peak has been assigned to transitions from the $2\pi^*$ level into Rydberg states. However, recent INDO-CI calculations on $Mo(CO)_6$ show that the excitation of the carbonyl complex is split into a triplet and a singlet parent coupled state of the $2\pi_b$ to $2\pi_a$ transition due to exchange interaction of the open shells [3.129]. The singlet state appears at 6 eV above the main line. It is therefore possible that some of the intensity in this region for the adsorbate has this origin and may overlap with the Rydberg transitions.

The third peak in the shake-up spectra nearly coincides in energy with the first C 1*s* and N 1*s* shake-up states in free CO and N_2. This transition has been assigned to a 1π–$2\pi^*$ excitation related to the intramolecular relaxation [3.89].

In Figs. 3.37 and 38, we compare the C 1*s* and O 1*s* spectra for CO chemisorbed on Ni(100), Cu(100) and Ag(110) [3.126]. The spectra show the so-called *giant satellites* in the case of adsorption on Cu and Ag. The main line in the C 1*s* spectrum for CO on Ag(110) is only a tiny structure preceding the totally dominating satellites. A lot of fine structure is observed, resulting in rather complicated spectra. We will concentrate on the features in C 1*s* spectra in relation to the discussion of CO and N_2 on Ni; a more detailed description has been given by *Tillborg* et al. [3.126].

The $2\pi_b$ states for NO are due to the hybridization with the metal *d*-states. In Ni, the *d*-band is not completely filled and the $2\pi_b$ states are found around 1 eV below the Fermi level. In Cu and Ag, the *d*-band is filled and positioned about 2 eV and 4 eV below the Fermi level. It can therefore be anticipated that the $2\pi_b$ states shift downwards in energy when NO is chemisorbed on Cu and Ag. The $2\pi_a$ states as probed by XAS, are located just above the Fermi level on all the substrates [3.30]. The $2\pi_b$–$2\pi_a$ shake-up transition energy will therefore increase going from Ni to Cu to Ag. This is observed in the C 1*s* shake-up spectrum in Fig. 3.37 where the corresponding excitation energy goes from 2 eV in Ni to 3 eV in Cu and 4.2 eV in Ag. Furthermore, the intensities of the shake-up transitions increase dramatically as the *d*-band is shifted to higher energies. Recent studies for CO on Au(110) confirm this picture [3.73]. The *d*-band in Au is located between the *d*-bands in Cu and Ag. The shake-up spectrum has an appearance which places it between CO on Cu and CO on Ag both in terms of energies and intensities.

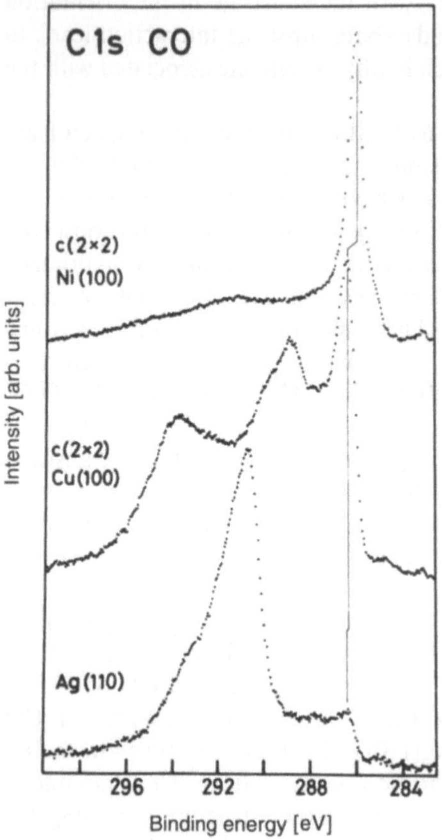

Fig. 3.37. C 1*s* region from CO adsorbed on Ni(100), Cu(100) and Ag(110). For the two former substrates, the overlayer structures involve purely on top adsorption sites. For CO/Ag(110) the adsorption geometry and site is not known

The satellite intensity is related to the difference in electronic structure between the initial and lowest energy final states. A large shake-up intensity indicates that the lowest final state deviates strongly from (is close to orthogonal to) the Koopmans' state. For adsorbed CO, it can be argued that the shake-up intensity should relate to the population of the molecular $2\pi_b$ orbital in the ground state (adsorbed CO without core-hole). This orbital is populated by approximately one electron in the core-ionized molecule, because of charge transfer screening from the metal (Sect. 3.4.1). The relative change between the initial and final states, and thus the satellite intensity, is larger if the ground-state population is low. This indicates that the ground state CO-$2\pi^*$ population in the $2\pi_b$ state is lower for CO on Cu and Ag than on Ni, which can be expected because of smaller interaction strengths.

The same arguments can be used in the NO-like final states for CO and N_2 on Ni(100) in Fig. 3.35. The fact that the shake-up intensity is lower in the C 1*s* spectrum in CO than in the N 1*s* spectrum in N_2 may be taken as a sign of a larger occupation of the $2\pi^*$ orbital in CO. Due to the difference in electronega-

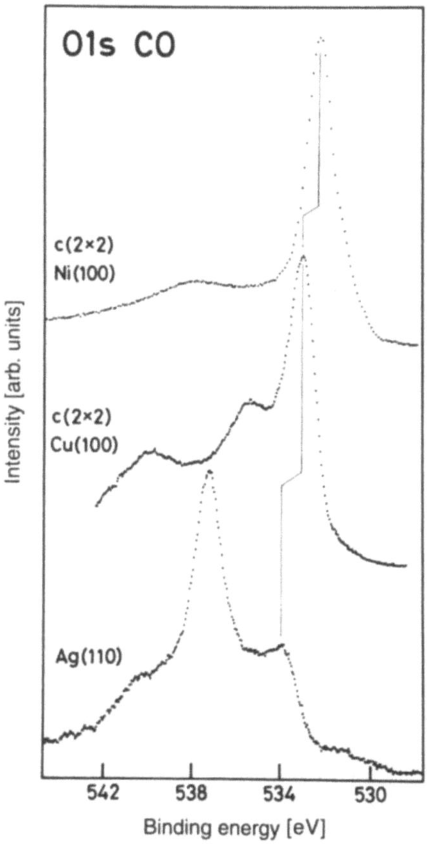

Fig. 3.38. O 1*s* region from CO adsorbed on Ni(100), Cu(100) and Ag(110). For the two former substrates, the overlayer structures involve purely on top adsorption sites. For CO/Ag(110) the adsorption geometry and site is not known

tivity between C and O the $2\pi^*$ orbital is more localized on the carbon atom allowing for a larger overlap with metal states in the substrate than for N_2.

In order to make a correct estimation of the satellite intensity, relative to the main line, it is necessary to include the complete shake-off continuum. The extended satellite spectra for the C 1*s* lines from CO adsorbed on Ni(100) (lower) and Cu(100) (upper) are depicted in Fig. 3.39 [3.126]. It is seen that a substantial part of the whole intensity extends up to 70 eV above the main lines. Similar shake-off intensities are observed in the O 1*s* spectrum from CO/Ni(100) and in N_2/Ni(100). Two structures are clearly seen in the CO/Cu(100) spectrum at about 23 and 35 eV above the main line. A broad feature is centered at 35 eV in the spectrum from CO/Ni(100). States at these energies have been observed in the C1*s* spectrum from $Cr(CO)_6$. The assignment of these states is not straightforward but can be regarded as shake-off resonances, as discussed in [3.31]. The C1*s* main line/total spectrum intensity ratios are estimated to be 0.12 and 0.29 for CO on Cu(100) and Ni(100), respectively. These ratios may be compared to the corresponding ratio of 0.67 for the free molecule. The observation of satellite

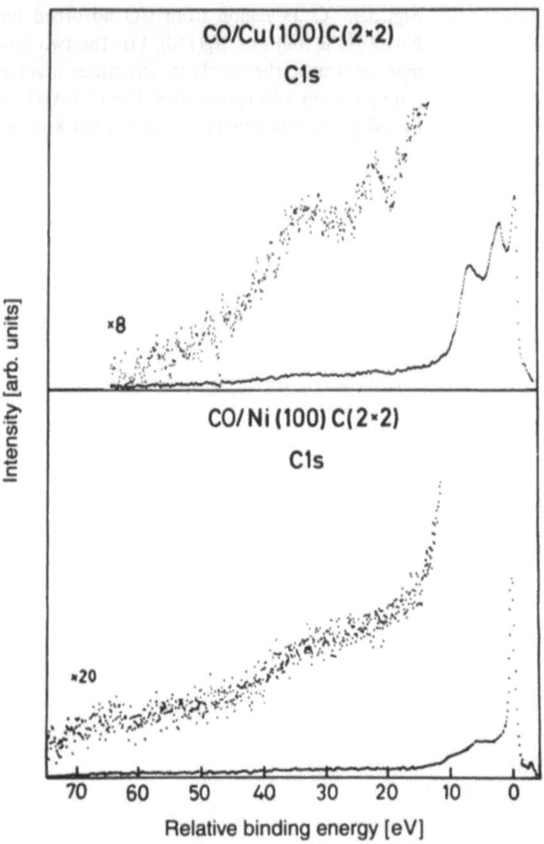

Fig. 3.39. Extended C 1*s* shake-up and shake-off spectra from CO adsorbed in a $c(2 \times 2)$ structure on Ni(100) and Cu(100). The spectra are aligned according to the main line positions and normalized to give the same main line peak height. Note the different enlargement of the two shake-off continua

intensity up to more than 70 eV above the main line is important not only in the use of core-level photoemission for quantitative analysis but also for the fundamental understanding of the photoionization process in adsorbates.

Acknowledgements. We gratefully acknowledge the people involved in the various projects on which this chapter is based, especially the former students Helena Tillborg, Anders Sandell, Olle Björneholm, Anders Stenborg and Erik Zdansky and the present students Bo Hernnäs and Peter Bennich. We also acknowledge the many fruitful discussions with Svante Svensson, Hans Ågren, Sten Lunell, Ulrik Gelius, Jesper Andersen, Ralf Nyholm, Börje Johansson, Richard Palmer, Jo Stöhr and Paul Brühwiler. The technical assistance of Jan-Olof Forsell has been invaluable.

References

3.1 K. Siegbahn, C. Nordling, G. Johansson, J. Hedman, P.F. Hedén, K. Hamrin, U. Gelius, T. Bergmark, L.O. Werme, R. Manne, Y. Baer: *ESCA Applied to Free Molecules* (North-Holland, Amsterdam 1969)

3.2 C.S. Fadley: *Electron Spectroscopy: Theory, Techniques and Applications* (Academic Press, New York 1978) Vol. 2
3.3 B. Feuerbacher, B. Fitton, R.F. Willis (eds.): *Photoemission and the Electronic Properties of Surfaces* (Wiley, New York 1978)
3.4 D. Menzel: In *Chemistry and Physics of Solid Surfaces*, ed. by R. Vanselow. (CRC Press, Boca Raton, FL 1979) Vol. II, p. 277
3.5 H. Siegbahn, L. Karlsson: In *Corpuscles and Radiation in Matter*, ed. by W. Mehlhorn, Handbuch der Physik, Vol. 31 (Springer, Berlin, Heidelberg 1982), vol. XXXI
3.6 M. Scheffler, A.M. Bradshaw: In *The Chemical Physics of Solid Surfaces and Heterogenous Catalysis*, ed. by D.A. King, D.P. Woodruff, (Elsevier, Oxford 1983) Vol. 2, p. 165
3.7 E.W. Plummer, C.T. Chen, W.K. Ford, W. Eberhardt, R.P. Messner, H. J. Freund: Surf. Sci. **158**, 58 (1985)
3.8 W.F. Egelhoff: Surf. Sci. Reps. **6**, 253 (1986)
3.9 N. Mårtensson, A. Nilsson: J. Elec. Spec. and Relat. Phenom. **52**, 1 (1990)
3.10 J. Stöhr: *NEXAFS Spectroscopy* Springer Ser. Surf. Sci., Vol. 25 (Springer, Berlin, Heidelberg 1992)
3.11 A. Nilsson, E. Zdansky, H. Tillborg, O. Björneholm, N. Mårtensson, J.N. Andersen, R. Nyholm: Chem. Phys. Lett. **197**, 12 (1992)
3.12 J.C. Fuggle: In *Electron Spectroscopy IV* ed. by C.R. Brundle, A.D. Baker, (Academic London 1981) p. 85
3.13 O. Björneholm, A. Sandell, A. Nilsson, N. Mårtensson, J.N. Andersen: Physica Scripta **T41**, 217 (1992)
3.14 J. Nordgren, N. Wassdahl: Physica Scripta T **31**, 103 (1990)
3.15 N. Wassdahl, A. Nilsson, T. Wiell, H. Tillborg, L.-C. Duda, J.H. Guo, N. Mårtensson, J. Nordgren, J.N. Andersen, R. Nyholm, Phys. Rev. Lett. **69**, 812 (1992)
3.16 C.S. Fadley, In *Synchrotron Radiation Research: Advances in Surface and Interface Science*, ed. by R.Z. Bachrach (Plenum, New York 1992) p. 421
3.17 A. Nilsson, A. Stenborg, O. Björneholm, N. Mårtensson, J.N. Andersen, C. Wigren (to be published)
3.18 N. Mårtensson, A. Stenborg, O. Björneholm, A. Nilsson, J.N. Andersen: Phys. Rev. Lett. **60**, 1731 (1988)
3.19 A. Nilsson, B. Eriksson, N. Mårtensson, J.N. Andersen, J. Onsgaard: Phys. Rev. B **38**, 10357 (1988)
3.20 T.A. Carlsson: Phys. Rev. **156**, 142 (1967)
3.21 N. Mårtensson, S. Svensson, U. Gelius: J. Phys. B **20**, 6243 (1987)
3.22 S. Svensson, B. Eriksson, N. Mårtensson, U. Gelius, G. Wendin: J. Elec. Spec. and Relat. Phenom. **47**, 327 (1988)
3.23 L.J. Saethre, T.D. Thomas, L. Ungier: J. Elec. Spec. and Relat. Phenom. **33**, 381 (1984)
3.24 L. Asplund, U. Gelius, S. Hedman, K. Helenelund, K. Siegbahn, P.E.M. Siegbahn: J. Phys. B **18**, 1569 (1985)
3.25 U. Gelius, S. Svensson, H. Siegbahn, E. Basilier, Å. Faxälv, K. Siegbahn: Chem. Phys. Lett. **28**, 1 (1974)
3.26 A. Nilsson, N. Mårtensson: Solid State Commun. **70**, 923 (1989)
3.27 A. Nilsson, N. Mårtensson: Phys. Rev. Lett. **63**, 1483 (1989)
3.28 N. Mårtensson: P. Baltzer, P.A. Brühwiler, J. Forsell, A. Nilsson, A. Stenborg, B. Wannberg, J. Elec. Spec. and Relat. Phenom., In press
3.29 P. Baltzer, L. Karlsson, M. Lundqvist, B. Wannberg: Rev. Sci. Instrum. **64**, 2174 (1993)
3.30 J.N. Andersen, O. Björneholm, A. Sandell, R. Nyholm, J. Forsell, L. Thånell, A. Nilsson, N. Mårtensson: Synchrotron Radiation News **4**, 15 (1991)
3.31 A. Nilsson, N. Mårtensson: Phys. Rev. B **40**, 10249 (1989)
3.32 U. Gelius, L. Asplund, E. Basilier, S. Hedman, K. Helenelund, K. Siegbahn, Nucl. Instrum. and Meth. B **1**, 85 (1984)
3.33 B. Johansson, N. Mårtensson: Phys. Rev. B **21**, 4427 (1980)
3.34 A. Nilsson, H. Tillborg, N. Mårtensson, Phys. Rev. Lett. **67**, 1015 (1991)

3.35 W.F. Egelhoff Jr.: Surf. Sci **141**, L324 (1984)
3.36 H. Antonsson, A. Nilsson, N. Mårtensson, I. Panas, P.E.M. Siegbahn: J. Elec. Spec. and Relat. Phenom. **54/55**, 601 (1990)
3.37 A. Nilsson, H. Antonsson, A. Sandell, N. Mårtensson: In *The Structure of Surfaces III* Springer Ser. Surf, Sci, Vol. 24 (Springer, Berlin, Heidelberg 1991) p. 467
3.38 H. Basch: J. Elec. Spec. and Relat. Phenom. **5**, 463 (1974)
3.39 U. Gelius: J. Elec. Spec. and Relat. Phenom. **5**, 985 (1974)
3.40 A. Nilsson, R.E. Palmer, H. Tillborg, B. Hernnäs, R.J. Guest, N. Mårtensson: Phys. Rev. Lett. **68**, 982 (1992)
3.41 M. Larsson, P. Baltzer, S. Svensson, B. Wannberg, N. Mårtensson, A. Naves de Brito, N. Correia, M.P. Keane, M. Carlsson-Göthe, L. Karlsson: J. Phys. B **23**, 1175 (1990)
3.42 R.M. Lambert, M.E. Bridge: In *The Chemical Physics of Solid Surfaces and Heterogeneous Catalysis*, ed. by D.A. King, D.P. Woodruff, (Elsevier, New York 1984), Vol. 3B
3.43 A. Sandell, A. Nilsson, N. Mårtensson: Surf. Sci. **251/252**, 971 (1991)
3.44 P. Uvdal, P.A. Karlsson, C. Nyberg, S. Andersson, N.V. Richardsson: Surf. Sci. **202**, 167 (1988)
3.45 H. Tillborg, A. Nilsson, N. Mårtensson: Surf. Sci. **273**, 47 (1992)
3.46 L. Westerlund, L. Jönsson, S. Andersson: Surf. Sci. **199**, 109 (1988)
3.47 P.R. Norton, J.W. Goodale, E.B. Selkirk: Surf. Sci. **83**, 189 (1979)
3.48 H. Tillborg, A. Nilsson, N. Mårtensson: To be published
3.49 A. Sandell, A. Nilsson, N. Mårtensson, Surf. Sci. **241**, L1 (1991)
3.50 J. Stöhr, R. Jaeger: Phys. Rev. B **26**, 411 (1982)
3.51 J.L. Gland, B.A. Sexton, G.B. Fisher: Surf. Sci. **95**, 587 (1980)
3.52 W. Wurth, J. Stöhr, P. Feulner, X. Pan, K.R. Bauchspiess, Y. Baba, E. Hudel, G. Rocker, D. Menzel: Phys. Rev. Lett. **65**, 2426 (1990)
3.53 R.J. Guest, O. Björneholm, B. Hernnäs, A. Nilsson, R.E. Palmer, P. Bennich, N. Mårtensson: Surf. Sci. **278**, 239 (1992)
3.54 G. Odörfer, R. Jaeger, G. Illing, H. Kuhlenbeck, H.J. Freund: Surf. Sci. **233**, 44 (1990)
3.55 R.E. Palmer, P.J. Rous: Rev. Mod. Phys. **64**, 383 (1992)
3.56 R.J. Guest, A. Nilsson, O. Björneholm, B. Hernnäs, A. Sandell, R.E. Palmer, N. Mårtensson: Surf. Sci. **269/270**, 432 (1992)
3.57 H. Tillborg, A. Nilsson, B. Hernnäs, N. Mårtensson, R.E. Palmer: Surf. Sci. **295**, 1 (1993)
3.58 T.-C. Chiang, G. Kaindl, T. Mandel: Phys. Rev. B **33**, 695 (1986)
3.59 G. Kaindl, T.C. Chiang, D.E. Eastman, F.J. Himpsel: Phys. Rev. Lett. **45**, 1808 (1980)
3.60 K. Jacobi, H.H. Rotermund: Surf. Sci. **116**, 435 (1982)
3.61 T. Mandel, M. Domke, G. Kaindl, C. Laubschat, M. Prietsch, U. Middelman, K. Horn: Surf. Sci. **162**, 453 (1985)
3.62 K. Jacobi: Phys. Rev. B **38**, 6291 (1988)
3.63 D. Nordfors, H. Ågren, K.V. Mikkelsen: Chem. Phys. Lett. **164**, 173 (1992)
3.64 B. Johansson, N. Mårtensson: Helvetica Physica Acta **56**, 405 (1983)
3.65 P.H. Citrin, G.K. Wertheim: Phys. Rev. B **27**, 3176 (1983)
3.66 D. Spanjard, C. Guillot, M.C. Desjonquéres: Surf. Sci. Rep. **5**, 1 (1985)
3.67 A. Flodsröm, R. Nyholm, B. Johansson: In *Synchrotron Radiation Research: Advances in Surface and Interface Science*, ed. by R.Z. Bachrach. (Plenum, New York 1992) Vol. 1, p. 199
3.68 W.F. Egelhoff Jr.: Crit. Rev. of Solid State and Mathematical Sci. **16**, 213 (1990)
3.69 N. Mårtensson, H.B. Saalfeld, H. Kuhlenbeck, M. Nuemann: Phys. Rev. B **39**, 8181 (1989)
3.70 D. Tománek, V. Kumar, S. Holloway, K. H. Bennemann: Solid State Commun. **41**, 273 (1982)
3.71 A. Rosengren, B. Johansson: Phys. Rev. B **22**, 3706 (1980)
3.72 A. Nilsson, O. Björneholm, A. Sandell, P. Bennich, B. Hernnäs, C. Puglia, N. Mårtensson: To be published
3.73 A. Sandell, P. Bennich, A. Nilsson, B. Hernnäs, O. Björneholm, N. Mårtensson: Surf. Sci., **310**, 16 (1994)
3.74 O. Björneholm, A. Nilsson, H. Tillborg, P. Bennich, A. Sandell, B. Hernnäs, C. Puglia, N. Mårtensson: Surf. Sci., In press

3.75 M. Tüshaus, E. Schweizer, P. Hollins, A.M. Bradshaw: J Elec. Spec. and Relat. Phenom. **44**, 305 (1987)
3.76 M.O. Krause, J.H. Oliver: J. Phys. and Chem. Ref. Data **8**, 329 (1979)
3.77 S. Svensson, N. Mårtensson, E. Basilier, P.-Å. Malmquist, U. Gelius, K. Siegbahn: Physica Scripta **14**, 141 (1976)
3.78 G. Wendin, M. Ohno: Physica Scripta **14**, 148 (1976)
3.79 S. Doniach, M. Sunjic: J. Phys. C **3**, 285 (1970)
3.80 S. Tougaard: Phys. Rev. B **34**, 6779 (1986)
3.81 H. Wiedemann: *Particle Accelerator Physics* (Springer, Berlin, Heidelberg 1993)
3.82 M. Bader, C. Ocal, B. Hillert, J. Haase, A.M. Bradshaw: Phys. Rev. B **35**, 5900 (1987)
3.83 L. Wenzel, D. Arvanitis, W. Daum, H.H. Rotermund, J. Stöhr, K. Baberschke, H. Ibach: Phys. Rev. B **36**, 7689 (1987)
3.84 J. Stöhr, R. Jaeger, T. Kendelewicz: Phys. Rev. Lett. **49**, 142 (1982)
3.85 W. Daum, S. Lehwald, H. Ibach: Surf. Sci. **178**, 528 (1986)
3.86 M. Rocca, S. Lehwald, H. Ibach, T.S. Rahman: Phys. Rev. B **35**, 9510 (1987)
3.87 M. Rocca, S. Lehwald, H. Ibach: Surf. Sci. **163**, L738 (1985)
3.88 N. Mårtensson, A. Nilsson: Surf Sci. **211/212**, 303 (1989)
3.89 J. Schirmer, G. Angonoa, S. Svensson, D. Nordfors, U. Gelius: J. Phys. B **20**, 6031 (1987)
3.90 A.P. Hitchcock, C.E. Brion: J. Elec. Spec. and Relat. Phenom. **18**, 1 (1980)
3.91 A. Nilsson, N. Mårtensson, S. Svensson, L. Karlsson, D. Nordfors, U. Gelius, H. Ågren: J. Chem. Phys. **96**, 8770 (1992)
3.92 S. Andersson, Solid State Commun. **21**, 75 (1977)
3.93 R. Berndt, J.P. Toennies, C. Wöll: J. Elec. Spec. **44**, 183 (1987)
3.94 N.V. Richardson, A.M. Bradshaw: Surf. Sci. **88**, 255 (1979)
3.95 A.M. Lahee, J.P. Toennies, C. Wöll: Surf. Sci. **177**, 371 (1986)
3.96 G. Doyen, G. Ertl: Surf. Sci. **69**, 157 (1977)
3.97 R.L. Martin, D.A. Shirley: Phys. Rev. A **13**, 1475 (1976)
3.98 R. Arneberg, J. Müller, R. Manne: Chemical Physics **64**, 249 (1982)
3.99 T. Åberg: Ann. Acad. Sci. Fennicae (Ser. A) **IV**, 308 (1969)
3.100 R. Manne, T. Åberg: Chem. Phys. Lett. **7**, 282 (1970)
3.101 B.I. Lundqvist: Phys. Kondens. Materie **9**, 236 (1969)
3.102 A. Sandell, A. Nilsson, P. Bennich, N. Mårtensson: Phys. Rev. Lett. **72**, 2604 (1994)
3.103 A. Hoffman, G.L. Nyberg, J. Liesegang: Phys. Rev. B **45**, 5679 (1992)
3.104 A. Nilsson, O. Björneholm, A. Sandell, B. Hernnäs, N. Mårtensson: Surf. Sci. **293**, L835 (1993)
3.105 N. Mårtensson, R. Nyholm, B. Johansson: Phys. Rev, B **29** (1984)
3.106 C.W. Bates, G.K. Wertheim, D.N.E. Buchanan, Phys. Lett. A **72**, 178 (1979)
3.107 R.A. Pollak, L. Ley, F.R. McFeely, S.P. Kowalczyk, D.A. Shirley: J. Elec. Spec. and Relat. Phenom. **3**, 381 (1974)
3.108 D. Lovric, B. Gumhalter: Phys. Rev. B **38**, 10323 (1988)
3.109 A. Nilsson, N. Mårtensson: Chem. Phys. Lett. **182**, 147 (1991)
3.110 E. Zdansky, A. Nilsson, H. Tillborg, O. Björneholm, N. Mårtensson, J.N. Andersen, R. Nyholm: Phys. Rev. B **48**, 2632 (1993)
3.111 H. Tillborg, A. Nilsson, T. Wiell, N. Wassdahl, N. Mårtensson, J. Nordgren: Phys. Rev. B **47**, 16464 (1993)
3.112 C.E. Moore: *Atomic Energy Levels* (National Bureau of Standards, Washington, DC 1949 Vol. 1
3.113 C.W. Bauschlichter: J. Chem. and Phys. **84**, 250 (1986)
3.114 A. Bosch, H. Feil, G.A. Sawatzky, N. Mårtensson, Solid State Commun. **41**, 355 (1982)
3.115 J.C. Fuggle, U.E, D. Menzel, K. Wandelt, C.R. Brundle: Solid State Commun. **27**, 65 (1978)
3.116 E. Umbach: Surf. Sci. **117**, 482 (1982)
3.117 K. Schönhammer, O. Gunnarsson: Solid State Commun. **23**, 691 (1977)
3.118 O. Gunnarsson, K. Schönhammer: Phys. Rev. Lett. **41**, 1608 (1978)
3.119 P.S. Bagus, M. Seel: Phys. Rev. B **23**, 2065 (1981)

3.120 H.-J. Freund, E.W. Plummer: Phys. Rev. B **23**, 4859 (1981)
3.121 R.P. Messmer, S.H. Lamson, D.R. Salahub: Phys. Rev. B **25**, 3576 (1982)
3.122 H.J. Freund, F. Greuter, D. Heskett, E.W. Plummer, Phys. Rev. B **28**, 1727 (1983)
3.123 E. Umbach: Solid State Commun. **51**, 365 (1984)
3.124 H.-J. Freund, R.P. Messmer, C.M. Kao, E.W. Plummer: Phys. Rev. B **31**, 4848 (1985)
3.125 D. Lovric, B. Gumhalter, K. Wandelt: Surf. Sci. **278**, 1 (1992)
3.126 H. Tillborg, A. Nilsson, N. Mårtensson: J. Elec. Spec. and Relat. Phenom. **62**, 73 (1993)
3.127 O. Björneholm, A. Nilsson, E. Zdansky, A. Sandell, H. Tillborg, B. Hernnäs, N. Mårtensson, J.N. Andersen: Phys. Rev. B **47**, 2308 (1993)
3.128 P. Avouris, N.J. Dinardo, J.E. Demuth: J. Chem. and Phys. **80**, 491 (1984)
3.129 J. Burstad, C. Engkvist, S. Lunell, H. Tillborg, A. Nilsson, S. Osborne, N. Mårtensson, S. Svensson: Chem. Phys. **179**, 303 (1994)
3.130 O. Björneholm, A. Nilsson, E. Zdansky, A. Sandell, H. Tillborg, N. Mårtensson, J.N. Andersen, R. Nyholm: Phys. Rev. B **46**, 10353 (1992)

4. Structure Determination of Molecular Adsorbates Using Photoelectron Diffraction

A.M. Bradshaw and *D.P. Woodruff*

This chapter is concerned with photoelectron diffraction in both its scanned-energy and scanned-angle forms. The phenomenon itself and its possible use as a structural tool was first described by *Liebsch* in 1974 in a theoretical study [4.1]. The first experimental studies were performed independently by three groups in 1978 [4.2–4]. As in Surface EXtended Absorption Fine Structure (SEXAFS) the technique is based on the interference of photoemitted electron waves and does not have the requirement of long-range order, but there are distinct differences between the two experiments, as will be discussed below. The scanned energy form (referred to here as PhD) also requires the use of synchrotron radiation, whereas the scanned angle variant at higher photoelectron kinetic energies can be performed with laboratory sources. The latter is often referred to as XPD (X-ray Photoelectron Diffraction), but in the case of adsorbed molecules is largely restricted to the determination of bond directions. The scanned energy form of the method, on the other hand, has the potential of delivering much more quantitative geometric information and, in some cases, even providing an almost complete structural characterisation [4.5, 6]. The present chapter therefore concentrates on the scanned-energy experiment and the reader is referred to two recent review articles for more exhaustive accounts of XPD [4.7, 8].

4.1 Introductory Comments

Despite considerable interest over many years in the adsorption complexes formed by molecules on metal surfaces, there is surprisingly little quantitative information available on their microscopic structure. In view of the important role played by adsorbed molecules and molecular fragments in heterogeneous catalysis, structural details such as adsorption site, orientation, substrate-adsorbate separation, change of intramolecular bond lengths and angles as well as the occurrence of adsorbate-induced reconstruction are extremely important for understanding reaction mechanisms and predicting selectivity. Another feature of considerable chemical interest is the analogy between the surface situation and the properties of the same molecule as a ligand in various co-ordination compounds [4.9]. Recent work on a number of molecular adsorption

Springer Series in Surface Sciences, Vol. 35
Applications of Synchrotron Radiation Ed.: W. Eberhardt

systems [4.10–12] has shown that these similarities extend well beyond the traditional examples of adsorbed CO and NO, where the internal stretching frequency has often been used, perhaps somewhat too arbitrarily [4.13], as an indication of the surface site.

The accurate determination of structural parameters in molecular chemisorption has recently become important for a further reason: Because of advances in theoretical techniques and concomitant vast increases in computing power, the determination of total energies in *ab initio* cluster calculations has recently become possible in an almost routine way [4.14]. This, in turn, means that predictions concerning geometry and bonding need to be tested with accurate structural parameters from model systems obtained under well-defined conditions. These data have hitherto been lacking because the most wide-spread technique for surface structural analysis – Low-Energy Electron Diffraction (LEED) – suffers from two major disadvantages in the present context. Firstly, the majority of molecular adsorption systems do not give rise to ordered overlayers and therefore cannot be investigated by conventional crystallographic techniques such as LEED or X-ray diffraction. An important exception here is adsorbed CO which appears to form ordered overlayers on most single-crystal metal surfaces. Ordered structures have also been reported for NO, NH_3, PF_3, C_2H_2 and C_6H_6 on some surfaces but this may be the exception rather than the rule. In particular, long-range order is very rarely observed in the case of molecular fragments formed as intermediates in heterogeneous reactions. In the future, *diffuse* LEED [4.15] may offer the possibility of structural determination in the absence of long-range order, but so far results for only one or two-molecular systems have been published. Secondly, a simple practical consideration has so far limited the applicability of LEED: molecular adlayers are particularly prone to electron beam-induced effects such as fragmentation and/or desorption as well as disordering. In fact, this might well be one reason why relatively few examples of studies of ordered molecular overlayers can be found in the literature. This problem can actually be solved by the use of more sophisticated LEED equipment with microchannel plate displays which allow considerably lower incident beam currents, although as yet little advantage has been taken of this possibility.

What are the other structural characteristics of chemisorbed molecules apart from a possible tendency to form ordered overlayers at finite coverage? Binding energies on metal surfaces can be as high as 2 eV/molecule which, in the case of predominantly covalent bonding, might be expected to be indicative of strongly directional bonds. Indeed, molecules and molecular fragments do tend to adsorb on surfaces sites of high symmetry (for example, atop a metal atom or bridging two metal atoms) and display a specific, fixed orientation relative to the surface. The latter should actually be expressed in terms of a mean angle of inclination since the soft external modes of vibration (the frustrated rotations and translations) give rise to large displacements parallel to the surface [4.16]. Very little is known about the intramolecular parameters of molecular adsorbates – bond lengths and bond angles – although the evidence that does exist suggests that normally no large structural perturbations take place relative to the free molecule. On the contrary, any changes are small and subtle in nature, which again is similar to the situation in coordination chemistry and underlines the need for more accurate and detailed information on structural geometry.

This general information on molecular adsorbates comes from a variety of direct and indirect structural-sensitive methods (including LEED). The technique of core-level absorption spectroscopy, often referred to as NEXAFS (Near-Edge X-ray Absorption Spectroscopy), has, for instance, proved to be particularly useful in the determination of orientation [4.17]. The spectrum, for example, at the carbon K-edge of an adsorbed hydrocarbon molecule is characterised by absorption resonances which may be identified with excitations into antibonding orbitals. In the case of an unsaturated system these can be both σ and π in nature. Since the molecule has a fixed geometry on the surface and the excitation is of the electric dipole type, the transitions are polarised. Although an electron-yield technique is usually employed to register the absorption spectrum, molecular orientation can still be determined by varying the direction of the electric vector of the incident radiation, as in the conventional linear dichroism experiment. The method thus requires polarised soft-X rays of variable wavelength which, in practice, can only be provided by a synchrotron-radiation source. Chapter 5 contains more details. In general, the method allows molecular orientation to be determined to within 10°, provided that the symmetry is sufficiently high and care is taken in interpretation [4.18].

A further technique providing geometric information is ESDIAD (Electron-Stimulated Desorption Ion Angular Distributions) [4.19]. Incident electrons produce electronic excitation which result in desorption of the molecular ion or of a fragment ion. The desorbing ions, e.g. H^+ from NH_3, leave the surface along the direction of the fragmented bond, giving rise to an angular distribution which contains orientational information. The method has been used with great success in several cases, but suffers from the fact that it is restricted to specific molecules (CO, NH_3, H_2O and PF_3 appear to work particularly well) and from the possibility that minority species on the surface may dominate the measured signal due to a high cross-section. Angle-resolved photoemission [4.20] has also been used as a probe of molecular orientation, although the possibilities here are limited if low symmetry pertains. By measuring the emission from adsorbate-induced levels either normal to the surface or in a symmetry plane and appropriately orienting the electric vector of the incident radiation, photoemission selection rules can (sometimes) decide for or against a particular high-symmetry configuration. The angle of inclination of a molecular symmetry axis cannot, however, be determined, unless of course it is perpendicular to the surface. (Adsorbed diatomics provide an exception here: the strong directional properties of the core-to-σ^* excitation often enables the direction of the molecular axis to be determined). As in the case of NEXAFS, a necessary prerequisite is the symmetry assignment of the adsorbate-induced spectral features.

Vibrational spectroscopy is a further rich, but indirect source of structural information on adsorbed molecules and should be considered in somewhat more detail. Of the available techniques both EELS (Electron Energy Loss Spectroscopy) [4.21] and, to a lesser extent, IRAS (Infrared Reflection-Absorption Spectroscopy) [4.22, 23] have proved extremely useful, in general, surface characterisation (Chap. 7). First, the so-called surface selection rule can be utilised in IRAS as well as in the dipole scattering mode of EELS (specular conditions). This simply states that only vibrations with a dynamic dipole, or a component thereof, perpendicular to the surface will give rise to a measurable absorption (or loss) signal due to the screening effect of the metal conduction electrons. As in the case of photoemission, there is thus the possibility of using the effect for orientation determination but in a rather limited way. One can decide, for example, whether a symmetry axis is perpendicular or inclined relative to the surface normal. If the selection rule indicates that it is inclined, however, the angle of inclination cannot be determined. Further assistance is provided by measuring under non-specular conditions in EELS which may give access to modes for which the dynamic dipole is parallel to the surface via so-called *impact scattering*. The second way in which vibrational spectroscopy provides structural information is through the change in frequency of internal modes as a result of adsorption. The most notable example is that of adsorbed CO for which the various frequency ranges of the C–O stretch have been correlated with different adsorption sites in analogy to the transition metal carbonyls. As already mentioned above, this rather qualitative approach has been applied somewhat uncritically in the past.

The above discussion makes clear that none of these techniques is capable of providing the accurate structural information required for a detailed understanding of the chemisorption of molecules and molecular fragments on metal surfaces. Even the most obvious, truly-structural technique, namely LEED, suffers from certain disadvantages. Attention has therefore focussed recently on two further methods in which core electron excitation provides the structural probe: SEXAFS (Surface EXtended Absorption Fine Structure) and photoelectron diffraction [4.24]. SEXAFS consists of weak oscillations in the absorption coefficient well above the absorption edge of an adsorbate core level. It occurs beyond the core-to-valence excitations (NEXAFS) and extends for several hundred eV photon energy above the edge. The oscillations are due to interference between the emitted photoelectron wave and the waves backscattered from the surrounding atoms. Since the frequency of the oscillations as a function of photon energy is determined by the distances to surrounding substrate atoms, an appropriate analysis can provide structural information. SEXAFS studies of adsorbates on single-crystal surfaces began in 1978 at Stanford with the pioneering work of two groups [4.25, 26]. The method has been applied with success to a large number of atomic adsorbates, both in the soft and hard X-ray regions [4.27, 28]. So far, there are few examples of its application to adsorbed molecules.

4.2 Basis of the Technique and Its Potential

The basic physics of the photoelectron-diffraction technique is the coherent interference of a directly emitted photoelectron wave field emerging from a surface atom with components of this same wave field elastically scattered by surrounding atoms. This interference gives rise to angle-resolved photoemission intensities which vary with emission angle and energy (and thus electron wavelength) in a way which is characteristic of the local geometrical arrangement of the scatterer atoms around the emitter; i.e., which is characteristic of the surface structure. Although this basic physics is common to all of the various techniques which are referred to as photoelectron diffraction, the ease of interpretation of the associated experimental data and the character of the dominant information content depends on the kinetic energy of the photoelectrons investigated. The qualitative difference between high- and low-kinetic-energy photoelectron diffraction experiments arises from the difference in the atomic scattering factor for electrons. At high energies, the scattering is concentrated in the near-forward direction, and the phase shift (δ) introduced in this forward scattering [for a complex scattering factor written as $f\exp(i\delta)$] is small compared with π. At low energies, by contrast, the scattering factor is more nearly isotropic in angle, although it most typically shows peaks in the near-forward and near-backward (i.e., close to 0° and 180°) scattering directions, and has associated phase shifts which can lie anywhere between 0 and 2π. The actual values of the energies which correspond to these 'high'- and 'low'-energy regimes are not uniquely defined, and depend on the scattering species, but typically energies above about 500 eV show characteristic dominance for forward scattering. Backscattering becomes significant for energies much lower than this value.

A quantitative illustration of these effects is provided by the data depicted in Figs. 4.1 and 2 for Ni scatterers. Figure 4.1 exhibits the modulus of the relative electron scattering factor for Ni atoms as a function of scattering angle at several energies (normalised to the same values for 0° scattering) in the 'low'- and 'high'-energy ranges [4.29]. Figure 4.2 provides added phase information by showing the results of a simple photoelectron-diffraction calculation in which an emitter atoms is placed 2.5 Å from a Ni atom and the scattered electron intensity resulting from the interference of the directly emitted and scattered components is shown as a function of emission angle [4.30]. Note, particularly, the effect in the forward scattering (0°) direction along the interatomic axis. In the direction there is no scattering path difference between the direct and scattered components of the wave field, so in the absence of a scattering phase shift the

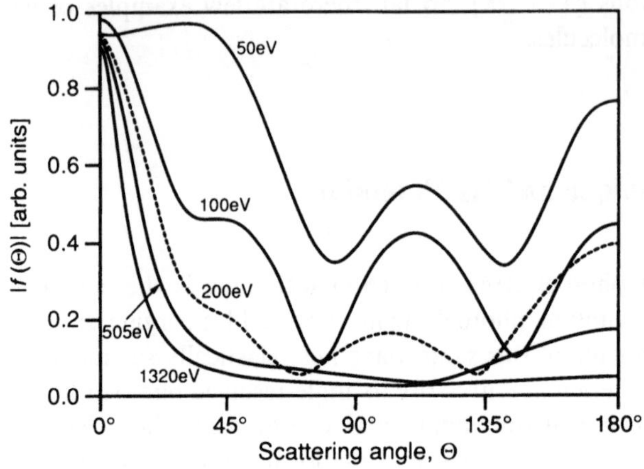

Fig. 4.1. Modulus of the electron scattering of a Ni atom as a function of scattering angle for several different electron energies. *Fadley* [4.29].

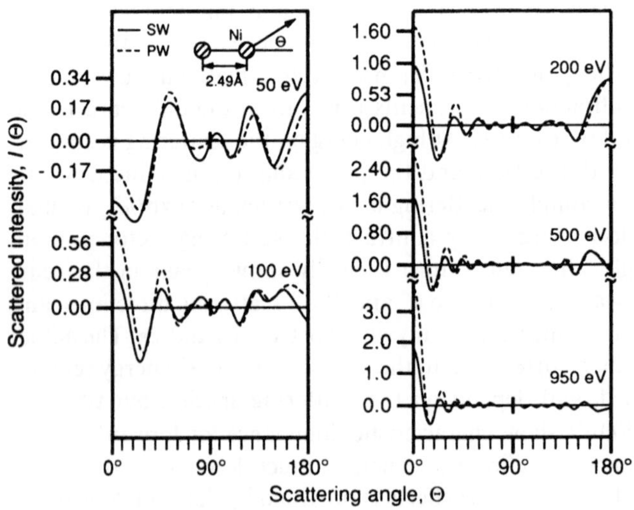

Fig. 4.2. Results of simple two-atom electron scattering calculations in which electrons are scattered from a Ni atom having been emitted from an atom 2.49 Å away (a distance corresponding to the Ni–Ni nearest-neighbour distance). The result of the interference of the directly emitted wave field with that scattered from the single Ni atom is shown as a function of angle relative to the interatomic forward scattering direction, intensities being normalised to that of the directly emitted flux, for several different electron energies. In each case two plots are shown corresponding to plane wave (PW) and spherical wave (SW) scattering approximations. *Fadley* [4.29]

intensity reflects directly the strength of the forward scattering. At high energies, at which the forward scattering strength is high, a clear peak is seen in the angular dependence along this interatomic, forward-scattering, direction. At lower energies, however, at which the forward-scattering strength is still large, it is possible to observe either a maximum or a minimum in this direction. The minimum results from destructive interference between the direct and scattered wave field components, clearly signalling a scattering phase shift of about π. Notice, also, that at low energies there is strong, and richly structured, emission in the backscattering direction.

The angular peak to be seen in the forward-scattering interatomic direction at high photoelectron kinetic energies provides the simplest form of the photoelectron-diffraction experiment. Because this peak corresponds to a zero path length difference between the interfering components (Fig. 4.3), it is actually a *zero-order* diffraction feature; i.e., it is akin to the white-light straight-through (as though geometrically shadowed) beam seen from a transmission diffraction grating. However, while in an optics experiment this zero-order beam provides no useful information, in photoelectron diffraction it provides a direct real-space measurement of an interatomic bond direction in the crystal surface. One obvious example of the value of this information, which is especially pertinent to the topic of this volume, is the case of photoemission from a core-electron energy level of an atom within an adsorbed molecular species. In this case a search for angular peaks in the emission pattern can reveal important information on the orientation and internal geometry of the molecular adsorbate. This is the XPD experiment mentioned in the previous section. Because the basic information of the interatomic directions is carried by the zero-order diffraction features (which are independent of photoelectron energy), there is little to be gained from the use of the variable energy capability of a synchrotron-radiation beam line.

Figure 4.4a illustrates the way in which the XPD technique can be applied to this problem of internal molecular geometry and orientation on a surface, but shows only the forward-scattered contributions to the total photoelectron wave field resulting from emission from an atom within the molecule. Figure 4.4b, on the other hand, includes the role of scattering from the substrate in which the scattering angles are typically large (i.e., in the range of about 90° to 180°). In this case the scattering paths involve substantial path-length differences relative to the directly emitted wave-field component. These path-length differences carry information about the position of the emitter atom(s) within the adsorbed

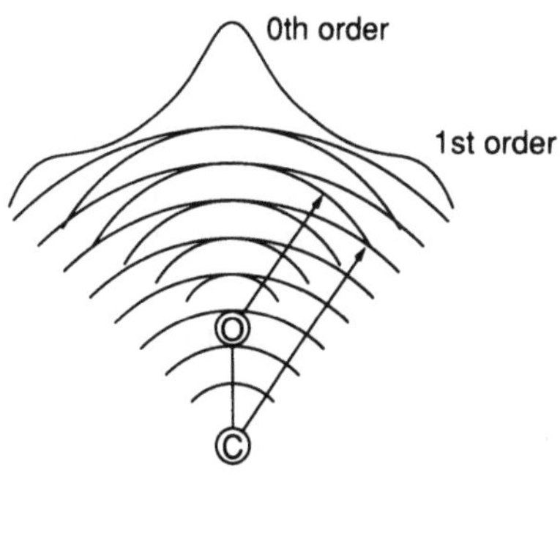

Fig. 4.3. Schematic diagram showing the effect of forward scattering photoelectron diffraction in a simple diatomic (CO) molecule. The spherical wave fronts shown correspond to the direct emission from the C atom, and the scattered wave from the O atom. The resulting polar plot of the intensities resulting from the interference of these two wave components shows the zero and first-order diffraction features

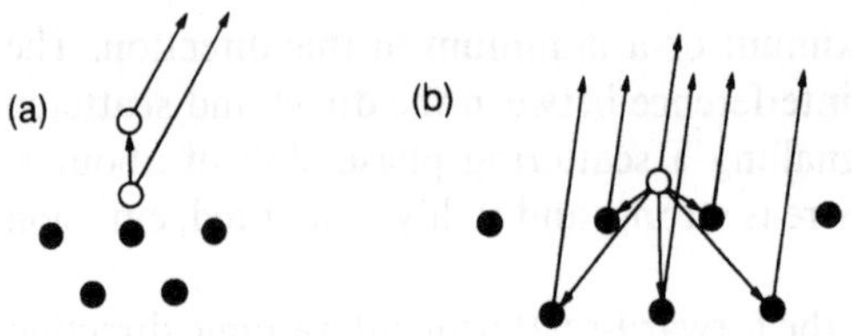

Fig. 4.4a, b. Schematic diagram showing the directly emitted and principal singly-scattered paths contributing to photoelectron diffraction (**a**) in the forward-scattering geometry from a simple diatomic molecule (Fig. 4.3) and (**b**) in backscattering from an atom adsorbed on a surface

molecule relative to the substrate. Specifically, they depend on the adsorbate-substrate atom bond-lengths and on the adsorbate-substrate registry or adsorption site. In general, this important adsorption-structure information can only be obtained by performing experiments at photoelectron kinetic energies which are low enough to guarantee strong backscattering. One obvious consequence of this condition, however, on the basis of the previous discussion, is that the interpretation of the data is more complex than in XPD. In particular, true (non-zero order) diffraction occurs, so there is no simple general correspondence between peaks in the angular distribution of the photoelectrons and the interatomic bond directions, and quantitative modelling must take account of the influence of many scatterers (because of the more nearly isotropic scattering), and of the scattering phase shifts involved. One might also anticipate that as the energies associated with strong electron backscattering are precisely those which are used in low energy electron diffraction (LEED). We might also expect to encounter the sample problems of multiple scattering which have led to the computational complexities associated with LEED [4.31, 32]. We shall show, however, that these problems can be surmounted, and indeed are not always as severe as this qualitative assessment would suggest.

A slightly more quantitative description highlights the main features of PhD in general. Specifically, consider the case of an emitter atom surrounded by scatterers at relative positions defined by the vectors $\mathbf{r}_j$. If we first take into account only the consequences of single scattering of the outgoing photelectron wave field by these surrounding atoms, the total amplitude in a direction defined by the electron wave vector $\mathbf{k}$ has the form

$$A(\mathbf{k}) \approx 1 + \Sigma_j a_j(\Theta_j, k) f_j(\Theta_j, k) \exp\{\mathrm{i}[\delta_j(\Theta_j, k) + kr_j(1 + \cos\Theta_j)]\} \qquad (4.1)$$

where Θ_j is the scattering angle at the jth scatterer, i.e. the angle between $\mathbf{r}_j$ and $\mathbf{k}$. The first term, unity, represents the directly emitted wave amplitude and the second term is the sum of scattered amplitudes. Notice that the scattering factor for each atom has been included explicitly as $f \exp(\mathrm{i}\delta)$ in order that the phase shift experienced in the scattering event can be added to the phase shift associated with the scattering path-length difference which has the form $r(1 + \cos\Theta)$. The amplitude term $a_j(\Theta_j, k)$ is itself comprised of several terms which describe: (a) the photoemitted wave-field amplitude at each scatterer relative to the directly emitted component in the collection direction (which depends on the photoemission initial-state angular-momentum quantum number, on the direction of the scatterer at $\mathbf{r}_j$ and the direction of $\mathbf{k}$, and on the distance r_j), (b) the influence of thermal vibrations of both emitter and scatterer through a Debye-Waller factor, and (c) the influence of inelastic electron scattering (which removes coherent photoelectron flux) through an exponential damping over the scattering pathlength. One important feature of the component terms (a) and (c) is that they introduce damping in the scatterer waves relative to the directly emitted wave, via the inelastic exponential damping, and by a $1/r_j$ term in the wave-field amplitude at the scatterer, both of which conspire to localise the scattering information. The dominant scatterer information is therefore local to the emitter. Clearly this description also highlights the fact that long-range order is irrelevant in photoelectron diffraction.

The emitted intensity, as measured at the detector, is given by the square modulus of the amplitude $|A(\mathbf{k})|^2$ and the dominant modulation terms (i.e., the terms other than the unity term coming from the direct emission) are essentially a cosine of the phase factor. This phase factor thus contains the local structural information through the components $r(1 + \cos\Theta)$. It is tempting to compare this situation to that in EXAFS (EXtended X-ray Absorption Fine Structure [4.27, 28]) in which one measures the *total* photo-ionisation cross-section as a function of photon energy and observes modulations in this cross-section due to the interference of the directly emitted photoelectron wavefield with components backscattered onto the emitter itself (see Fig. 4.5a), causing a modulation of the final-state wave-function amplitude and thus of the photo-ionisation matrix elements. In EXAFS, therefore, one separates out of the modulation term (by subtracting, and dividing by, the atomic cross-section in the absence of adjacent scatterers) to give

$$X(k) = \Sigma_j a_{\mathrm{E}j}(k) \sin[2kr_j + \delta_j(\pi, k)] \tag{4.2}$$

whereas the equivalent approximate expression for photoelectron diffraction [the Angle Resolved Photoemission Extended Fine Structure (ARPEFS) function [4.33] is

$$X(\mathbf{k}) = \Sigma_j a_{\mathrm{P}j}(\Theta_j, k) \cos[kr_j(1 + \cos\Theta_j) + \delta_j(\Theta_j, k)]\,. \tag{4.3}$$

Superficially, it would therefore seem that if one measures photoelectron

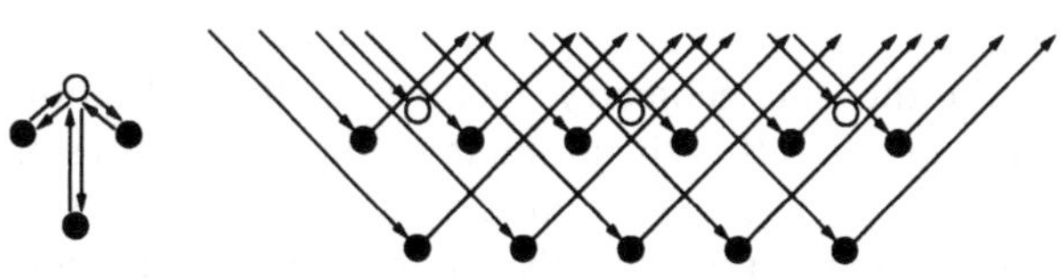

Fig. 4.5. Schematic diagram showing the single scattering paths involved in the EXAFS (*left*) and LEED (*right*) experiments

diffraction in a fixed direction as a function of photon, and hence photoelectron, energy (scanned-energy mode), one should produce a technique for which much of the methodology of EXAFS can be used.

There are, however, some important differences in these two techniques, which can be extracted from these two expressions. One rather obvious difference relates to the fact that in EXAFS (as in PhD) the k-periodicity of the fine-structure function can only yield the structural values r_j after correction for the scattering phase shifts δ_j. In EXAFS, however, only 180° backscattering enters the single-scattering expression, so the phase factors depend only on the atomic species of the emitter and scatterer, and not on the geometry. This leads to the very valuable 'phase shift transferability' of EXAFS which allows empirical correction of the phase shifts from measurements on model compounds of known structure. Thus, the basic idea of Fourier transforming of the data is especially valuable in EXAFS, because with this empirical correction, peaks do occur at distances $2r_j$, giving a direct determination of spatial parameters. Of course, the spherical averaging of EXAFS, such that the sum in (4.2) is over *shells* of equivalent scatterers around the emitter, means that the direct information obtained is only the *distances* to the scatterers, and not the spatial coordinates. Thus, adsorption site information needs secondary information, either via neighbour directions obtainable from the dependence of the EXAFS amplitudes on the X-ray polarisation direction or from the combination of several shell distances.

In the case of PhD one measures a derivative or angle-resolved photoionisation cross-section, as indicated by the fact that $X(\mathbf{k})$ depends on the vector $\mathbf{k}$, not the scalar k. Although this might appear to be a more difficult measurement because the angle-resolved cross-section is smaller, there are a number of important advantages. Firstly, from an experimental point of view, the fact that one measures a smaller quantity is offset by the fact that one has a means of measuring it rather directly (which is much less true, in the case of surface structural measurements, for the total photo-ionisation cross-section) and that the amplitude of the modulations in PhD is typically a factor of ten larger than in EXAFS which involves spherical averaging (tens of percent rather than percent). From a theoretical standpoint, as is clear from (4.3), another distinction is that the bond-length and site information are inexorably interlinked in photoelectron-diffraction because the scattering path lengths depend on both of these components. Of course, one consequence of this is that the Fourier-transform device which proves so valuable in EXAFS is of more limited value in photoelectron diffraction. Note, for example, that the phase-shift-correction problem is more severe because the scattering phase shifts depend on the scattering angle and thus on the site geometry relative to the collection direction. The net consequence of this is that the ultimate method of analysis of photoelectron-diffraction data is by the use of model calculations for test structures using phase shifts calculated theoretically, rather than the method often used in EXAFS, which is based on simple Fourier transforms and empirical phase shifts. The greater computational complexity of photoelectron diffraction is, however, more than offset by the greater site sensitivity and large amplitude of the modulations which eases many of the detailed practical problems (such as background subtraction and normalisation) of surface EXAFS.

The trial-and-error route to structure determination using model calculations has much in common with the methodology of LEED, and indeed there are many parallels in the basic physics and the theoretical formulation. This similarity, however, highlights one potential complexity which is known to be very important in LEED, namely multiple scattering. The problem of providing an accurate description of multiple scattering in LEED is that it leads to the use of rather demanding computer codes which has been a major factor in discouraging more wide-spread use of LEED. Of course, LEED, photoelectron diffraction and EXAFS are all based on the same basic physical effect of coherent interference of elastically scattered low-energy electrons, yet there is a marked contrast between the practical data-analysis methods of LEED and EXAFS. The relative lack of importance of multiple-scattering effects in EXAFS is due, in part, to the fact that the technique uses a local electron source (the emitter atom) and a local detector (also the emitter atom), leading to a $1/r_j^2$ weighting of the scattering contribution from the atomic shell at a distance r_j. This clearly attenuates the role of long-distance scattering paths such as those in multiple scattering. Indeed, if the dominant interest is in the nearest-neighbour shell (as it often is in surface EXAFS) then multiple-scattering features are inevitably outside the first shell peak in the Fourier transform and thus can be filtered out. In fact, EXAFS can show multiple-scattering effects. In common with LEED, the relative strength of forward scattering relative to backscattering means that the important multiple-scattering contributions come from paths which involve only one backscattering event, but also one or more forward scatterings. In EXAFS this means that the main effect of multiple scattering is found in more distant shells of atoms which are almost directly behind ('shadowed' by) nearer-neighbour atoms. These higher shells are rarely investigated in surface EXAFS.

PhD is clearly intermediate in its degree of localisation between LEED and EXAFS; it has a local source (leading to a $1/r_j$ term in the scattering amplitudes), but a remote detector. For LEED both source and detector are remote. We might therefore surmise that multiple scattering is of intermediate importance, and this indeed is the case. Specifically, single-scattering calculations in photoelectron diffraction can reproduce much of the dominant structure observed in experiments, at least in geometries leading to the strongest modulations. By contrast, single scattering in LEED can, in the simplest surface structures, provide almost no useful structural information at all. For example, let us consider the case of simple chemisorption in an ordered superstructure which produces no distortion of the substrate. In this situation the fractional-order diffracted intensities, in single scattering, come only from the adsorbate atoms, and thus carry no structural information on the adsorbate-substrate registry. Although new information is carried by the adsorbate scattering contributions to the integral order beams, the intensities of these beams are typically dominated by substrate scattering. In fact, the single-scattering events relevant to PhD (Fig. 4.4b) correspond to the *double* scattering processes in LEED which start with scattering by the adsorbate atoms. These are the lowest-order scattering events producing site information in the fractional-order LEED beams. We can therefore view a single-scattering calculation in PhD as roughly equivalent to a double-scattering calculation of LEED fractional-order diffracted beams. This comparison helps to account for the relative success of simple single-scattering calculations in the PhD interpretation. Current methodology, however, normally involves final structural refinement using converged multiple-scattering calculations, as will be shown later in this chapter.

Although we have remarked already that Fourier-transform methods are likely to prove troublesome because of the problems of scattering phase shifts, which are themselves dependent on the detailed structure and measurement geometry, some effort has been invested in exploring these methods in photoelectron diffraction. The essential motivation for this is that *direct* structural analysis, such as that offered by Fourier transforms in EXAFS and in X-ray diffraction, are much more efficient than determining the structure by trial-and-error model calculations. Thus, although we can anticipate real problems in making Fourier-transform methods effective in photoelectron diffraction, it is common-argued that if it is possible to determine the *approximate* structure by a direct method, the scale of effort required to refine this structure by current trail-and-error methods is greatly reduced. The basic idea of using Fourier transforms in the analysis of PhD spectra has been explored since the early 1980s [4.34–36], but a more recent version of this idea which has recently received a great deal of attention, is the notion of *photoelectron holography* [4.37].

If we regard the directly emitted photoelectron wave field as a reference wave, it is clear that the angular dependence of the photoemission intensity resulting from the interference with wave-field components scattered by the surrounding atoms (i.e., the photoelectron-diffraction angular distribution) is effectively a photoelectron hologram. Clearly this hologram cannot usefully be directly (optically) reconstructed to give the original object (the juxtaposition of the surrounding atoms), but if the hologram can be inverted numerically (by some kind of Fourier-transform method), the result would be a direct image of this atomic environment, i.e. of the local structure. So far most of the effort devoted to this idea, at least for the more interesting backscattering geometry which can provide adsorption sites, has been through the use of theoretically simulated data. A number of problems have been identified, including the fact that the reference wave is not simply an outgoing spherical wave, but is structured (e.g., a *p*-wave for emission from an initial *s*-state), and the fact that the scatterers are complex and introduce (structure-dependent) scattering phase shifts. The possible disturbing influence of multiple scattering has also been debated. The visible effects of such difficulties in early attempts to invert simulated data of this kind [4.37] were poor resolution in the 'atomic images' (as poor as 1 Å uncertainties in position), systematic errors in positions (of similar magnitude), and the presence of many spurious artifacts which could not, *a priori*, be distinguished from the features attributed to real atomic images. There is no doubt that considerable progress has been made in reducing the scale of all of these problems, again based on simulated data; this has been achieved by some increase in the computational complexity of the analysis (including phase shift correction procedures), but also by a substantial increase in the amount of data used in the inversion. In particular, these schemes [4.38] may necessitate the use of complete angular scans in polar and azimuthal angle ('holograms') on quite a fine angular mesh, recorded at 10 or more distinct photoelectron energies. The fact that this data base is many times larger than that currently being used to produce successful structural analyses using current trial-and-error methods (see the following sections) indicates that if this is the only practical method of using photoelectron holography, it may not be a very fruitful approach.

However, alternative ideas for obtaining first-order structural information from photoelectron diffraction, by some kind of direct method, have also been explored. In particular, *Shirley* and his co-workers have used Fourier transformations of individual PhD spectra to try to identify all the principal near neighbour backscatterers. Initially [4.34], it was thought that peaks in the Fourier transforms could be related to adsorbate-substrate layer spacings, but it was quickly recognised that the peaks correspond to scattering path-length differences [4.35, 36]. Although there are certainly dangers in attempting to read too much into the detailed structure of these transforms, as has been discussed by *Fadley* and coworkers [4.39, 40], the main near-neighbour scattering peaks do appear to be useful in obtaining a first guide to the structure and are routinely used by *Shirley*'s group. More recent ideas, which seek to use wide-range angular measurements in a fashion somewhat akin to those of the holographic approach, concentrate on attempts to identify the directions of the nearest neighbour (substrate) scatterers alone [4.41, 42]. Clearly, such information gives the adsorption site and approximate adsorbate-substrate bond-length, and could be a good starting point for structural refinement. Compared to the full holographic inversion process, this less ambitious set of objectives can be achieved with very much less experimental data.

These methods rely on a feature of scanned-energy-mode photoelectron diffraction which has been noted in many experimental studies such as the examples from our own work presented in the next section, and in similar studies by other groups [4.35, 40]. This common feature is that if data are recorded in a direction which places a (substrate) scatterer almost directly behind the emitter, relative to the detector, then the photoelectron-diffraction spectrum is dominated by modulations attributable to scattering by this one near neighbour. An obvious consequence of this is that the scanned-energy-mode photoelectron-diffraction spectra recorded in these directions are particularly simple, with a single dominant modulation periodicity which, being associated with a short scattering path length, has a long periodicity in electron wave number. One route to first-order structure determination is therefore to perform a coarse search in emission angle for the directions leading to these dominant nearest neighbour backscattering directions. One possibility [4.41] for doing so is to use to the fact that the nearest-neighbour adsorbate-substrate distance is commonly predictable, to within about 0.1 Å, even before the experiment is performed. It is therefore possible, with only a very simple, two-atom scattering calculation, to predict the energies corresponding to the maxima and minima in the photoelectron diffraction modulation spectrum in the aligned backscattering geometry. Angular searches at these energies can then provide the experimental data base with which to identify the nearest-neighbour directions.

A more general version [4.42] of this idea is to collect scanned-energy-mode spectra at a range of emission angles, to Fourier transform these and to map the intensity of the shortest (nearest-neighbour backscattering path) contribution in each of these as a function of angle. The largest peak will correspond to the aligned nearest-neighbour directions. Notice that this experiment can be performed on a rather coarse mesh of both angular and energy parameters, as the short scattering path length will be preserved at the expense of more complex single and multiple-scattering paths. Moreover, as one is only concerned with identifying the *directions* of the nearest-neighbour scatterers, phase shift corrections are not required. Both model calculations with simulated data [4.42], and work with very recent experimental data [4.43] (Sect. 4.3.4), suggests that this method may be a rather effective and efficient route to first-order structure determinations. In the rest of this review, however, we will concentrate on quantitative structural analyses achieved with the aid of multiple scattering simulations. The particular formulation of the multiple scattering approximations and the computational codes are those developed and described by *Fritzsche* [4.44, 45], who has also offered comparisons [4.46] with alternative schemes used by others working in this field [4.47–51].

4.3 Examples

4.3.1 Forward-Scattering Photoelectron Diffraction (XPD): CO on Ni(110)

As explained in the last section, the dominance of forward scattering at electron energies greater than about 500 eV allows the direct real-space measurement of intramolecular bond directions from the angular distribution of the core-level photoemission signal. Of the dozen or so molecular chemisorption systems which have so far been investigated with the XPD method, the work of *Bonzel* and his group on the Ni(110)-CO system is perhaps one of the most illustrative [4.52]. At a coverage of $\theta = 0.9$–1.0 (CO molecules per surface Ni atom) a $(2 \times 1)p2mg$ overlayer is formed. Earlier investigations with vibrational spectroscopy [4.53] had indicated a mixture of two different adsorption sites (possibly atop and bridge) up to $\theta = 0.8$ and a single, rather broad loss peak at higher coverages. A LEED I-V analysis [4.54], photoemission [4.55] and ESDIAD [4.56] all pointed to a structure similar to that of Fig. 4.6. The latter two methods gave tilt angles of 17° and 19°, respectively. The LEED analysis suggested a bent species and also favoured the bridge site. Symmetry considerations in EELS [4.57] favoured, however, the atop site as shown in Fig. 4.6.

Figure 4.7 exhibits the normalised intensity of the C 1*s* photoelectron line at 120 K as a function of the polar angle of emission in the ⟨100⟩ azimuth. Mg$K\alpha$ radiation with a photon energy of 1254 eV has been used, so that the C 1*s* electrons have a kinetic energy of 964 eV. The various curves refer to increasing CO exposure; saturation coverage ($\theta = 1.0$) appears to be reached after 4×10^{-5} mbar s. At the lowest CO exposures a peak centered along the surface normal may be attributed to forward scattering at the oxygen atom, indicating that the C–O molecular axis is aligned perpendicular to the surface. The width of the peak is greater than that expected on the basis of single-scattering calculations due to the broadening effect of the soft, high-amplitude vibrational modes parallel to the surface. This has been confirmed by measurements at different temperatures [4.58]. On increasing the coverage, the peak along the surface normal is replaced by a double-peak structure with lobes at $\Theta = +21°$ and $-21°$, as expected from Fig. 4.6. The peak width decreases somewhat, probably as result of the constraints imposed on the frustrated translations by the high density of the layer. In the ⟨1 1 0⟩ azimuth (not shown) only a single

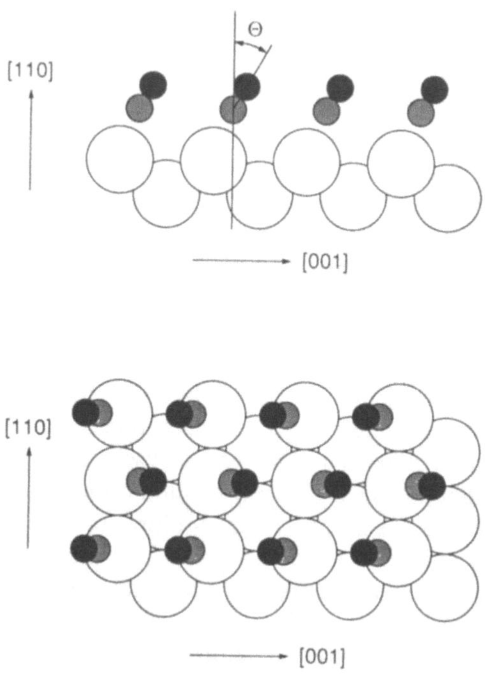

Fig. 4.6. Sectional (top) and plan (bottom) views of the Ni(110) (2 × 1)-CO (*p2mg*) structure. The large open circles denote Ni atoms, the small hatched (filled) circles denote C (O) atoms. The sectional view shows only one row of CO molecules. CO molecules are depicted occupying atop adsorption sites, although there remains some doubt as to whether atop or bridge sites are actually involved

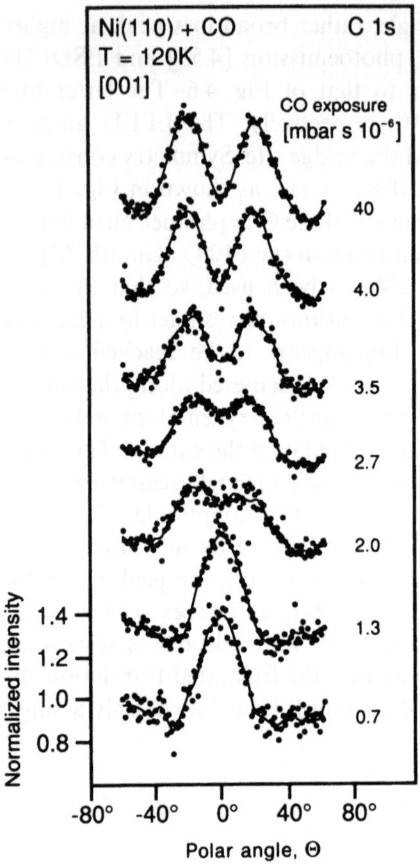

Fig. 4.7. XPD results taken from the Ni(110)/CO system. Polar-angle plots are shown of the C 1*s* signal taken using Mg *Kα* radiation as a function of CO exposure. The highest exposures correspond to the (2×1) *p*2*mg* saturation phase shown in Fig. 4.6. *Wesner* et al. [4.52]

peak along the surface normal is observed both at low and, with reduced intensity, at high coverages. The picture which emerges from the XPD data therefore supports the results of previous studies and gives a value of 21° for the mean angle of inclination of the C–O axis in the ⟨100⟩ azimuth.

In a more recent study of the (2×1) *p*2mg structure the same group has extended the polar angle range to $\pm 85°$ in four different azimuths [4.59]. Forward-scattering enhancement peaks in the C1*s* signal were detected at various angles due not only to intramolecular scattering but also to *inter*-molecular scattering. In particular, strong intermolecular peaks were found at $\pm 68°$ in the ⟨100⟩ azimuth. However, intermolecular peaks observed in the ⟨111⟩ azimuth at $\pm 65°$ indicated that the Ni–C–O configuration is linear, i.e., that the species is tilted and not bent at the carbon atom. Multiple-scattering calculations were also performed for different CO geometries and binding sites, although the calculated curves were not found to vary much for the different configurations. A comparison with the experimental data supported, however, the atop site and a linear Ni–C–O configuration, although the authors described the balance of evidence favouring the adsorption site as 'marginal'. In view of the very low backscattering contribution to the measured intensity distributions at these kinetic energies (Figs. 4.1 and 2), this lack of sensitivity to the adsorption site is not surprising.

4.3.2 Experimental Methodology and Qualitative Site Sensitivity in PhD

Before discussing the qualitative and quantitative interpretation of PhD spectra, some explanation of the method of performing the experiment and processing the data will prove of value. Figure 4.8 shows the key ingredients of the method. Having prepared the adsorbate-covered surface in the usual way, and set the soft X-ray incidence and photoelectron collection geometries to the desired values, an electron energy distribution around the adsorbate peak of interest is recorded by the angle-resolving electron spectrometer at a fixed photon energy, typically chosen to give a photoelectron kinetic energy of about 100 eV. In the present case (Fig. 4.8) the sample is a Ni(111) surface onto which ammonia has been adsorbed and the photoelectron peak of interest is therefore that originating from the N 1*s* core level of the adsorbed ammonia [4.60]. The binding energy of this peak is approximately 390 eV, so the first spectrum was recorded at a photon energy of 490 eV, giving rise to the photoelectron peak seen in Fig. 4.8 at approximately 100 eV kinetic energy. The photon energy is then increased by a fixed increment (3.5 eV in this case), and a new photoelectron-energy spectrum recorded over a kinetic energy range displaced to higher energy by the same increment. This procedure is then repeated in equal increments until a final

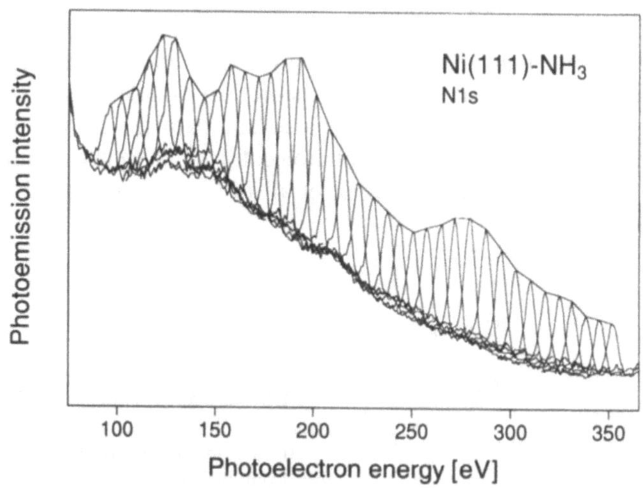

Fig. 4.8. Experimental N 1*s* photoelectron-energy spectra taken from NH_3 on Ni(111) in normal emission for a series of photon energies. Relative to the data used in full photoelectron-diffraction analysis of this system, only alternate spectra are shown [4.60]

photon energy, corresponding to a photoelectron kinetic energy of typically 300–400 eV, is reached.

Figure 4.8 depicts the succession of N 1*s* peaks which are produced, although for clarity only alternate spectra are shown. Clearly the lower envelope of these spectra is rather featureless, whereas the peaks reveal strong oscillations which are due to the photoelectron-diffraction effect. For the comparison with theory, a numerical routine is applied to extract the peak areas of the N 1*s* features in this spectrum, and the resulting plot of the peak area versus kinetic energy is normalised to the average values to obtain a modulation function. This latter stage is identical to that used in EXAFS. The modulation function is thus

$$X(E_{\text{kin}}) = \frac{I(E_{\text{kin}}) - I_0(E_{\text{kin}})}{I_0(E_{\text{kin}})} \tag{4.4}$$

in which $I(E_{\text{kin}})$ are the actual recorded peak areas, and $I_0(E_{\text{kin}})$ is the smooth fit though the average values. This final PhD spectrum (Fig. 4.9) is also sometimes referred to as the Angle-Resolved Photoemission Fine-Structure spectrum (ARPEFS) in analogy to EXAFS [4.35], as discussed in Sect. 4.2. The extraction of the fine-structure function in this way has the merit that it allows us to correct for both the smoothly varying monochromator output flux and the variation of the (total) photo-ionisation cross-section with energy in a simple fashion. It also provides a direct comparison for the results of theoretical calculations which calculate the scattering interferences but have no need to include the atomic cross-section variations.

Although aspects of this normalisation procedure have much in common with EXAFS, we have already remarked in Sect. 4.2 that there are important differences in the underlying physics. This means that PhD is expected to be significantly more sensitive to adsorption site than is EXAFS. The most direct method of illustrating this site sensitivity is to take an adsorbate species and to move it to different locations on a surface, measuring the PhD spectra for each geometry. This is generally only possible in model theoretical calculations. One way in which it *can* be achieved experimentally is to compare the results of experiments on the adsorption of several different molecules which each contain a common element. Providing that the bonding of all of these species to the

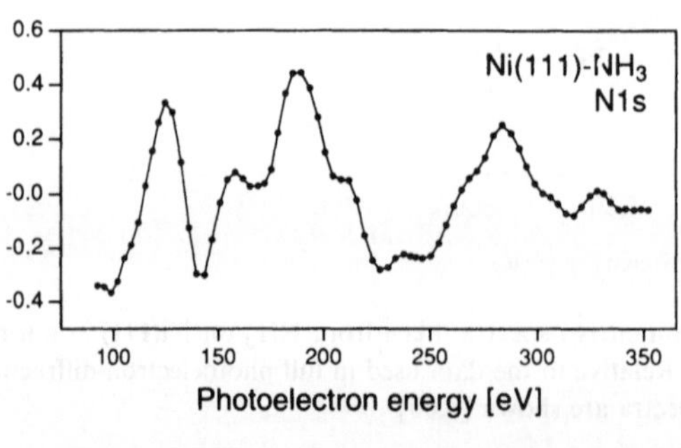

Fig. 4.9. Normal emission N 1*s* photoelectron diffraction (PhD) spectrum from NH_3 on Ni(111) resulting from the complete experiment summarised in Fig. 4.8 [4.60]

surface takes place through this common atom, we have the possibility of effectively moving this atom to different sites, since the bonding geometry may differ due to the influence of the remaining parts of the molecule. Notice, however, that if the adsorbate-substrate bonding is through the common atom, the local nearest neighbour bondlength to the surface is likely to be very similar in each case. A specific example of this idea is to compare the O 1*s* photoelectron diffraction from chemisorbed oxygen on Cu(100) with the same signal from two catalytic intermediates, the methoxy species (CH_3O–) and the formate species (HCOO–), on the same surface. The methoxy species is formed by the deprotonation of methyl alcohol (CH_3OH) on this surface by interaction with preadsorbed atomic oxygen (the alcohol hydrogen and excess oxygen being desorbed as water) [4, 61], while the formate species results from a similar deprotonation of formic acid (HCOOH) either by direct interaction with the clean Cu(1 0 0) surface (with H_2 desorption) or, more effectively, by a similar interaction with preadsorbed atomic oxygen [4.62]. Prior vibrational spectroscopy of these systems makes it clear that both of the molecular fragments are bonded to the surface through the O atoms, and that in the case of formate the two oxygen atoms are symmetrically equivalent in the adsorbed species [4.63–65].

Normal emission O 1*s* scanned-energy-mode photoelectron-diffraction spectra from these three systems [4.1, 66, 67] are compared in Fig. 4.10. Superimposed on these experimental spectra are the results of particularly simple theoretical calculations to simulate the effect of the dominant Cu substrate backscatterer atoms. The three experimental spectra clearly differ

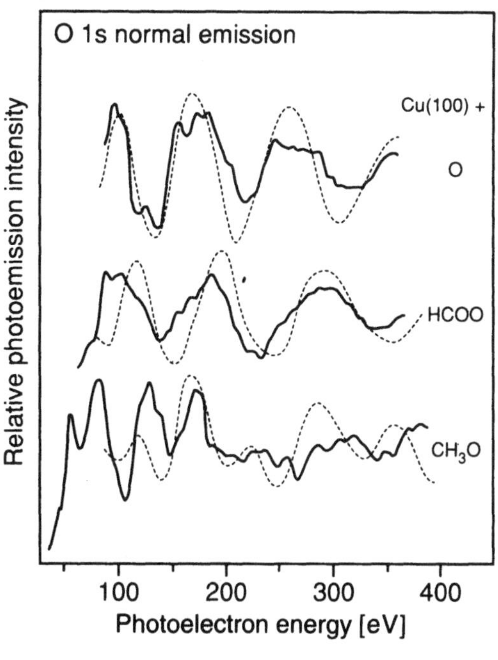

Fig. 4.10. Normal emission O 1*s* PhD spectra taken from chemisorbed O on Cu(100) (in the $(\sqrt{2} \times 2\sqrt{2})R45°$ phase), from the formate (HCOO–) and methoxy species (CH_3O–) on Cu(100). The dashed lines are the results of very simple model calculations as described in the text. Data taken from [4.12, 66 and 67]

substantially. In the case of the chemisorbed oxygen and formate, the spectra are dominated by three main peaks with slightly different periodicity. For these two species, we see that similar, but slightly different, dominant scattering path-length differences are involved, although the difference in periodicity is found, in a full structural analysis [4.66, 12], to result from different sites rather than different bond lengths (in fact, the Cu–O bondlength is *shorter* in the case of chemisorbed oxygen, although the shorter energy periodicity in Fig. 4.10 implies a longer dominant scattering pathlength). By contrast, the methoxy spectrum is quite different, with an apparent dominant peridocity leading to about 6 peaks in the same energy range (implying a substantially longer dominant scattering path-length difference).

Full quantitative analysis of these three adsorption systems shows that formate [4.12] bridges two nearest-neighbour top layer Cu atoms such that the O atoms of the formate are almost exactly atop top layer Cu atoms. Chemisorbed oxygen [4.66], on the other hand, occupies a site which is essentially the simple hollow of the Cu(100) surface, although the oxygen induces a $(\sqrt{2} \times 2\sqrt{2})R45°$ reconstruction in which one of the top-layer nearest neighbours of each oxygen atom is removed. The key feature of this structure is that the O atom lies directly above a second-layer Cu atom, which in normal emission produces strong 180° scattering. Because the chemisorbed oxygen is quite close to coplanar with the top Cu atom layer, the scattering path-length difference to this second-layer substrate atom is only slightly larger than the similar path length to the nearest-neighbour, top layer, Cu atom in the case of formate. The dashed lines in Fig. 4.10 are the computed spectra for scattering from these single Cu atoms only, in each of these two cases. The local bonding geometries are illustrated in Fig. 4.11, which also shows the local geometry for the methoxy species, which is found to adopt an off-hollow adsorption site displaced significantly towards the bridging site [4.67]. The dashed-line fit to the methoxy spectrum of Fig. 4.10 includes the

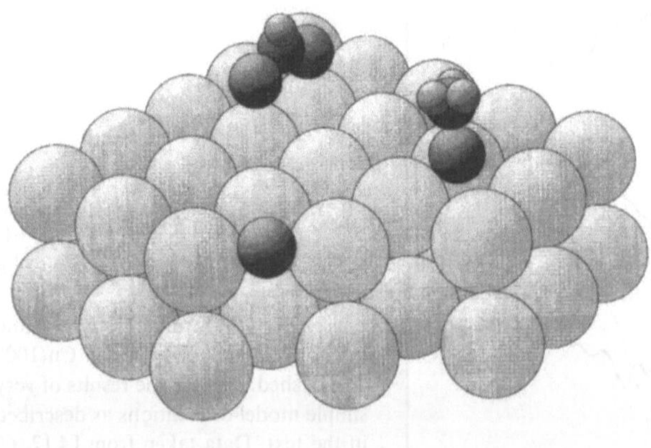

Fig. 4.11. Model of a Cu(100) surface showing the local adsorption sites adopted by chemisorbed O (in the $(\sqrt{2} \times 2\sqrt{2})R45°$ phase), and by formate and the methoxy species on this surface (as determined by the PhD experiments [4.12, 66, 67]

effect of the more distant second layer Cu atom below the O, and the four Cu nearest-neighbour atoms in the top layer, associated with this rather different site.

Although the full structural statements about these three adsorption systems can only be justified by recourse to more complete theoretical calculations, as described later in this section, the primary feature of importance illustrated by Fig. 4.10 is that PhD spectra do show a substantial sensitivity to adsorption site in cases where the nearest neighbour bond-lengths change very little. A second point is that, especially when the spectra show strong, dominant, long-energy-period oscillations, they can often be attributed to the scattering effect of a single substrate backscatterer placed close to the optimum 180° scattering geometry. One further illustration of the qualitative site sensitivity of the technique is provided by a comparison of the results of experiments on the formate species adsorbed on two different faces of copper – Cu(1 0 0) and Cu(1 1 0). Figure 4.12 shows the normal emission PhD spectra for both O 1*s* and C 1*s* photoemission from each of these systems [4.4]. It is clear that the results from the two surfaces are extremely similar, indicating that the local adsorption geometry is the same. This actually contrasts with conflicting evidence from SEXAFS at the time of these PhD measurements which seemed to imply quite different local adsorption sites for the two faces [4.68–70, 4]. Notice that although the geometry of the

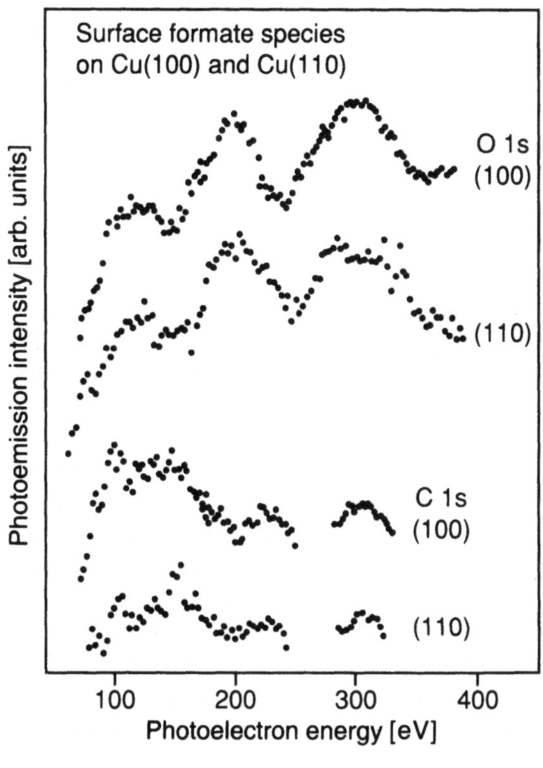

Fig. 4.12. Comparison of O *1s* and C *1s* normal emission PhD spectra taken from the formate species (HCOO–) on Cu(100) and Cu(110) [4.12]

Cu(100) and (110) faces is quite different, both faces do have nearest neighbour Cu atoms in the top layer (in ⟨110⟩ azimuths) so the same bridging configuration with the oxygen atoms almost atop is possible. Quantitative fits to the data of Fig. 4.12 confirm this picture, and also show that the small differences between the data from the two faces (e.g., the different fine structure on the three main peaks in the O 1*s* spectra) are attributable to second-layer Cu atom scattering which is, of course, different for the two different crystal faces. The results of Fig. 4.12 also serve to remind us that for adsorbed molecules containing more than one elemental species (in addition to H!), separate measurements of PhD spectra are possible for each species. This can provide not only reinforcing information about the molecular adsorption site, but also some details of possible changes in the internal shape of the molecule which might result from its interaction with the substrate, because we can effectively determine the exact location of each constituent atom independently.

A rather different example of a *qualitative application* of the method is given by a study of NO adsorption on Ni(1 1 1), in which some very important conclusions, of quite general significance, can be drawn even before any quantitative fitting is pursued. The properties of adsorbed NO on Ni(111) have already been studied extensively by vibrational spectroscopy [4.71–73]: both electron energy loss and infrared reflection-absorption spectroscopies. The key feature of these measurements is that three quite distinct N–O stretching frequencies are seen under different adsorption conditions. At low coverage on a clean Ni surface, this frequency is about 1500 cm^{-1}. For half-monolayer and saturation coverage a single frequency of approximately 1580 cm^{-1} is seen. Preadsorption of atomic oxygen leads to a much higher N–O stretching frequency of around 1800 cm^{-1}. By analogy with the extensive literature on the C–O stretching frequencies of CO in metal carbonyls and on different surfaces, these three different vibrational states of NO have been assigned to three different local adsorption geometries. The favoured explanation was that they corresponded to tilted bridge, linear bridge, and atop adsorption sites, respectively.

Figure 4.13 exhibits the normal emission N 1*s* PhD spectra obtained from these three different adsorption states [4.5]. Because NO is known to bond to the surface through the N atom, we can expect these spectra to show similar site sensitivity to the O 1*s* spectra of Fig. 4.10. What Fig. 4.13 actually demonstrates, is very little difference between the three NO spectra. There are differences in the relative amplitudes of peaks, especially relative to the background, but all the main features appear at the same energies. A clear implication of this is that the local adsorption geometries are actually the same for all three states with quite distinct vibrational properties. As further confirmation of this conclusion, we include in Fig. 4.13 the normal emission N 1*s* spectrum taken from ammonia (NH_3) adsorbed on the same surface (which we have already shown in Fig. 4.9) [4.60]. Clearly, this spectrum is completely different, and indeed we can anticipate, by comparison with the O 1*s* spectrum from formate on Cu(1 0 0) in Fig. 4.8, that the strong oscillations with three dominant peaks probably result from atop site adsorption of NH_3 (a conclusion confirmed in a full analysis

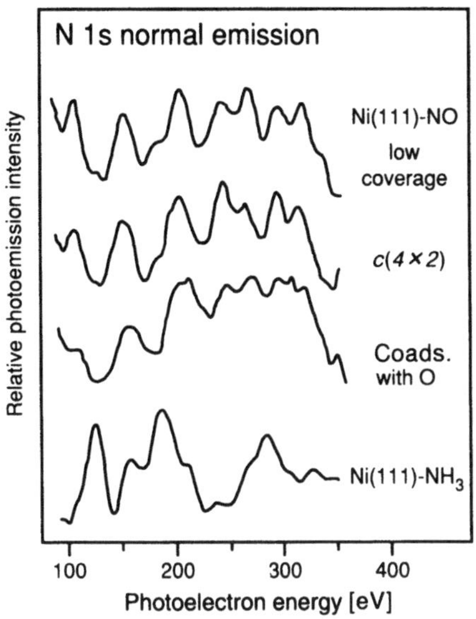

Fig. 4.13. N 1*s* PhD spectra taken from NO adsorbed on Ni(111) at low coverage, at the saturation coverage corresponding to the *c*(4 × 2) phase, and when coadsorbed with atomic oxygen, compared with the spectrum resulting from NH_3 adsorption on the same substrate (Fig. 4.9). Data are taken from [4.13, 60]

[4.60]). Quantitative fitting of the NO data of Fig. 4.11, on the other hand, together with other N and O 1*s* PhD spectra recorded in other emission geometries, indicates that NO actually occupies hollow sites in all three adsorption states [4.13, 74]. The most important conclusion of this study, however, is not what the exact adsorption site and geometry is, but rather that it is unchanged between states with quite different N–O stretching frequencies.

These results remind us that vibrational spectroscopy is not primarily a structural technique. Despite the wide-spread use of internal vibrational frequencies as a fingerprint of different adsorption sites in surface science, it is clear that this is potentially dangerous. In fact, there are probably two different reasons for the changing N–O frequencies in the different Ni(111) adsorption states. Firstly, in the case of O coadsorption, for which the observed frequency indicates a strong stiffening of the N–O bond, the change is most likely attributable to an electronic interaction. NO in the gas phase has a single electron occupying the antiboding 2π level, so any change in the occupation of this level can have a large effect on the N–O bond strength. Co-adsorption of the electronegative oxygen is likely to produce a net transfer of electron charge from the NO to the oxygen, relative to the adsorption on the clean substrate; this will reduce the degree of antibonding in the N–O bond, producing the observed frequency change. By contrast, the more subtle frequency change with increasing coverage on the clean metal may be attributed to dynamic coupling, in addition to a possible 'chemical' effect. Simple modelling of the formation of the observed 0.5 ML *c*(4 × 2) structure formed at saturation coverage shows that if hollow site occupation is involved, coverage beyond 0.25 ML is only possible by filling near-neighbour sites, leading to a sudden increase in dynamic coupling. This rationalisation highlights the fact that vibrational spectra can be sensitive to electronic and structural effects, including adsorbate-adsorbate interactions, which do not reflect changes in the local adsorbate-substrate registry. Of course, NO, with its partially occupied antibonding level, is especially susceptible to changes in internal bond strength due to electronic interactions, but there is growing evidence that even vibrational spectroscopy of adsorbed CO must be interpreted with care and not be used as too simplistic a fingerprint of local adsorption sites.

4.3.3 Quantitative Structural Analysis with PhD

a) PF_3 on Ni(111)

As noted in Sect. 4.2, the present authors and their respective research groups have recently applied the multiple-scattering codes of *Fritzsche* to the more rigorous quantitative structural analysis of various molecular adsorption systems. Two examples are provided by ammonia and phosphorus trifluoride on Ni(111) [4.60, 75] where in each case adsorption occurs in the atop site and the threefold molecular symmetry axis is oriented perpendicular to the surface. Only the PF_3 system will be discussed in detail here. NH_3 adsorption on Ni(111) was mentioned briefly in the last section (Fig. 4.13). The chemisorption of PF_3 has been the subject of particular interest in recent years because of the similarity between its surface properties and its behaviour as a ligand in coordination chemistry [4.9, 11, 76]. PF_3 forms zero-valent, mononuclear complexes with most transition metals and bonds only in a linear configuration [4.77]. Moreover, the bonding mechanism is similar to that of CO: σ-donation occurs via the $8a_1$ orbital with π back donation from the metal into the $7e$ orbital. Several theoretical studies (e.g., *Bagus* et al. [4.78]) supported the notion that the analogy to CO also pertains in surface chemistry. On Ni(111) *Alvey* and *Yates* [4.11] inferred from the sixfold symmetric ESDIAD pattern that PF_3 is adsorbed on the Ni(111) surface in atop sites and is azimuthally oriented at low temperature so that the individual P–F bonds are directed over next-nearest neighbour Ni atoms (Fig. 4.14). Since there are two such equivalent orientations the observed pattern contains six F^+ beams in the form of a hexagon. In the PhD study described here the adsorption site, the Ni–P bond length and the change in the Ni–Ni first layer separation were determined using the P $2p$ level at 136 eV binding energy. Before going into the details of the structural analysis, it is appropriate to describe briefly the multiple scattering calculations as well as to discuss the question of reliability (or R-) factors as an objective means of assessing the quality of fits between theory and experiment.

The multiple-scattering cluster calculations are based on a magnetic quantum-number expansion [4.44]. An optimised set of basis functions is used in the angular-momentum representation, with the result that all sums over magnetic quantum numbers converge rapidly and can be truncated after a few terms

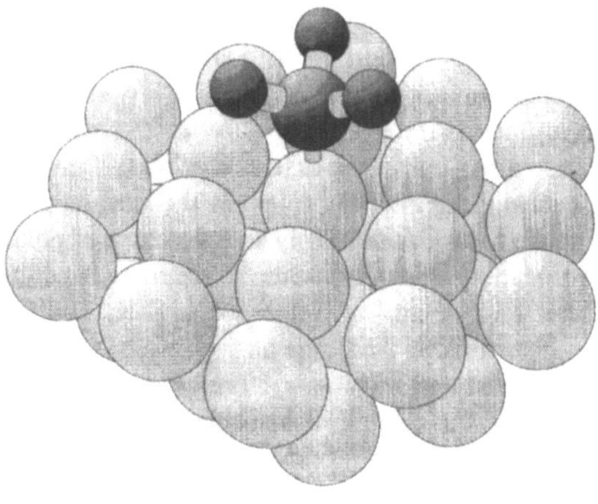

Fig. 4.14. Model showing the adsorption geometry of PF_3 on Ni(111)

without loss of accuracy. Further, by specifically taking into account the finite energy resolution in the experiment, additional damping factors arise in the theoretical expression for the intensity. These systematically suppress the contribution of long scattering path ways [4.45]. Thus, for the calculations reported in this chapter, third- and higher-order scattering events could largely be neglected due to their longer path length. The finite acceptance of the electron-energy analyser is also specifically taken into account and similarly reduces the importance of certain long scattering path ways. The reader is referred to the papers of *Fritzsche* [4.44–46] for a fuller account of the method, in particular for a comparison with the other approaches which may be used for multiple-scattering cluster simulations of photoelectron-diffraction spectra.

As in LEED, structural analysis in photoelectron diffraction proceeds by comparison between experiment and theoretical simulations for a series of trial structures. In this situation it is helpful to apply some form of *R*-factor, in order to put this matching procedure on an objective, quantitative basis. The structural parameters in the calculations are then varied in order to localise the global minimum in the *R*-factor. In the case of LEED, many years of experience have indicated that there is a typical numerical value of the *R*-factor to be expected for an acceptable fit. In the case of photoelectron diffraction it is not yet clear what *R*-factor values can be achieved. Moreover, there is some question as to which type of *R*-factor should be applied. In this context we note that the nature of the raw data is different: LEED spectra comprise a sum of Lorentzian peaks, whereas photoelectron-diffraction spectra comprise a sum of harmonic oscillations about a mean value. Furthermore, it is not clear how large a data set should be used in photoelectron diffraction. A single PhD spectrum is typically far more sensitive to the adsorption site than a single LEED spectrum because, as noted above, the dominant intensity variations in integral-order beam LEED spectra are due to scattering interferences within the substrate. By comparison, the source of electrons in photoelectron diffraction is localised on the adsorbate, so that *all* the structure in the spectrum is sensitive to the local adsorbate-substrate registry. In favourable cases it may, therefore, be possible that a single PhD spectrum is adequate for the determination of the adsorption site, although a larger data set is almost certainly necessary if a large number of structural parameters has to be varied. We also note that in the case of the few systems for which there are extensive PhD spectra (e.g., Cu(111)–C_2H_2 in the next subsection), the intensity modulations tend to vary quite slowly as a function of polar angle in a particular azimuth, indicating that there can be much redundant information in such large data sets. Two reliability factors, R_p and R_m, will be used in this and the following subsections in order to arrive at an objective quantitative comparison between experiment and theory. The former is essentially that introduced by *Pendry* [4.79] for structural analysis in LEED. The latter is based on that used by *Shirley* and co-workers in their photoelectron diffraction studies [4.80], but normalised such that the value of unity corresponds to zero correlation between theory and experiment, as in the case of R_p:

$$R_m = \frac{\Sigma(X_{th} - X_{exp})^2}{\Sigma(X_{th}^2 + X_{exp}^2)} \tag{4.5}$$

where $X(E)$ is the modulation function. Values of zero and two correspond to perfect agreement and to completely anticorrelated data, respectively. To avoid the problem of zeros and negative values of $X(E)$, which are not found in LEED intensities, R_p has been applied to $[X(E) + 1]$, rather than $X(E)$.

In the scanned-energy photoelectron-diffraction study of the Ni(111)-PF_3 system described here, the P 2*p* photoelectron peak was measured at photon energy intervals of 3 eV in the kinetic energy range 80–380 eV at polar emission angles of $\phi = 0°$, 30° and 50° in the ⟨110⟩ azimuth. In the simulations the height of the P atom above the first Ni layer was varied for the atop site (Fig. 4.14), as well as for the bridge, fcc hollow and hcp hollow sites. The F atoms were not included since experience shows that the scattering from other light atoms of a molecule is generally negligible, unless the configuration is close to the favoured 180° scattering geometry. Relatively small metal atom clusters were used initially in the calculations with scattering pathlengths of up to 14 Å and 9 Å for single and double scattering, respectively, and with all Ni–Ni spacings set at the bulk value. Larger clusters were then employed in a later stage of optimisation. Plots of the normal emission *R*-factors as a function of the Ni–P separation are shown for all four sites in Fig. 4.15. By far the best agreement between theory and experiment is obtained for the atop site at a Ni–P distance of 2.07 Å with $R_p = 0.24$ and $R_m = 0.15$. Particularly striking, however, are the oscillations in the *R*-factors due to the phenomenon of multiple 'site' coincidences which are also known in LEED. Thus, quite reasonable agreement is also observed for a Ni–P distance of 1.65 Å where R_p and R_m have minima of 0.50 and 0.29, respectively. The reason for the oscillatory behaviour of the *R*-factors can be seen from the calculated diffraction curves in Fig. 4.16. For the atop site the normal emission photoelectron-diffraction spectrum is dominated by strong modulations due to

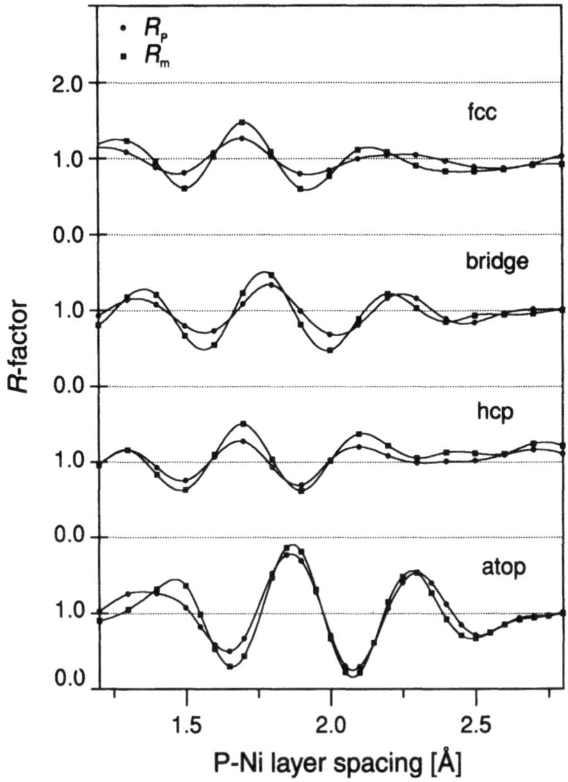

Fig. 4.15. Summary of the quality of experiment/theory agreement, as determined by two different R-factors, for the P $2p$ normal emission PhD spectrum from PF_3 adsorbed on Ni(111), when compared with theoretical simulations assuming adsorption in four different surface sites at a range of P–Ni top layer spacings [4.75]

the interference between the directly emitted wave and the wave 180° backscattered from the Ni atom directly 'behind' the P atom, i.e., the frequency of oscillations depends on the path-length difference. Good agreement with the experimental curve is obtained at 2.10 Å, but at 1.90 Å the two curves are almost completely out of phase. Relatively good agreement is, however, observed again at 1.7 Å, but the modulation frequency of the calculated diffraction spectrum is clearly lower due to the decreased path-length difference. Although the result is already given essentially by Fig. 4.15, any remaining ambiguity can be removed by considering the R-factor plots for comparisons with the experimental data taken at other emission angles (Fig. 4.17). The R-factor values at the minima are somewhat larger (because of the less favourable back-scattering geometry) but the only Ni–P spacing corresponding to a minimum in all three plots is 2.07 Å. We discuss multiple site coincidences in more detail elsewhere [4.81].

Further optimisation with a larger cluster (10 Å and 20 Å for single and double scattering, respectively) was carried out by varying the separation

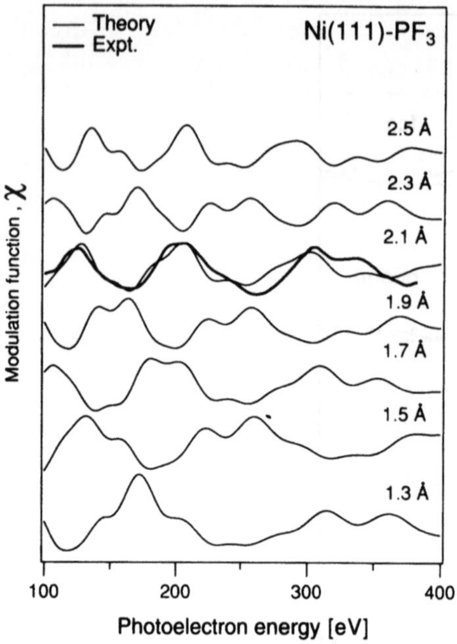

Fig. 4.16. Comparison of the experimental normal emission P 2*p* PhD spectrum from PF_3 adsorbed on Ni(111), with the results of theoretical calculations assuming atop site adsorption at various P–Ni top layer spacings as indicated [4.75]

between the first and second Ni layers, as well as the vibrational amplitude of the soft absorbate modes parallel to the surface [4.60, 82], the spline function was used for the derivation of $X(E)$ from the raw spectra and the exact photoelectron emission angle. A small, but perceptible 0.03 Å expansion of the Ni–Ni layer spacing produces a better fit, but the Ni–P separation remains at 2.07 Å as in the preliminary analysis. A mean square displacement of the molecule of 0.02 $Å^2$ parallel to the surface also had to be included. The resulting R_p and R_m values were then 0.15 and 0.09, respectively.

This PhD study thus shows conclusively that the PF_3 molecule is adsorbed in the atop site with a Ni–P separation of 2.07 ($\pm$ 0.03) Å and also indicates that an expansion of the Ni–Ni outermost spacing of 0.03 ($\pm$ 0.05) Å takes place. The precisions were calculated using a procedure derived for LEED by *Pendry* [4.79]. In a simultaneous structural study of this system with other synchrotron radiation-based techniques [4.83], it was also shown that the molecule adsorbs in an atop site with a Ni–P separation of 2.07 ($\pm$ 0.03) Å (SEXAFS) and 2.13 ($\pm$ 0.10) Å (X-ray standing-wave analysis), in remarkably good agreement with the present data. These investigations were also able to confirm that the threefold symmetry axis is perpendicular to the surface (using NEXAFS) and that the molecule is essentially undistorted relative to the gas phase. Finally, we recall the surface chemistry/coordination chemistry analogy and note that the Ni–P separation in $Ni(PF_3)_4$ is 2.099 ($\pm$ 0.003) Å [4.84].

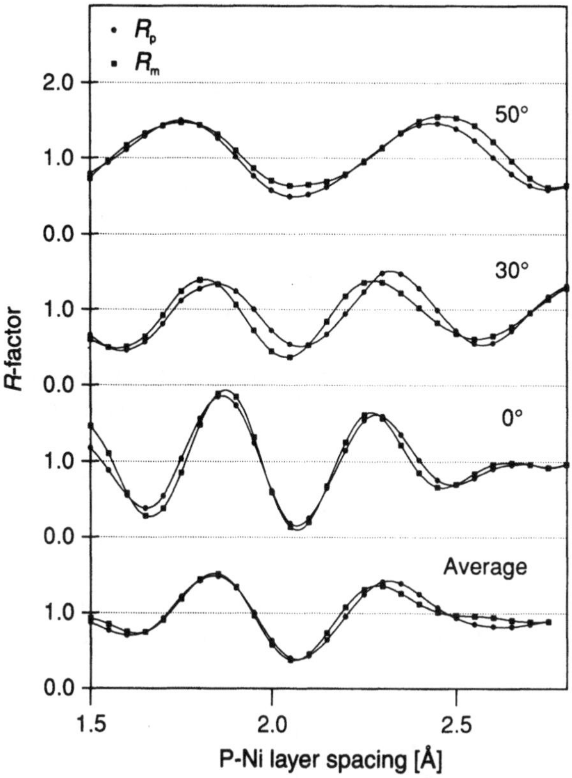

Fig. 4.17. Summary of the quality of experiment/theory agreement, as determined by two different R-factors, for P $2p$ PhD spectra taken at three different polar emission angles from PF_3 adsorbed on Ni(111), when compared with theoretical simulations assuming atop adsorption at a range of P–Ni top layer spacings. The lowest curve shows the average of the three individual spectral R-factors [4.81]

b) C_2H_2 on Cu(111)

Simple hydrocarbons form one class of adsorbates for which relatively little structural information is currently available. The reactivity of transition- and noble-metal surfaces to these molecules is of considerable importance for a whole range of catalytic reactions in the petrochemical industry. On the basic research side, the dehydrogenation of ethylene on a Ni(111) surface actually provided the first example of the direct observation of a catalytic reaction on a single-crystal surface using a spectroscopic technique (UV photoemission) [4.85]. Subsequently, vibrational spectroscopy, in particular EELS, has indicated that substantial rehybridsation occurs when unsaturated hydrocarbons, in particular acetylene, are adsorbed at low temperature on various metal surfaces. It was first suggested that the origin of this effect was structural in that the lattice parameters of Ni and Cu offered favourable adsorption sites for acetylene, especially on their (111) faces [4.86]. Now that further data has become available [4.87], showing similar low frequencies on other surfaces, it is clear that the real explanation must be somewhat more sophisticated. Nevertheless, the reduction in C–C bond order from three to somewhere between one and two, which occurs for acetylene on Cu(111), still suggests adsorption in a site which allows the molecule to co-ordinate to several Cu atoms. This can be achieved on the (111) surface if the two carbon atoms occupy the two nearest-neighbour, inequivalent, three-fold hollow sites which are separated by 1.47 Å. Assuming that the C–C axis is approximately parallel to the surface, such a bondlength would be compatible with the observed frequency of the C–C stretch of 1307 cm^{-1}.

A strong indication of the essential correctness of this model was provided by the application of Fourier-transform mapping [4.42, 43] to this problem [4.88]. As explained in Sect. 4.2, this is a direct method for obtaining adsorption site information in which PhD spectra, at a range of emission angles, are Fourier transformed to map the intensity contributions from the various (nearest-neighbour) backscattering paths as a function of angle. Using PhD spectra recorded at polar angles from − 10° to 60° in 10° increments in the ⟨110⟩ azimuth and in the two non-equivalent ⟨211⟩ azimuths, nearest-neighbour substrate peaks are found in the directions to be expected for both fcc and hcp

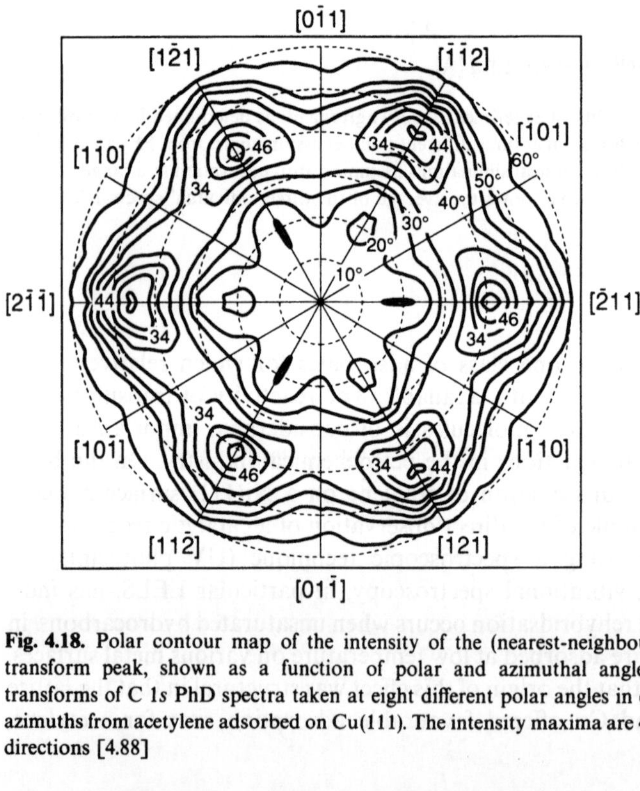

Fig. 4.18. Polar contour map of the intensity of the (nearest-neighbour scattering) 4 Å Fourier transform peak, shown as a function of polar and azimuthal angle, obtained from Fourier transforms of C 1*s* PhD spectra taken at eight different polar angles in each of the three principal azimuths from acetylene adsorbed on Cu(111). The intensity maxima are 44 and 46 in the two ⟨211⟩ directions [4.88]

sites. Although the measurements were only performed in three azimuths, the rotational symmetry of the crystal face and appropriate interpolation enable a contour plot of the Fourier amplitude for a back-scattering path-length difference of 4 Å to be made, as shown in Fig. 4.18. The sixfold symmetry is immediately apparent. The distance of 4 Å corresponds to roughly twice the nearest-neighbour separation which is the path-length difference for 180° scattering.

In order to confirm this conclusion, and to refine the associated structural parameters, extensive model calculations were performed using the multiple-scattering code in the same way as for Ni(111)-PF_3 in the last section. This optimisation process has been performed using six of the 24 different experimental spectra described above, specifically the normal emission and 40° emission data taken in each of the three principal azimuths. The 40° off-normal emission data are rather important because, as can be seen from Fig. 4.18, this angle places the nearest-neighbour Cu atoms almost directly behind the carbon atom emitters, a geometry leading to particularly strong modulations in the diffraction spectra. Three principal structural parameters have been investigated in detail: these are the average C–Cu top layer spacing, the difference in this layer spacing for the C atoms above the hcp and fcc hollow sites and the displacement of the C atoms from the exact threefold hollow sites parallel to

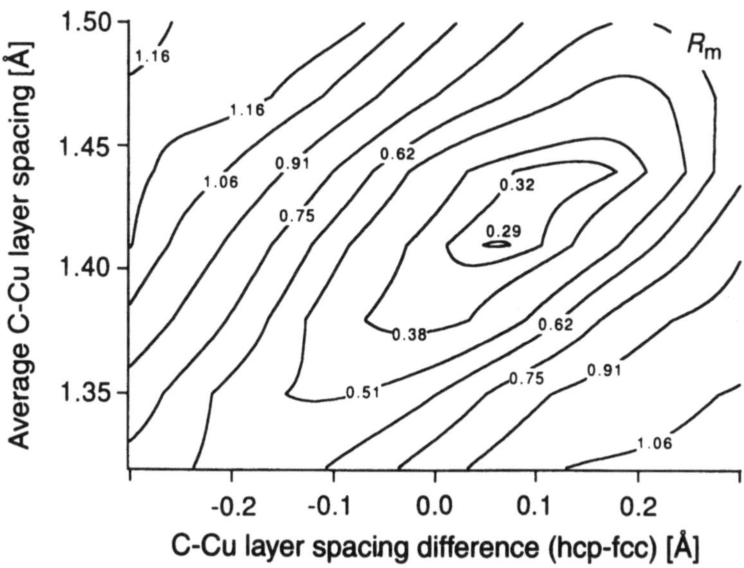

Fig. 4.19. Multi-spectral R-factor contour map based on the comparison of six different C 1s PhD spectra (taken at 0° and 40° polar angle in three different azimuths) from acetylene on Cu(111), with the results of calculations assuming that the two C atoms lie above the two different (fcc and hcp) hollow sites on this surface. The structural parameters varied in the calculation are the average C–Cu top layer spacing, and the difference between the C–Cu top layer spacings for the two different sites [4.88]

the surface. The influence of these parameters on the *R*-factor (R_p and R_m) for the six individual spectra as well as on the unweighted average value (multispectral *R*-factor) was investigated.

Figure 4.19 provides an example of the results of this process in the form of a contour map for the average value of R_m as a function of both the average C–Cu top layer spacing and the difference in height of the C atoms. A clear single minimum is seen corresponding to values of 1.41 ($\pm$ 0.03) Å and 0.06 ($\pm$ 0.03) Å, respectively. The carbon atom above the hcp site is marginally higher than that above the fcc site. The optimum value of the lateral displacement of C atoms from the threefold symmetry sites is 0.00 ($\pm$ 0.05) Å. The

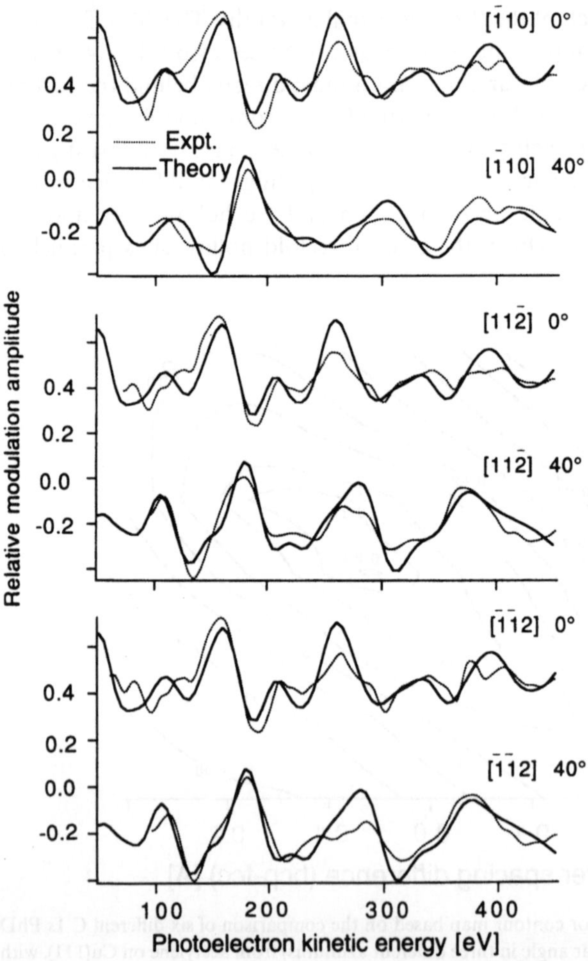

Fig. 4.20. Comparison of experimental C 1*s* PhD spectra (taken in six different emission geometries) from acetylene adsorbed on Cu(111), with the results of calculations for the optimised structure as shown in Fig. 4.21 [4.88]

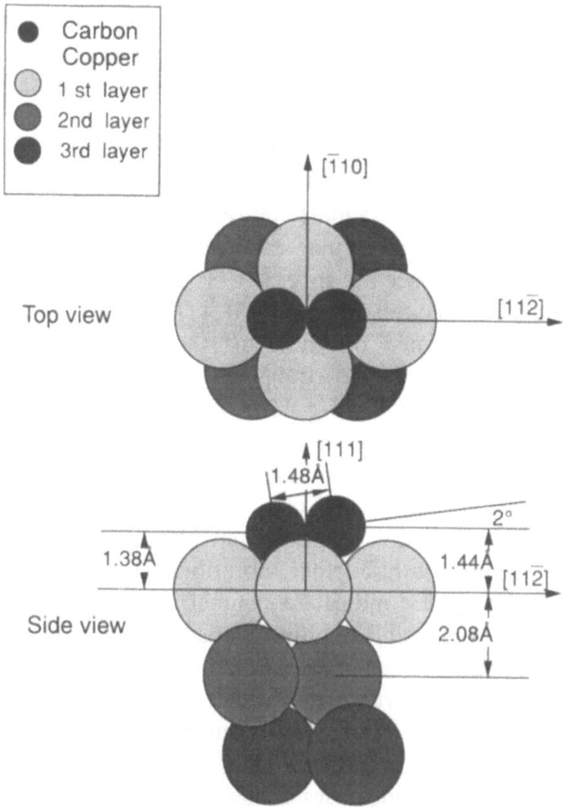

Fig. 4.21. Plan and sectional views of the structure of acetylene adsorbed on Cu(111) as determined by the PhD study [4.88] described in the text

calculated photo-electron-diffraction spectra for this optimal structure are compared with the six experimental spectra in Fig. 4.20. The optimal structure itself is shown schematically in Fig. 4.21. These results provide direct quantitative confirmation that the acetylene molecule adsorbed on Cu(111) is very substantially stretched, i.e. by 0.28 ($\pm$ 0.10) Å, consistent with the considerably reduced C–C stretching frequency, and that the C–C axis is very nearly parallel to the surface. Although this result confirms the original suggestion of *Bandy* et al. [4.86] concerning the adsorption site, the observation of similarly low C–C stretching frequencies on other Cu surfaces (and, indeed, on other transition metal surfaces) suggests, as noted above, that there is no special structural significance attached to Cu(111). Clearly, further quantitative studies are required to elucidate the binding site in other acetylene adsorption systems.

4.3.4 Chemical-Shift Photoelectron Diffraction

In the case of larger adsorbed molecules, or of different molecules co-adsorbed, several atoms of the same element may be present on the surface in different locations. If these atoms are in chemically distinct environments, there will be differences in their photoelectron binding energies, i.e., they will exhibit chemical shifts. If scanned-energy-mode photoelectron diffraction is performed on such systems, the modulations in the spectra will be different if the adsorption sites of the inequivalent atoms and/or their separations from the surface are different. Thus, we have the possibility of determining independently the surface coordinates of each atom. This has been referred to as chemical-shift photoelectron diffraction and is likely to prove particularly powerful in future for the structural analysis of more complicated molecular adsorption systems. Two examples are described here: the surface acetate species (CH_3COO–) adsorbed on Cu(110) [4.89] and PF_x fragments co-adsorbed on Ni(111) [4.90].

The surface acetate species is formed on the Cu(110) surface by the decomposition of acetic acid and contains two equivalent carbon atoms, as indicated by the C 1*s* photoelectron spectrum of Fig. 4.22a. From literature data [4.91] on acetic acid and acetates it is known that the carboxyl carbon atom (C1) has a 3.5–4.0 eV higher 1s photoelectron binding energy than the methyl carbon atom (C2). The polarisation dependence of the oxygen and carbon *K*-edge absorption spectra [4.92] reveals that the molecular plane is perpendicular to the Cu surface in the ⟨110⟩ azimuth. Further, vibrational spectra of the same species on the Cu(100) surface [4.93] indicate a symmetric configuration with two equivalent oxygen atoms. A plausible bonding geometry based on that of the corresponding formate species described in Sect. 4.3.2 is shown as an inset to Fig. 4.22. Note that in this proposed structure, which is confirmed by the study reported here, the so-called *aligned bridge site* is occupied. On the same surface the corresponding trifluoroacetate species (CF_3COO–) can also be formed; the binding energy of the C2 atom increases in this case by ca. 6 eV and the corresponding peak shifts to the low-kinetic-energy side of the C1 peak.

Figure 4.22b exhibits the photoelectron diffraction spectra from the C1 and C2 carbon atoms in the acetate species at a polar emission angle of $\theta = 20°$ in the ⟨110⟩ azimuth. It is immediately obvious that the modulations in the two spectra are completely different, indicating clearly that the two carbon atoms have different coordinates relative to the surface. The same result is obtained at other polar emission angles. As above, quantitative determination of the sites of the two C atoms requires, however, simulations based on multiple-scattering cluster calculations. Results obtained for the best fit between theory and experiment are shown in Fig. 4.23 for the carbon atoms ($\theta = 20°$) and in Fig. 4.24 for the O atoms ($\theta = 0°$) [4.89]. These two detection angles place a nearest-neighbour Cu atom directly behind the appropriate emitter atom, a geometry which leads to the strongest diffraction effects, as we have seen already several times in this article. The experimental acetate and trifluoroacetate spectra are very similar for the C1 atom and have a strong resemblance in the case of the C2 atom, although here the modulations are weaker for the fluoroacetate. Further, the theory curves for the two species are virtually identical, indicating that the influence of the F atoms on the diffraction spectra is minimal. The same was found to be true for the O 1*s* spectra (Fig. 4.24). The relative insensitivity of photoelectron-diffraction spectra to the presence of other light atoms turns out

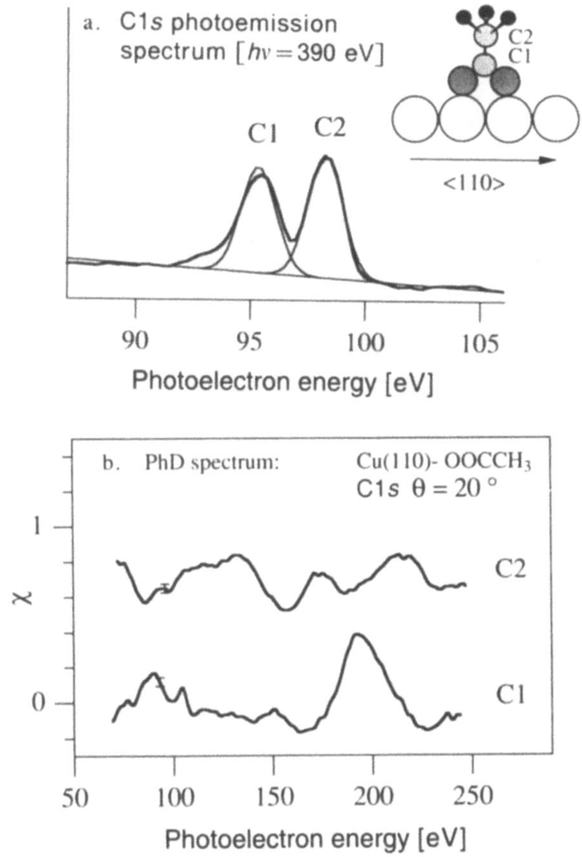

Fig. 4.22a, b. Results of a PhD investigation of the acetate species adsorbed on Cu(110). (**a**) shows the C 1*s* photoelectron energy spectrum recorded at a photon energy of 390 eV; the two chemically shifted C 1*s* peaks are related to the carboxyl (C1) and methyl (C2) carbons as labelled in the sectional model of the adsorption geometry shown in the inset. (**b**) shows the C 1*s* PhD spectra taken at a polar emission angle of 20° for these two chemically distinct atoms [4.89]

to be particularly useful in this situation. The quantitative determination of the position of each atom in the molecule would require, in principle, a complicated optimisation procedure in multiparameter space. To a very good approximation, however, the simulations can be performed for just the emitter atom alone on the substrate. Inclusion of the other atoms of the molecule at a later stage only shifts the position of the minimum in the *R*-factor by ca. 0.05 Å and marginally reduces its value.

The best fits to the experimental data were found for the aligned bridge site which gives significant minima in both R_p and R_m for all emission angles measured. The normal emission acetate data of Fig. 4.24 give best fit values of $z(O) = 1.90$ Å, where z is the O–Cu top layer separation, d(O–O) = 2.25 Å

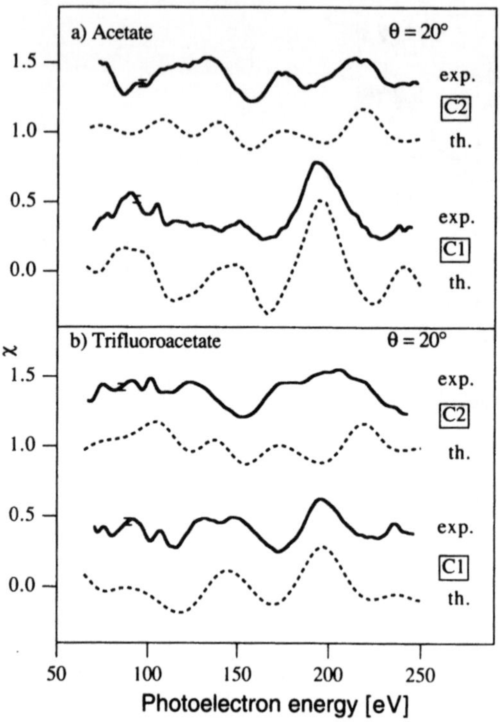

Fig. 4.23. Comparison of the two chemically shifted C 1*s* PhD spectra taken at a polar emission angle of 20° from the (a) acetate and (b) trifluoroacetate species on Cu(110), with the results of model calculations for the optimum adsorption geometry [4.89]

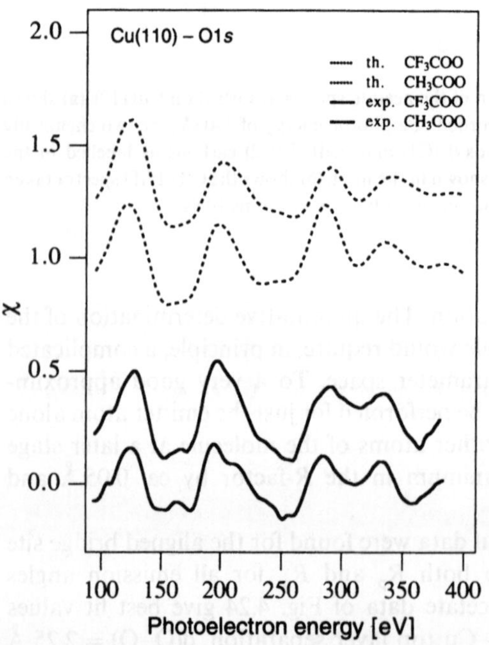

Fig. 4.24. Comparison of the normal emission O 1*s* PhD spectra from the acetate and trifluoroacetate species on Cu(110) with the results of model calculation for the optimum adsorption geometry [4.89]

and a contraction of the outermost Cu–Cu layer spacing relative to the bulk of 0.03 Å ($R_p = 0.19$, $R_m = 0.18$). The corresponding off-normal emission data gave essentially the same values (± 0.05 Å) but somewhat higher R-factors. From the data of Fig. 4.23a, z(C1) = 2.44 Å ($R_p = 0.35$, $R_m = 0.15$) and z(C2) = 3.98 Å ($R_p = 0.36$, $R_m = 0.30$). The normal emission data gave identical values with comparable R-factors. The data at 27° and 34° gave essentially the same values (± 0.05 Å) but with R-factor values that were generally somewhat larger. Thus for the surface acetate species it may be concluded that the O–Cu bond length is 1.91 Å, the O–C–O bond angle 129°, the C–O bondlength 1.25 Å, and the C–C bond length 1.54 Å. The precision of these determinations was estimated to be ± 0.04 Å for the bond lengths and ± 5° for the bond angle. For inorganic acetate compounds the bond angle and the C–O, C–C bond lengths are typically 119°, 1.26 and 1.50 Å, respectively [4.94]. From the analysis of the trifluoroacetate data z(O) and d(O–O) appear to remain as in the acetate species; z(C1) and z(C2) shift to 2.47 and 4.02 Å, respectively. The R-factors are somewhat worse, as might be expected from the poorer agreement seen in Fig. 4.23b. Since the measured contraction in the O–C–O bond angle (ca. 3°) is smaller than the estimated precision it may be concluded that structure parameters of the surface acetate species do not change significantly upon fluorination.

In the second example of chemical-shift photoelectron diffraction we are concerned with elucidating the structure of co-adsorbed molecular fragments. If PF_3 adlayers on metal surfaces are subject to irradiation by photons or electrons, rapid fragmentation occurs [4.11], giving rise to a mixture of adsorbed PF_x species, where $x = 0$–3. In the case of PF_3 on Ni(111), encountered in Sect. 4.3.3a, electron bombardment at 200 eV produces the P 2p photoelectron spectra shown in Fig. 4.25. The chemically shifted, but still overlapping, spin-

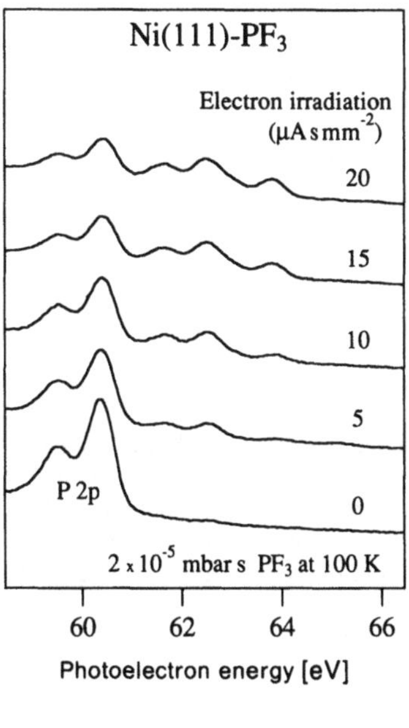

Fig. 4.25. P 2p photoelectron energy spectra taken from PF_3 adsorbed on Ni(111) after different exposures to electron beam irradiation [4.90]

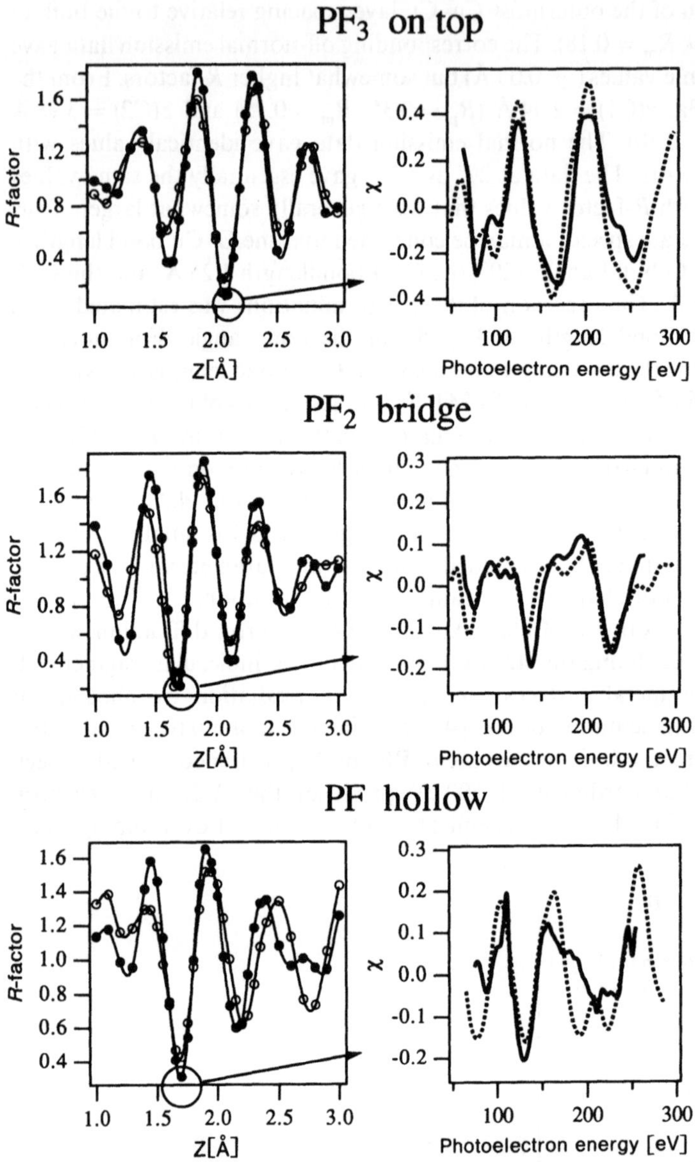

Fig. 4.26. Results of a PhD study of the three different chemically shifted P 2*p* photoemission (spin-orbit split doublet) peaks associated with the PF_3, PF_2 and PF species produced by electron irradiation of PF_3 on Ni(111) as seen in Fig. 4.25. On the left-hand side are shown *R*-factor plots (for the two different *R*-factors) resulting from comparisons between the experimental spectra and the results of theoretical computations in which the P–Ni top layer spacing was varied but the site fixed as atop for PF_3, bridge for PF_2 and hollow for PF. On the right-hand side are shown the comparisons of the experimental PhD spectra with the results of the computations corresponding the optimal layer spacings given by the *R*-factor minima [4.90]

orbit split doublets of PF_3, PF_2 and PF can be clearly distinguished; electrons, as opposed to photons, do not give rise to adsorbed P atoms. Phosphorus $2p$ PhD spectra were measured in normal emission as well as at a polar angle of 30° in the ⟨110⟩ azimuth [4.90]. The normal emission data are shown on the right-hand side of Fig. 4.26. Note that the energy range of the spectra is unusually short (70–260 eV) because the monochromator resolution worsens with increasing photon energy and the separation of the overlapping doublets thus becomes increasingly difficult. Comparison with theoretical simulations using the multiple-scattering codes shows that PF_3 adsorbs in the atop site, as might be expected from the analysis of the pure adlayer presented in Sect. 4.3.3a. On the other hand, PF_2 adsorbs in the twofold bridge site and PF in the threefold hollow site which confirms the conclusions drawn from the corresponding ESDIAD patterns [4.11]. The best fit calculated curves are presented as dashed lines on the right-hand side of Fig. 4.26. The corresponding R-factor plots as a function of the P–Ni top layer spacing are on the left. Note that oscillations in the R-factor due to multiple site coincidences (Sect. 4.3.3a) occur not only for PF_3 but also for PF_2 and PF. For PF_3 the Ni–P bondlength is 2.05 ($\pm$ 0.04) Å in excellent agreement with the value for PF_3 in the pure layer (2.07 Å). The P–Ni layer spacing for PF_2 in the bridge site is 1.70 ($\pm$ 0.04) Å, giving al nost exactly the same value for the Ni–P bond length as for PF_3 in the atop site. The layer spacing for PF could not be determined with the same precision. The best agreement between theory and experiment was obtained with a mixture of hcp and fcc sites with a slightly greater layer spacing in the hcp site. Whereas the 30° diffraction spectra gave the same quantitative result for PF_3 and PF_2 as in normal emission, for PF the minimum in both R-factors was 0.1 Å lower. At present we can, therefore, only conclude that the P–Ni layer spacing for PF is in the range 1.50–1.70 Å. Despite this minor problem, which could probably be eliminated by a more extensive data set (spectra at further emission angles), the structural analysis of this co-adsorbed layer represents an interesting example of a further area of application of chemical-shift photoelectron diffraction.

4.4 Concluding Remarks

Although the number of adsorbed molecular systems so far investigated with photoelectron diffraction in the scanned-energy mode is limited, it should be clear from the material presented in this chapter that it is a potentially very powerful probe of surface structure that has not yet been exploited to the full. The technique is not only able to provide information on the adsorption site and molecular orientation but also on adsorbate-substrate layer spacing, intramolecular parameters and changes in substrate geometry. One of its major disadvantages is that it requires monochromatised synchrotron radiation in a particular spectral region which, for most molecules, corresponds to the excitation energies of the C, N and O 1*s* levels. Whereas various grating monochromators are now available for this part of the soft X-ray region at several synchrotron-radiation sources, few of them have sufficiently high flux to measure adsorbate PhD spectra within a reasonable space of time. The experiments described here were all carried out on the high-energy TGM at BESSY which was specifically designed as a high-flux, albeit low-resolution, monochromator [4.95]. Further improvements in signal intensity, which can be utilised for better data quality, shorter measuring times and/or more extensive data sets, can also be attained by the use of more efficient parallel detection devices on electron energy analysers. Perhaps the most significant future development, however, is the next generation of XUV synchrotron-radiation sources which will enable PhD experiments to be carried out on insertion devices with appropriately tailored monochromators. The very large increase in spectral brilliance will not only provide higher flux for these experiments, but will also enable them to be performed at higher spectral resolution [4.96].

The two most recent advances in the technique itself are the application of direct methods and the utilisation of the chemical shift; both are likely to influence considerably future developments. As we have seen in Sects. 4.2 and 4.3.4 it is already possible, at least with the Fourier-mapping technique, to obtain reasonably directly the symmetry of the adsorption site and the approximate nearest-neighbour separation. A pre-requisite is, however, a sufficiently large data set, in this case, a set of energy scan spectra at several polar emission angles in at least two azimuths (depending on the symmetry of the surface). A far larger data set is required to obtain a single-photoelectron hologram, which consists of a complete angular scan in polar and azimuthal angle on quite a fine angular mesh. This places unreasonable demands on experimenters working with currently available resources and, in addition, the inversion method has yet to prove capable of providing a realistic image of the surface. Indeed, the measurement of up to a dozen such holograms combined with more efficient inversion techniques may be necessary in order to produce structural information of the desired precision. This does not mean that holography will never be used for adsorption site determination, but rather that the techniques presently available will need some improvement, in particular simplification, before they prove practical. The second development – the use of the chemical shift-promises to be particularly significant for the analysis of larger C-containing molecules, but only if the binding-energy difference in the photoelectron spectrum is sufficiently large (relative to the natural linewidth) and the peaks are easily resolved with a minimum of curve analysis. At present there is no large data base of high-resolution photoelectron spectra of adsorbed molecules which could answer this question.

Lastly, we should remark that so far PhD studies of molecular adsorbates have only involved a small range of species on nickel and copper substrates. Not only are there many more molecular adsorbates of interest, but also it will be interesting to explore the applicability of the method to other materials.

Acknowledgements. We would like to thank our many colleagues who have worked with us over the last five years to produce many of the results described in this chapter. Their contributions are clearly identifiable in the references below. We also thank K. Hermann for providing us with Figs. 4.11 and 14. Further, we acknowledge not inconsiderable financial support from the German Federal Ministry of Research and Technology, the British Science and Engineering Research Council as well as the SCIENCE and Large Installations programmes of the European Community.

References

4.1 A. Liebsch: Phys. Rev. Lett. **32**, 1203 (1974)
4.2 S. Kono, C.S. Fadley, N.F.T. Hall, Z. Hussain: Phys. Rev. Lett. **41**, 117 (1978)
4.3 D.P. Woodruff, D. Norman, B.W. Holland, N.V. Smith, H.H. Farrell, M.M. Traum: Phys. Rev. Lett. **41**, 1139 (1978)
4.4 S.D. Kevan, D.H. Rosenblatt, D. Denley, B.-C. Lu, D.A. Shirley: Phys. Rev. Lett. **41**, 1565 (1978)
4.5 A.M. Bradshaw: Faraday Discuss. Chem. Soc. **89**, 1 (1991)
4.6 D.P. Woodruff: In *Angle-Resolved Photoemission: Theory and Current Applications*, ed. by S.D. Kevan (Springer, Berlin, Heidelberg 1992) p. 243
4.7 C.S. Fadley: In *Synchrotron Radiation Research: Advances in Surface and Interface Science*, ed. by R. Z. Bachrach (Plenum, New York 1992) Vol. 1
4.8 H.P. Bonzel: Progr. Surf. Sci., **42**, 219 (1993)
4.9 M.R. Albert, J.T. Yates Jr.: *The Surface Scientist's Guide to Organometallic Chemistry* (Am. Chem. Soc., Washington, DC 1987)
4.10 C.E. Anson, B.J. Bandy, M.A. Chesters, B. Keiller, I.A. Oxton, N. Sheppard: J. Elect. Spect. **29**, 315 (1983)
4.11 M.D. Alvey, J.T. Yates, Jr.: J. Am. Chem. Soc. **110**, 1782 (1988)
4.12 D.P. Woodruff, C.F. McConville, A.L.D. Kilcoyne, Th. Lindner, J. Somers, M. Surman, G. Paolucci, A.M. Bradshaw: Surf. Sci. **201**, 228 (1988)
4.13 M.C. Asensio, D.P. Woodruff, A.W. Robinson, K.-M. Schindler, P. Gardner, D. Ricken, A.M. Bradshaw, A.R. Gonzalez-Elipe: Chem. Phys. Lett. **192**, 259 (1992)
4.14 G. Pacchiani, P.S. Bagus (eds): *Cluster Models for Surface and Bulk Phenomena* (Plenum, New York 1991)
4.15 J.B. Pendry, D.K. Saldin: Surf. Sci. **145**, 33 (1984)
4.16 N.V. Richardson, A.M. Bradshaw: Surf. Sci.: **88**, 255 (1979)
4.17 J. Stöhr: *NEXAFS Spectroscopy*, Springer Ser. Surf. Sci., Vol. 25 (Springer, Berlin, Heidelberg 1992)
4.18 A.M. Bradshaw, J. Somers: Phys. Scripta **T31**, 189 (1990)
J. Somers, A. Robinson, Th. Lindner, A.M. Bradshaw: Phys. Rev. B **40**, 2053 (1989)
4.19 E.g. T.E. Madey: J. Vac. Sci. Tech. **A4**, 257 (1986)
4.20 E.g. A.M. Bradshaw: In *Interaction of Atoms and Molecules with Solid Surfaces*, ed. by V. Bortolani, N.H. March, M.P. Tosi (Plenum, New York 1990) p. 477
4.21 H. Ibach, D. Mills: *Electron Energy Loss Spectroscopy and Surface Vibrations* (Academic, New York 1982); H. Ibach: Surf. Sci. (Stockholm)
4.22 A.M. Bradshaw, E. Schweizer: Adv. Spectroscopy **16**, 413 (1988)
4.23 Y. Chabal: Surf. Sci. Rep. **8**, 211 (1988)
4.24 The use of X-ray standing waves and Auger electron angular distributions should also be mentioned at this point, although so far little work has been done with adsorbed molecules
4.25 P.H. Citrin. P. Eisenberger, R.C. Hewitt: Phys. Rev. Lett. **41**, 309 (1978)
4.26 J. Stöhr, D. Denley, P. Perfetti: Phys. Rev. B **18**, 4132 (1978)
4.27 J. Stöhr: X-ray absorption in *Principles, Applications Techniques of EXAFS, SEXAFS and XANES*, ed. by D. C. Koningsberger, R. Prins (Wiley, New York 1988) p. 443

4.28 J. Haase, A.M. Bradshaw: In *Synchrotron Radiation Research: Advances in Surface and Interface Science*, ed. by R. Z. Bachrach (Plenum, New York 1992) Vol. 2 p. 55
4.29 C.S. Fadley: Physica Scripta T17, 39 (1987)
4.30 M. Sagurton, E.L. Bullock, R. Saiki, A.P. Kaduwela, C.R. Brundle, C.S. Fadley, J.J. Rehr: Phys. Rev. B **33**, 2207 (1986)
4.31 J.B. Pendry: *Low Energy Electron Diffraction* (Academic, New York, 1974)
4.32 M.A. Van Hove, W.H. Weinberg, C.-M. Chan: *Low Energy-Electron Diffraction*, Springer Ser. Surf. Sci. Vol. 6 (Springer, Berlin, Heidelberg, 1986)
4.33 J.J. Barton, S.W. Robey, D.A. Shirley: Phys. Rev. B **34**, 778 (1986)
4.34 D.H. Rosenblatt, J.G. Tobin, M.G. Mason, R.F. Davis, S.D. Kevan, D.A. Shirley, C.H. Li, S.Y. Tong: Phys. Rev. B **23**, 3828 (1981)
4.35 J.J. Barton, C.C. Bahr, Z. Hussain, S.W. Robey, J.G. Tobin, L.E. Klebanoff, D.A. Shirley: Phys. Rev. Lett. **51**, 272 (1983)
4.36 J.J. Barton, C.C. Bahr, Z. Hussain, S.W. Robey, L.E. Klebanoff, D.A. Shirley: J. Vac. Sci. Technol. A**2**, 847 (1984)
4.37 J.J. Barton: Phys. Rev. Lett. **61**, 1356 (1988)
4.38 e.g. J.J. Barton: Phys. Rev. Lett. **67**, 3106 (1991):
S.Y. Tong, H. Li, H. Huang: Phys. Rev. Lett. **67**, 3102 (1991); see also B.P. Tonner, J. Zhang, X. Chen, Z.-L. Han, G.R. Harp, D.K. Saldin: J. Vac. Sci. Technol. **B10**, 2082 (1992):
L.J. Terminello, J.J. Barton, D.A. Lapiano-Smith: J. Vac. Sci. Technol. **B10**, 2088 (1992)
4.39 E.L. Bullock, C.S. Fadley, P.J. Orders: Phys. Rev. B **28**, 4867 (1983)
4.40 P.J. Orders, C.S. Fadley: Phys. Rev. B, **27**, 781 (1983)
4.41 R. Dippel, D.P. Woodruff, X.-M. Hu, M.C. Asensio, A.W. Robinson, K.-M. Schindler, K.-U. Weiss, P. Gardner, A.M. Bradshaw: Phys. Rev. Letters **68**, 1543 (1992)
4.42 V. Fritzsche, D.P. Woodruff: Phys. Rev. B **46**, 16128 (1992)
4.43 K.-M. Schindler, Ph. Hofmann, V. Fritzsche, S. Bao, S. Kulkarni, A.M. Bradshaw, D.P. Woodruff: Phys. Rev. Lett. **71**, 2054 (1993)
4.44 V. Fritzsche: J. Phys. Condens. Matter **2**, 9735 (1990)
4.45 V. Fritzsche: Surf. Sci. **265**, 187 (1992)
4.46 V. Fritzsche: J. Elec. Spec. Rel. Phenom., **58**, 299 (1992)
4.47 J.J. Barton, D.A. Shirley: Phys. Rev. B **32**, 1906 (1986)
4.48 V. Fritzsche, P. Rennert: Phys. Status Solidi B **135**, 49 (1986)
4.49 J.J. Barton, D.A. Shirley: Phys. Rev. A **32**, 1019 (1985)
4.50 J.J. Rehr, R.C. Albers: Phys. Rev. B **41**, 8139 (1990)
4.51 A.P. Kaduwela, D.J. Friedman, C.S. Fadley: J. Elec. Spec. Rel. Phenom. **57**, 223 (1991)
4.52 D.A. Wesner, F.P. Coenen, H.P. Bonzel: Phys. Rev. Lett. **60**, 1045 (1988). Phys. Rev. B **39**, 10770 (1989)
4.53 B.A. Gurney, W. Ho: J. Vac. Sci. Technol. **A3**, 1541 (1985)
4.54 D.J. Hannaman, M.A. Passler: Surf. Sci. **203**, 449 (1988)
4.55 H. Kuhlenbeck, M. Neumann, H.-J. Freund: Surf. Sci. **173**, 194 (1986)
4.56 M.D. Alvey, M.J. Dresser, J.T. Yates Jr.: Surf. Sci. **165**, 447 (1988)
4.57 B. Voigtländer, D. Bruchmann, S. Lehwald, H. Ibach: Surf. Sci. **225**, 151 (1990)
4.58 D.A. Wesner, F.P. Coenen, H.P. Bonzel: Surf. Sci. **199**, L419 (1988)
4.59 O. Knauff, U. Grosche, H.P. Bonzel, V. Fritsche: Mol. Phys. **76**, 787 (1992)
4.60 K-M. Schindler, V. Fritzsche, M.C. Asensio, P. Gardner, D. E. Ricken, A.W. Robinson, A.M. Bradshaw, D.P. Woodruff, J.C. Conesa, A.R. González-Elipe: Phys. Rev. B. **46**, 4836 (1992)
4.61 I.A. Wachs, R.J. Madix: J. Catalysis **53**, 208 (1978)
4.62 D.H.S. Ying, R. Madix: J. Catalysis **61**, 48 (1980)
4.63 B. Sexton: Surf. Sci. **88**, 299 (1979)
4.64 B. Sexton: Suf. Sci. **88**, 319 (1979)
4.65 B.E. Hayden, K.C. Prince, D.P. Woodruff, A.M. Bradshaw: Surface Sci. **133**, 589 (1983)
4.66 M.C. Asensio, M.J. Ashwin, A.L.D. Kilcoyne, D.P. Woodruff, A.W. Robinson, Th. Lindner, J.S. Somers, D.E. Ricken, A.M. Bradshaw: Surf. Sci. **236**, 1 (1990)

4.67 Th. Lindner, J. Somers, A.M. Bradshaw, A.L.D. Kilcoyne, D.P. Woodruff: Surf. Sci. **203**, 333 (1988)
4.68 D. Outka, R.J. Madix, J. Stöhr: Surf. Sci. **164**, 235 (1985)
4.69 M.D. Crapper, C.E. Riley, D.P. Woodruff, A. Puschmann, J. Haase: Surf. Sci. **171**, 1 (1986)
4.70 M.D. Crapper, C.E. Riley, D.P. Woodruff: Surf. Sci. **184**, 121 (1987)
4.71 S. Lehwald, J.T. Yates, Jr., H. Ibach: in *Proc. IVC-8, ICSS-4, ECOSS-3, Carnes*, Suppl. Le Vide Les Couches Minces, No 201, 221 (1980)
4.72 W. Erley: Surf. Sci. **205**, L771 (1988)
4.73 J.G. Chen, W. Erley, H. Ibach: Surf. Sci. **224**, 215 (1989)
4.74 M.C. Asensio, D.P. Woodruff, A.W. Robinson, K.-M. Schindler, P. Gardner, D. Ricken, A.M. Bradshaw, J.C. Conesa, A.R. González-Elipe: J. Vac. Sci. Technol. A **10**, 2445 (1992)
4.75 R. Dippel, K.-U. Weiss, K.-M. Schindler, P. Gardner, V. Fritzsche, A.M. Bradshaw, M.C. Asensio, X.M. Hu, D.P. Woodruff, A. R. Gonzalez-Elipe: Chem. Phys. Lett. **199**, 625 (1992)
4.76 F. Nitschke, G. Ertl, J. Küppers: J. Chem. Phys. **74**, 5911 (1981)
4.77 J.F. Nixon: Adv. Inorg. Radiochem. **29**, 62 (1985)
4.78 P.A. Bagus, K. Hermann, C.W. Bauschlicher: Ber. Bunsenges. **88**, 302 (1984)
4.79 J.B. Pendry, J. Phys. C **13**, 937 (1980)
4.80 L.-Q. Wang, A.E. Schach von Wittenau, Z.G. Li, L.S. Wang, Z. Q. Huang, D.A. Shirley: Phys. Rev. B **44**, 1292 (1991)
4.81 R. Dippel, K.-U. Weiss, K.-M. Schindler, D.P. Woodruff, P. Gardner, V. Fritzsche, A.M. Bradshaw, M.C. Asensio: Surf. Sci. **287/288**, 465 (1993)
4.82 V. Fritzsche, K.-M. Schindler, P. Gardner, A.M. Bradshaw, M. C. Asensio, D.P. Woodruff: Surf. Sci. **269/270**, 35 (1992)
4.83 M. Kerkar, D.P. Woodruff, J. Avila, M.C. Asensio, M. Fernandez-Garcia, J.C. Conesa: J. Phys.: Condens. Matter **4**, 6509 (1992)
4.84 A. Almenningen, B.A. Andersen, E.E. Estrup: Acta Chem. Scand. **24**, 1579 (1970)
4.85 J.E. Demuth, D.E. Eastman: Phys. Rev. Lett. **32**, 1123 (1974)
4.86 B.J. Bandy, M.A. Chesters, M.E. Pemble, G.S. McDougall, N. Sheppard: Surf. Sci. **139**, 87 (1984)
4.87 N. Sheppard: Ann. Rev. Phys. Chem. **39**, 589 (1988)
4.88 S. Bao, K.-M. Schindler, Ph. Hofmann, V. Fritzsche, A. M. Bradshaw, D.P. Woodruff: Surf. Sci. **291**, 295 (1993)
4.89 K.-U. Weiss, R. Dippel, K.-M. Schindler, P. Gardner, V. Fritzsche, A.M. Bradshaw, D. A.L. Kilcoyne, D.P. Woodruff: Phys. Rev. Lett. **69**, 3196 (1992)
4.90 K.-U. Weiss, R. Dippel, M. Schindler, P. Gardner, V. Fritzsche, A.M. Bradshaw, D.P. Woodruff, M.C. Asensio, A.R. Gonzalez-Elipe: Phys. Rev. Lett. **71**, 581 (1993)
4.91 S.R. Thomas, T.D. Thomas: J. Am. Chem. Soc. **100**, 5459 (1978)
4.92 J. Somers, Th. Lindner, A.M. Bradshaw: Unpublished (1988)
4.93 B.A. Sexton: Chem. Phys. Lett. **65**, 469 (1979)
4.94 N.W. Alcock, V. L. Tracy: Acta Crystallogr. Sect. B **35**, 80 (1974)
4.95 E. Dietz, W. Braun, A.M. Bradshaw, R. Johnson: Nucl. Instr. Methods A **239**, 359 (1985)
4.96 H. Wiedemann: *Particle Accelerator Physics* (Springer, Berlin, Heidelberg 1993)

5. Near Edge X-Ray Absorption and Decay Dynamics of Adsorbed Molecules

W. Wurth and *D. Menzel*

In this survey we will discuss the evolution of core-excited states of adsorbed molecules. The chapter will be organised as follows. After giving a short introduction (Sect. 5.1), we will start with a discussion of the primary core-excitation process (Sect. 5.2); followed by a brief review of results on decay dynamics (Sect. 5.3); finally we will touch on the question of core-excitation-induced dissociation and desorption (Sect. 5.4).

Special emphasis will be given to the coupling of core-excited states to internal (electronic and nuclear) and external (polarization, charge transfer) degrees of freedom. Hence, we will subdivide each subject in two parts, namely the discussion of weakly-coupled adsorbates (physisorbates) and the discussion of more strongly coupled species (chemisorbates). We will try to base our argumentation on the most recent experimental results obtained with new high-resolution soft X-ray monochromators at various synchrotron facilities.

5.1 Background

The detailed investigation of the creation and evolution of core excited states in atoms and molecules, both free and bound to a substrate as an adsorbate, gives access to an understanding of the intra- and extramolecular couplings or, in the time-dependent picture, of the dynamics of the coupled excitations. Conversely, information about the hierarchy of these couplings can be derived from such experiments, together with other information on the investigated systems. Therefore, these processes have received considerable attention in the past decade [5.1]. Recently, this research has seen a rapid development which is mainly due to significantly improved experimental conditions in near edge X-ray absorption studies which in turn are leading to an increase of theoretical efforts. A key improvement in the experimental sector is the introduction of new concepts in the design and realisation of soft X-ray monochromators [5.2–5]. Together with the realisation of second- and third-generation synchrotron radiation sources providing low-emittance beam profiles often in connection with high-power insertion devices, this has lead and will lead to high-flux, high-resolution beam lines in the soft X-ray regime. For isolated molecules the study of core-excited states with high resolution, i.e., where the energetic width of the

Springer Series in Surface Sciences, Vol. 35
Applications of Synchrotron Radiation Ed.: W. Eberhardt

excitation is on the order of the lifetime widths of the respective core holes, has been possible in the past using sophisticated inner-shell electron energy loss spectrometers [5.6–8]. On the other hand, for adsorbed and condensed molecules for which EELS (Electron Energy Loss Spectrosopy) is not usable due to the high background of unspecific secondaries, these developments have introduced a new quality to the investigation of the evolution of their core-excited states (Chap. 3). Paralleling this development of the excitation sources, improvements in detection efficiencies and state resolution of the decay products of the core-hole states are also significant. Rotatable electron and ion analyzers allow an independent choice of detection and polarization angles which, in turn, makes an independent analysis of the angular anisotropies of the core excitation and decay processes possible [5.9–11]. Furthermore, the development of liquid-He cooled manipulators providing sample temperatures down to 5 K has introduced the possibility of preparation of low-temperature states of adsorbates [5.12]. This allows us to access weakly adsorbed molecular states which often act as precursors for more strongly bound adsorbate states and to prepare well-defined spacer layers for variable adsorbate-substrate coupling.

The improved experimental conditions have fostered an improved understanding of the basic physical mechanisms. First, we are now able to study the development of a surface chemical bond starting from a physisorbed state of a particular molecule up to chemisorption and sometimes even finally to dissociation of the molecule, i.e., we can see changes in the ground-state properties of the molecule. Second, we can follow trends in extramolecular mechanisms for the excited states of the adsorbates with increasing or decreasing adsorbate-substrate coupling. These extramolecular processes are particularly important for metallic substrates because there the substrate often acts as an efficient excitation sink leading to fast delocalization of the excitation energies and/or charge [5.13–15]. Third, we can study intramolecular processes for high symmetry configurations because the surface very often tends to orient the molecules. Since the example of the isolated molecule always serves as point of comparison for the additional interactions experienced by the adsorbate, the parallel development of detailed investigations for free molecules is very important (Chap. 6) [5.16, 17].

Recently the investigation of the intramolecular dynamics of core-hole decay processes has received particular attention (Chap. 6). The attention has focused on the question of time scales of core-hole decay via autoionization and Auger decay in comparison to vibrational motion of the atoms in the core-excited state of the molecule [5.18–23]. Studies for isolated molecules show that there is significant interference between the electronic decay and nuclear motion and that in contrast to conventional wisdom nuclear motion may sometimes be fast on the timescale of the electronic decay [5.18–23]. The knowledge of these timescales is especially important when tertiary processes such as dissociation and desorption are considered, because the gain of kinetic energy during the lifetime of excited states will be essential for the breaking of intra- and/or extramolecular bonds [5.15].

In the following we will first discuss the primary core-excitation process via absorption of X-ray photons of adsorbates. In the second part we will review results on the decay dynamics of core-excited states. Both parts will be organized in the way that we will start with a discussion of physisorbed, i.e. weakly coupled, weakly perturbed, molecules and then move on to more strongly coupled chemisorbed molecules. In the third part we briefly survey results on core-induced dissociation of surface molecules, observed by the desorption of ions. The emphasis will be on recent results which are meant to exemplify the improved understanding of basic mechanisms due to the significant improvements on the experimental side.

5.2 X-Ray Absorption

It is well known that the fine structure of the X-ray absorption cross-section for molecules close to threshold contains a lot of information about the intramolecular properties [5.24]. Whereas for isolated molecules this has been extensively studied by using inner-shell electron energy loss spectrometry, thus indirectly obtaining the dipole excitation cross sections [5.25], for adsorbed molecules Near Edge X-ray Absorption Fine Structure (NEXAFS) spectra have almost exclusively been obtained using synchrotron radiation sources (with the exception of a few adsorbate ISEELS studies [5.26, 67]). The energetic positions of these resonances have been used to obtain chemical as well as structural and orientational information for the adsorbed molecules [5.24]. Polarization-dependent measurements where the orientation of the electric field vector is changed with respect to the surface often allow the determination of the orientation of molecular symmetry axes with respect to the surface using the known symmetry of certain resonances; or vice versa, if the orientation is fixed the symmetry of resonant channels can be deduced [5.24]. Due to the availability of high-resolution monochromators detailed line-shape studies have recently become feasible, thus, extending the potential of X-ray absorption studies to the detailed understanding of dynamic effects for core-excited states. This is particularly interesting for very weakly coupled molecules, e.g., physisorbed or weakly chemisorbed molecules, for which substrate-induced line-shape broadening effects are less pronounced. We will start with a short survey of this field.

5.2.1 Physisorbed Molecules

a) Resonances Below Threshold

We first discuss the unoccupied states below threshold for physisorbed molecules. These states can, to first order, be subdivided into two groups: the first group consisting mainly of antibonding valence states, the second group consisting of molecular Rydberg states converging to the ionisation threshold. This subdivision is somewhat artificial because the lowest molecular Rydberg levels certainly will locally possess some valence character which can be clearly seen when the desorption of fragments and, hence, the antibonding character is considered [5.28, 29]. For the sake of simplicity, and in order to keep contact with the literature we will, nevertheless, use the term *Rydberg states* for the states close to the ionisation threshold, keeping in mind that this is certainly not justified in the strict sense of hydrogenic Rydberg levels.

b) Energetics

Most molecules show strong bound resonances several eV below threshold which represent transitions into antibonding valence states related to inner-molecular double or triple bonds between adjacent atoms. These resonances are often termed π-resonances reflecting the fact that they result from a π-type molecular bond, although strictly speaking group theoretical considerations validate this nomenclature only for molecules belonging to the symmetry groups $C_{\infty v}$ or $D_{\infty h}$. The energetic position of these resonances is found to be essentially unchanged when comparing van der Waals coupled physisorbed molecules to the isolated molecules. This reflects the fact that the overlap between the molecular valence levels and the substrate valence bands is virtually nonexistent, so that no chemical shift is observed. In addition extramolecular relaxation is very small because the resonantly excited state is a neutral core excited state in which the excited electron is sufficiently spatially confined to the molecule to screen the core hole completely. Figure 5.1 shows a comparison of NEXAFS spectra for molecular oxygen illustrating these findings. The topmost spectrum is obtained for isolated O_2 molecules [5.16], in the middle a spectrum for condensed O_2 on Pt(111) is depicted and the bottom spectrum shows results for a physisorbed monolayer of O_2 on Pt(111). It can be clearly seen that the position of the $1s^{-1}\,1\pi_g^{+1}$ resonance is unaffected by condensation or physisorption [5.30]. This result has so far been confirmed by all other studies on small physisorbed molecules (e.g., CO/Al(111) [5.31], CO/Ag(111) [5.32], N_2/graphite [5.33], N_2/Pt(111) [5.13], O_2/graphite [5.34], N_2O/Pt(111) [5.35], C_2H_4/Cu(111) [5.36]).

The situation is less clear when the position of the Rydberg-type states is considered. It is well known that the ionisation potential of a physisorbed molecule on a metal surface is shifted to lower energies by an image charge screening contribution from the substrate as well as by extramolecular polarization of neighbouring atoms in an adlayer or condensate. In an extended system there is, however, a second important threshold; namely, the inner ionisation threshold corresponding to an excitation to the conduction band minimum. For a monolayer on a metal this could even be the metal Fermi level. Hence, for an extended molecular solid one could expect the Rydberg states to converge towards the new threshold upon condensation. For a molecular solid (either in two or three dimensions) the Rydberg states should develop excitonic

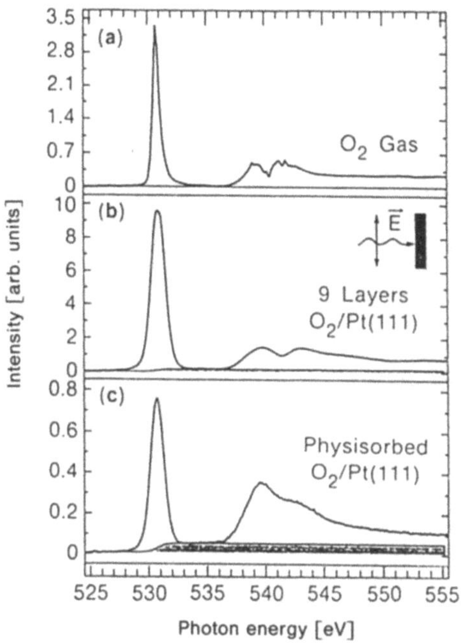

Fig. 5.1. O–K edge NEXAFS spectra for isolated O_2 molecules [5.16], condensed O_2 and physisorbed O_2 on Pt(111). In the adsorbate cases the electric field vector is oriented parallel to the surface [5.24]

character leading to the formation of exciton states which may be of Frenkel or Wannier type. In the molecular solid those excitons would converge towards the conduction band minimum.

Core excitonic states of a molecule, representing to first order (in the equivalent core approximation) an excited impurity, should rather correspond to a fairly immobile, localized excitonic state. However, for an adsorbate, depending on the spatial extension of the Rydberg (or excitonic) states, these excited molecular states should have considerable overlap with the substrate and therefore show an energetic shift and a significant line-shape broadening upon adsorption. The situation might in fact be comparable to the formation of image states in front of a metallic surface, now however for a $(Z + 1)$ impurity. Depending on the position relative to the empty substrate bands the Rydberg-type states would be pure surface states, i.e. spatially confined, or surface resonances, i.e. degenerate with delocalized substrate bands. Figure 5.2 shows a comparison for isolated N_2 [5.37], condensed N_2, and molecular nitrogen physisorbed on Pt(111) and Xe/Pt(111), respectively. For the condensate a shift of 0.5 eV to higher excitation energies and a significant broadening of the lowest-lying Rydberg levels is observed [5.38]. The results derived for the molecular solid are representative for the top-most layer because they were obtained using fragment ion detection [5.38] which is probably only sensitive to the first layer if desorption is not induced by secondary electrons. Comparison to electron yield spectra does, however, not show detectable differences for the peak positions thus implying that the bulk levels are within the detection limits

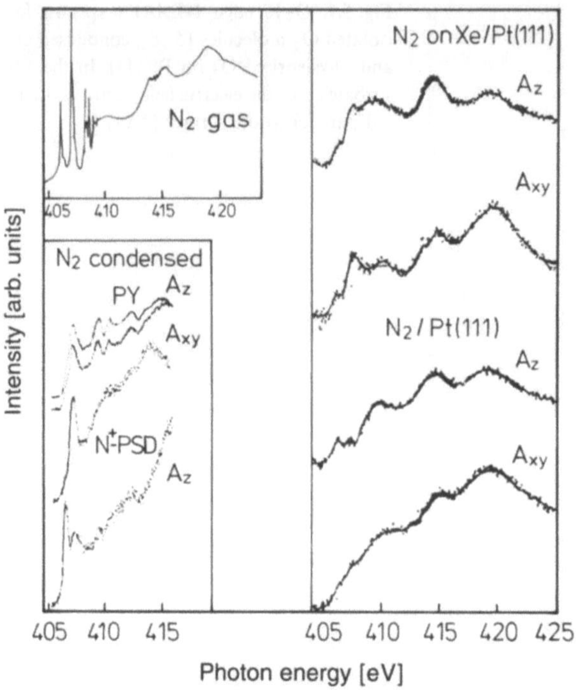

Fig. 5.2. A comparison of the higher-energy part of N–K-edge NEXAFS spectra for isolated N_2 [5.37] and condensed N_2 (left side) with spectra for N_2 molecules adsorbed on Pt(111) and on a monoatomic Xe spacer layer on Pt(111) (right side). The dominant π-resonance at 401 eV is not included. For the adsorbed molecules (condensed or monolayer) two orientations of the electric field vector with respect to the surface are shown (A_{xy}: E vector parallel to the surface; A_z: E vector perpendicular to the surface). Note that for the condensate the two orientations are equivalent for the electron yield spectra (PY) while large differences are observed in the case of N^+-ion detection perpendicular to the surface due to alignment effects [5.38]

identical to the surface levels, in contrast to recent results for rare-gas solids [5.39]. By exploiting alignment effects, using ion detection normal to the surface, it is possible to determine the symmetry of the Rydberg states. The assignment agrees well with the previously postulated assignment for the gas phase results based upon equivalent core considerations (relating the N_2 1s-*nl* energies to valence excited states of NO (2π-*nl*)) [5.37]. It is interesting to compare these results to the case of the physisorbed molecules. It should be noted that the observed transition energies for the adsorbates agree very well with the results for the isolated molecule. However, the symmetry of the states appears to be in contradiction to the assignment derived above. (Similar results have recently been reported for C_2H_4 physisorbed on Cu [5.36], while in the case of weakly chemisorbed N_2 on Fe no differences to the gas phase have been observed although there the levels are less well resolved [5.40]). For all levels a significant broadening (or quenching) is observed. For the $3p\pi$-levels (407.2 eV) this quenching is less pronounced for the molecules adsorbed on a Xe spacer layer. One can therefore conclude that the quenching is related to the overlap with empty density of states above the Fermi level of the metallic substrate resulting in a significant broadening of the respective levels and possibly even in a broken symmetry. The effect appears to be stronger the larger the size of the corresponding Rydberg level and represents a pure final state effect, i.e., the excited electron has a large probability for delocalization, the localized core exciton is energetically degenerate with a polarization-screened core ion plus an electron in the conduction band and the overlap is sufficient to lead to such a charge transfer within the core-hole lifetime. This leads to broadening of the absorption line. Consequences for decay spectra will be discussed below. No shift between the molecules on the bare metal surface and on the monoatomic rare-gas spacer layer is observed indicating that the energetic distance of the levels from the external ionisation limit, i.e., the term value is not fixed. However, due to the significant lifetime broadening it is not possible to determine the convergence limit of the Rydberg states. Hence, no definite answer concerning the degree of perturbation of the states upon physisorption can be given yet. However, the existence of well defined (though broadened) absorption resonances is indicative of weak coupling.

c) Line Shape

Coming back to the bound valence resonances we will, in the following, discuss the line shape of the resonances and try to deduce information on the vibrational motion of physisorbed molecules in the excited state. So far high-resolution spectra of bound valence resonances yielding vibrational resolution have been obtained only for a few adsorbate molecules such as N_2 (which is the prototype molecule for all high-resolution studies) [5.13], CO [5.31], and C_2H_4 [5.36, 41]. A comparison for N_2 on Pt(111), and Xe/Pt(111)[5.32], with gas-phase results [5.37] as shown in Fig. 5.3 indicates that the potential energy surface of the core-excited state is not noticeably disturbed upon physisorption. Within the limits of the experimental resolution no changes of the dynamical linewidth are observed either. This is apparently different in the case of CO adsorbed on Al(111) [5.31] and for C_2H_4 physisorbed on Cu [5.36]. For CO adsorbed on Al(111) a coverage dependent shift of the vibrational energy in the core-excited state has been observed [5.31]. Figure 5.4 shows results for the coverage dependence of the vibrational energy as well as the linewidths of the $\Delta v = 0$ and 1 transitions. The changes have been attributed to the different potential energy surfaces of the $1s$-2π excited states for CO molecules adsorbed parallel to the surface (low coverage) and those oriented straight up or tilted (high coverage). In a recent study for physisorbed C_2H_4 on Cu(100) [5.36] a dependence of the π-resonance line shape on molecular orientation was also observed, which in this case appears to be related to the orientation of the π-system with respect to the surface. These studies indicate that the substrate

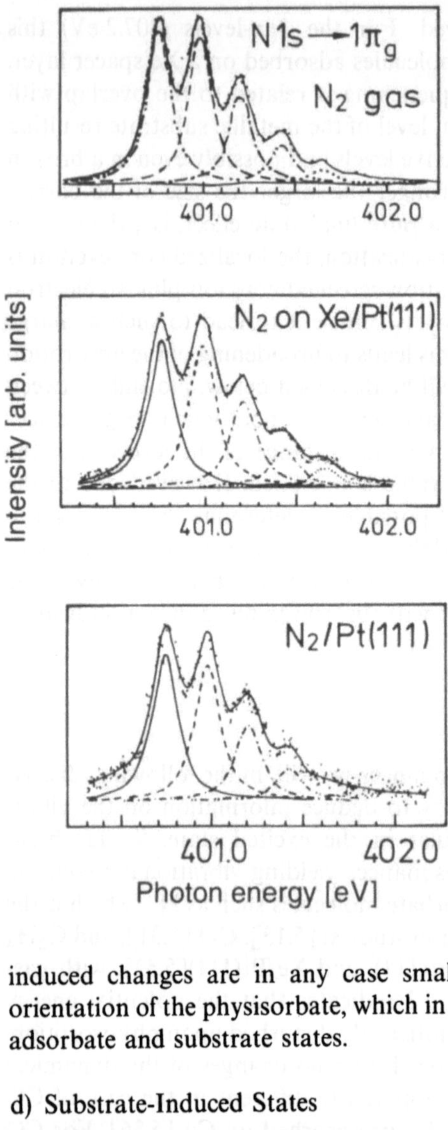

Fig. 5.3. Vibrationally resolved spectra of the $1\pi_g$-resonance of molecular nitrogen for physisorbed N_2 on Pt(111) and on a Xe spacer on Pt(111) are shown in comparison with the gas phase spectrum [5.37]

induced changes are in any case small and may generally depend on the orientation of the physisorbate, which in turn will influence the overlap between adsorbate and substrate states.

d) Substrate-Induced States

A subject of lively discussion is the question of the appearance of substrate-induced absorption below the ionisation threshold for weakly coupled molecules. It has been postulated that the empty conduction-band states of a metallic substrate should manifest themselves in a steplike absorption feature representing mainly the overlap of the spatially less confined *sp*-density of states with the adsorbate core level [5.42]. Although steplike features have been used quite frequently to deconvolute experimental NEXAFS spectra, recent calculations for ammonia on a Ni_{10}-cluster [5.43] indicate that, even in the case of a chemisorbed molecule, substrate induced states are fairly localized in energy and that the overlap depends strongly on the symmetry of the empty substrate states.

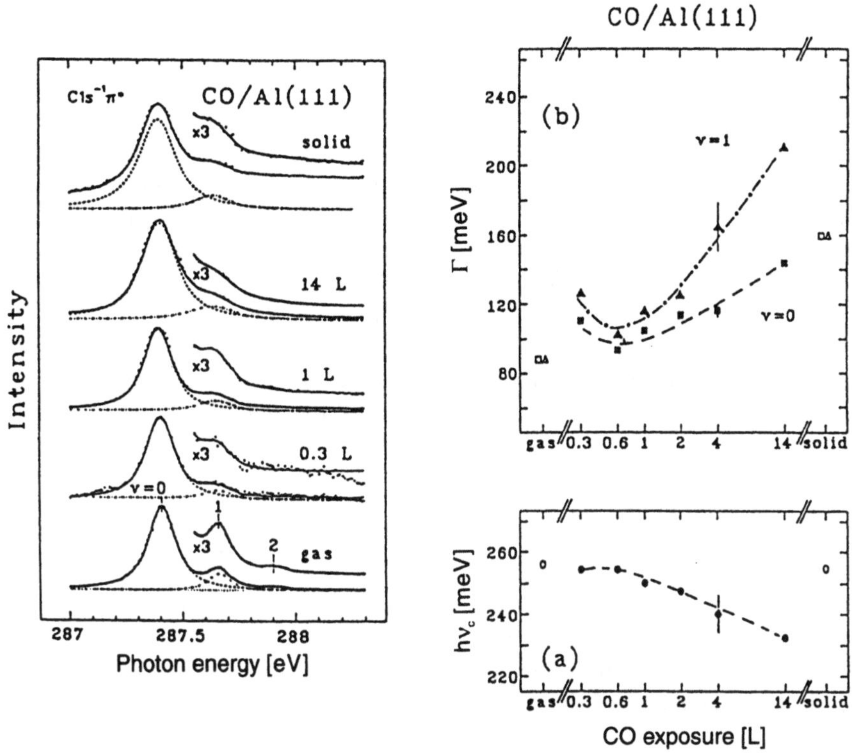

Fig. 5.4. Vibrationally resolved spectra of the 2π-resonance of CO taken at the C–K edge for gas phase CO, condensed CO, and CO adsorbed on Al(111) are shown for comparison on the left side. On the right side the resonance position and the linewidth for the different states are plotted. The spectra are taken from [5.31] (with permission)

Studies for physisorbed molecules on the bare metal substrate in comparison with molecules adsorbed on spacer layers of different thickness could probably help in settling this question.

e) Continuum (Shape) Resonances

Energetics

We will now briefly discuss resonant core-excited states above the outer ionisation threshold. The resonances above threshold can to first order be divided into shape resonances and resonances due to multi-electron excitations. For isolated molecules, an empirical relation between shape resonance position and intramolecular bond length has been established [5.44]; it basically extends the correlation between structure and absorption cross section used in interpreting extended X-ray absorption fine structure, based on single-scattering theory, into the multiple-scattering regime close to threshold. Recent calculations for diatomic molecules seem to confirm this picture [5.45]. Physisorbed molecules provide an interesting test ground for the reference level problem inherent in the application of the empirical relationship. In the case of isolated molecules the ionisation threshold has been exclusively used as reference level for shifts in the shape-resonance position [5.44]. It is well known and has already been discussed above that in the case of an adsorbed molecule the ionisation threshold is shifted to lower energies by extramolecular screening of the core hole. For a single physisorbed molecule this is simply due to image charge screening of the core ion from the metallic substrate. This relaxation shift is typically of the order of 1 eV. Hence, one might expect a similar shift of the shape resonance position assuming no change of the innermolecular bond distance upon physisorption. However, a shift of the absolute shape resonance position is usually not observed indicating that the ionisation threshold is not an appropriate reference level for the scattering problem. In contrast, the energetic difference to the bound valence resonance stays constant and is therefore a more reliable intrinsic reference level. This is not unexpected because the scattering of the outgoing electron is already pretty well described by a potential constructed from overlapping atomic potentials without explicitly taking into account the rearrangement of valence electrons. Furthermore, the scattering state is essentially a self screened state and, hence, does not experience the polarization screening of the core ionic state that develops when the excited electron leaves.

Line Shapes

The question of shape resonance line shapes has been addressed experimentally only recently because it needs a very careful treatment of the background intensity and of the normalization problem. One approach which is borrowed from nuclear and atomic physics is to treat the shape resonance in the *giant-resonance* framework [5.46]. However, this is still hampered by a lack of theoretical justification for the fitting procedure. The results seem to be in conflict with predictions from scattering theory and from ab-initio calculations for isolated molecules [5.47, 48]. The latter theoretical results predict a Feshbach-Fano type resonance profile resulting from the interaction of the virtual, quasi-bound, discrete levels with continuum states. Originally it was proposed that the asymmetry of the shape resonance line shape should reflect non-adiabatic coupling to the vibrational degrees of freedom [5.49]. So far reliable fits of the shape resonance line shape profile are still lacking. High-flux high-resolution data for a number of different, oriented physisorbed molecules are therefore needed to come to a better understanding of the underlying physics.

f) Multi-Electron Excitations

So far, we have only discussed single-electron excitations. However, it is well known that many features in the absorption cross section close to threshold result from multi-electron excitations. Zero-energy electron spectroscopy for isolated molecules has recently made excitation thresholds for satellite transitions observable [5.50]. For adsorbed molecules such experiments are very complicated due to the large background of secondary electrons and the complex geometry of the electrostatic potential in front of the surface. Multi-electron resonances are, however, also observable for adsorbed molecules. These features are especially important when the dynamics of core-excited states is considered, as will be discussed below. For physisorbed molecules multi-electron excitations are clearly visible in the absorption spectra and their energetic positions compare well with results for isolated molecules [5.33, 36, 51], as would be expected in view of the molecular nature of these transitions. Orientation of the molecule on the surface provides the opportunity to determine the symmetry of these transitions. For example in N_2 the double excitation regime close to 415 eV (Fig. 5.2) appears to be dominated by π-type final states indicating that the dominant configuration is most likely a $1s^{-1}\,1\pi_u^{-1}\,1\pi_g^{+1}$ $(\mathrm{nl}\pi)^{+1}$ configuration. Very often multi-electron excitation thresholds observed in X-ray absorption can be directly related to shake-up satellites observed in X-ray photoemission [5.52]. Therefore, symmetry analysis by means of polarization dependent measurements may also help to identify dominant shake-up contributions. This procedure can also be used in photofragmentation studies, as will be discussed below.

5.2.2 Chemisorbed Molecules

a) Resonances Below Threshold

In the following we will move on to the discussion of more strongly coupled adsorbate systems, i.e. chemisorbed molecules. Upon increase of the adsorbate-substrate coupling strength, adsorbate levels tend to broaden significantly. In part this is a consequence of fast electron transfer between adsorbate and substrate in the final excited state, similar to the situation for the extended Rydberg states of physisorbed atoms or molecules. However, in contrast to the situation discussed above there is also a significant broadening of electronic states in the initial ground state due to the bonding interaction between adsorbate and substrate. This interaction will also lead to a shift in the energetic position of the initial and final state of the excitation process. Hence, resonances in the X-ray absorption cross section will generally be expected to be energetically shifted and their line shape to be altered upon chemisorption.

b) Energetics and Line Shape

To illustrate these effects, Fig. 5.5 exhibits a comparison of the bound 2π-resonance for CO chemisorbed on Pt(111) and physisorbed on a monoatomic Xe spacer layer on Pt(111). It is obvious that the bound resonance is strongly perturbed upon chemisorption. The shift in the energetic position of the bound valence resonances can be used as indicative for the chemical state of the adsorbed molecule, although these shifts are sometimes quite small. Recently it has been shown that even molecules in different adsorption geometries lead to measurable shifts in the energetic position of the bound resonance [5.53]. Figure 5.6 shows site dependent changes in the resonance position for CO on Ni(100). The topmost spectrum is characteristic for bridge-bonded CO, the spectrum at the bottom represents the situation for on top adsorption. In the middle a mixed situation (2/3 bridge, 1/3 on top) is depicted.

Changes in the linewidth of bound resonances can also be indicative of the formation of a chemical bond. In Fig. 5.7 two spectra for chemisorbed molecular oxygen on Pt(111) are shown for two different polarizations of the exciting photons with respect to the surface normal. When comparing the width of the $1\pi_g$-resonance for the electric field vector $\bar{E}$ parallel to the surface to one close to perpendicular, a significant broadening becomes obvious. This can be interpreted as a broadening introduced by bond formation via the π_g-orbital perpendicular to the surface while the π_g-system parallel to the surface remains essentially unperturbed [5.30]. Besides line-shape broadening introduced by bond formation (e.g. hopping of electrons between adsorbate and substrate in the ground state), one could also imagine additional vibrational broadening upon chemisorption. This implies coupling of the electronic transitions to low energy vibrations, e.g., substrate phonons or frustrated translations, via Franck-Condon transitions to a final state with substantially different adsorbate-substrate equilibrium bond distance [5.54]. There is, however, so far no clear evidence for the importance of this effect which is known to be small for physisorbed molecules, as has been discussed above. Recent temperature-dependent measurements for NO on Pd(110) [5.55] seem to indicate that

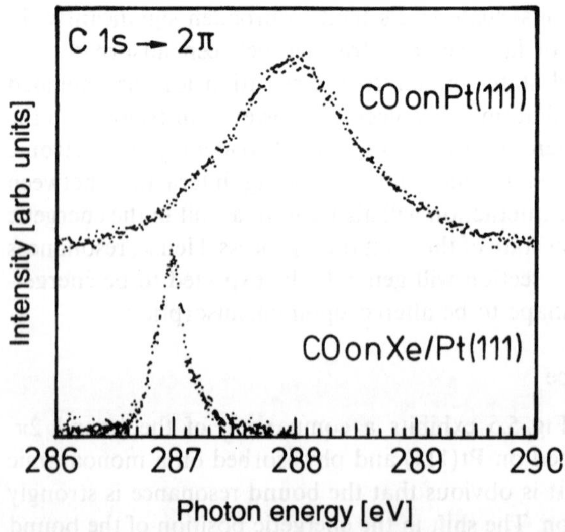

Fig. 5.5. C–K edge absorption spectra of the 2π-resonance for chemisorbed CO on Pt(111) and physisorbed CO on Xe/Pt(111) [5.11]

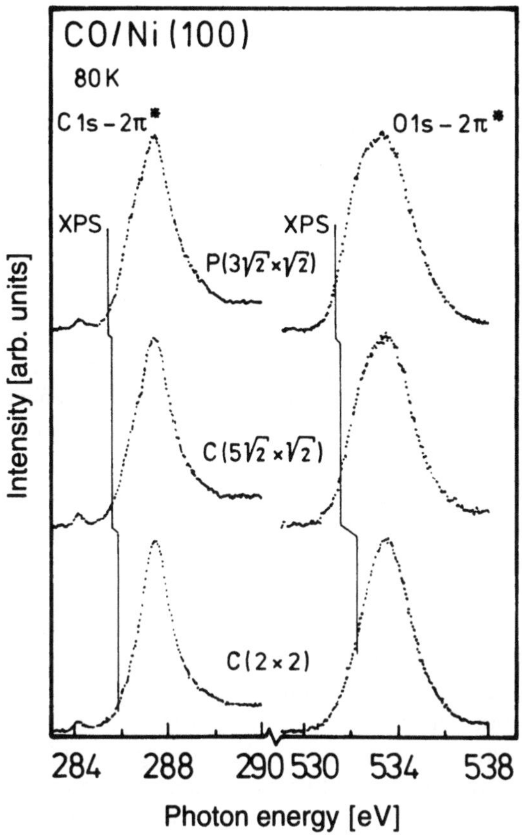

Fig. 5.6. C–*K* edge absorption spectra of the 2π-resonance for chemisorbed CO on Ni(100) for three different adsorption phases taken from [5.53] (with permission)

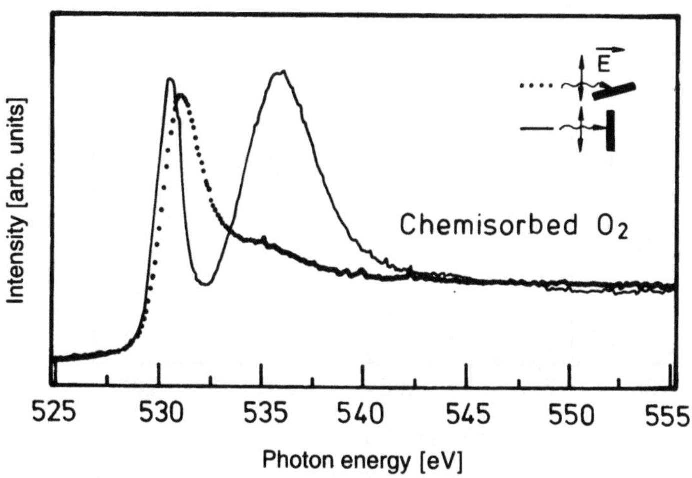

Fig. 5.7. NEXAFS spectra for chemisorbed molecular oxygen on Pt(111) for two different orientations of the electric field vector with respect to the surface

thermal excitation of frustrated translations in the chemisorption ground state might also induce additional broadening in specific cases. In summary, depending on the strength of the adsorbate-substrate interaction bound core valence resonances for chemisorbed molecules are shifted and significantly broadened when compared to the isolated or physisorbed molecule.

c) Rydberg States and Substrate Induced Features

The question of the existence of Rydberg type states for chemisorbed species is closely related to the question of substrate induced density of unoccupied states below the ionisation threshold. The absorption cross section on the high-energy side of the bound valence resonances tends to be non-zero in an energy regime where the isolated molecule does not have any oscillator strength. This effect appears to increase with bond strength [5.24]. Hence, especially for chemisorbed hydrocarbons the so-called *Fermi-level steps* representing the inner ionisation threshold have been invoked to explain the observed intensity below the ionisation threshold (an example of a fit to experimental data for ethylene is shown in Fig. 5.8 [5.41]). Arguments have also been given that this intensity is due to the hybridization of molecular Rydberg states with unoccupied density of states above E_F [5.56]. 'Rydberg-type' resonances which are not quenched upon chemisorption are very often explained as due to a large valence admixture of these final states leading to a stronger spatial confinement of these levels [5.57]. This seems to be especially important for bonds involving hydrogen [5.58]. The hydrogenic levels are often energetically degenerate with Rydberg-type states and therefore couple efficiently to these states. While the understanding of the degree of quenching of Rydberg-type levels is still incomplete, recent theoretical work on NH_3 adsorbed on a Ni_{10} cluster [5.43] at least shows that the picture of

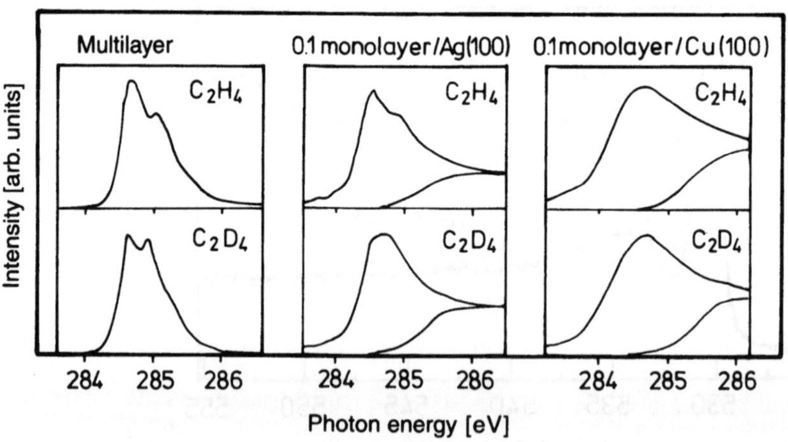

Fig. 5.8. A comparison of X-ray absorption spectra close to the π-resonance for ethylene and deuterated ethylene in the condensed phase and adsorbed on Ag(100) and Cu(100). The spectra are taken from [5.41] (with permission)

substrate induced states being very well represented by *sp*-density of states above the Fermi edge, i.e. a step-like function, is too simplistic and that one has at least to take into account the symmetry of the interacting adsorbate-substrate levels, – as one might have expected.

d) Continuum Resonances

Approaching the ionisation threshold brings us back to the quasi-bound shape resonances and their energetic position with respect to some reference level. There is strong empirical evidence that the large energetic shifts of the shape-resonance position of molecules upon chemisorption is a prime indicator for a change in intramolecular bond length [5.44] and can even be used as a quantitative measure (although there are probably more promising approaches using photoelectron diffraction techniques, see Chap. 7). Figure 5.9 exhibits an example for the shape-resonance position in O_2 for different isolated molecules in comparison with adsorbate systems. The bond length vs. energy relationship seems to be very well described by an almost linear relationship [5.44]. This appears to be a general finding for diatomic molecules and, as mentioned above, has recently been related to an EXAFS like treatment of the shape resonance

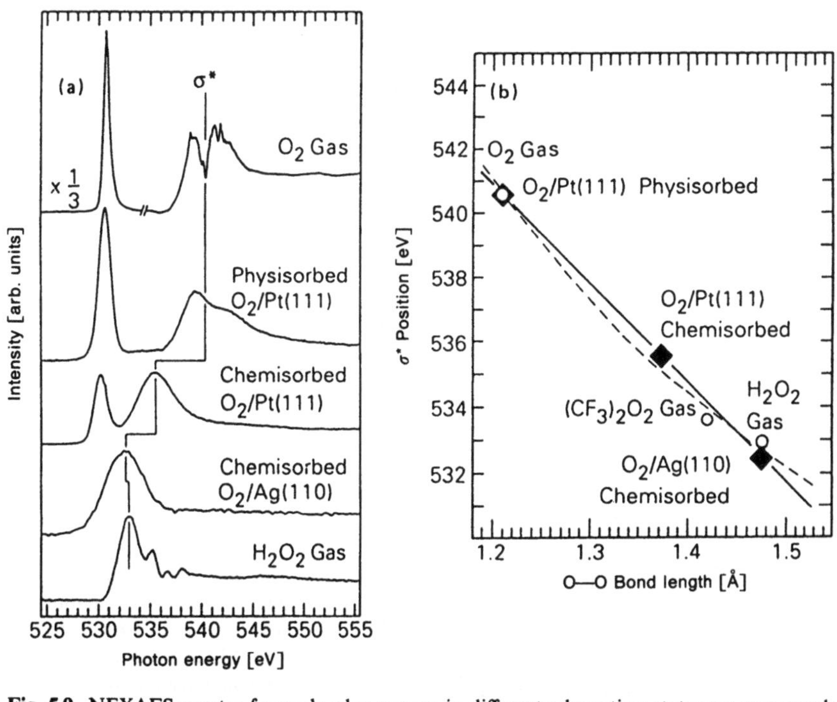

Fig. 5.9. NEXAFS spectra for molecular oxygen in different adsorption states are compared with gas phase spectra of O_2 and H_2O_2 (left side). On the right side the absolute position of the σ- shape resonance is correlated with the O–O bond length. The figure is taken from [5.24] (with permission)

[5.45], provided a proper choice of reference level is used. The ionisation threshold is certainly a poor choice because, as has been pointed out above for the case of physisorbed molecules, sizeable shifts of the ionisation threshold due to extramolecular screening are expected even without a change in intramolecular bond length in the ground state. This is even more important in chemisorption systems where screening via charge transfer has been shown to be so efficient that the adsorbate complex remains essentially neutral upon creation of a core hole [5.59]. The same arguments given above for physisorption concerning why the ionisation threshold is an inappropriate reference also apply here. Possible solutions to this problem are the use of the relative position of the shape resonance with respect to a bound valence resonance or the use of the absolute position (as in Fig. 5.9) if equivalent molecules in different environments are considered.

e) Multi-Electron Transitions

Finally, before considering dynamic effects of the evolution of core-excited states we will briefly discuss multi-electron excitations for chemisorbed molecules. An important aspect of multi-electron transitions is the localisation of holes after the decay of the multiply excited states which, in turn, makes these states

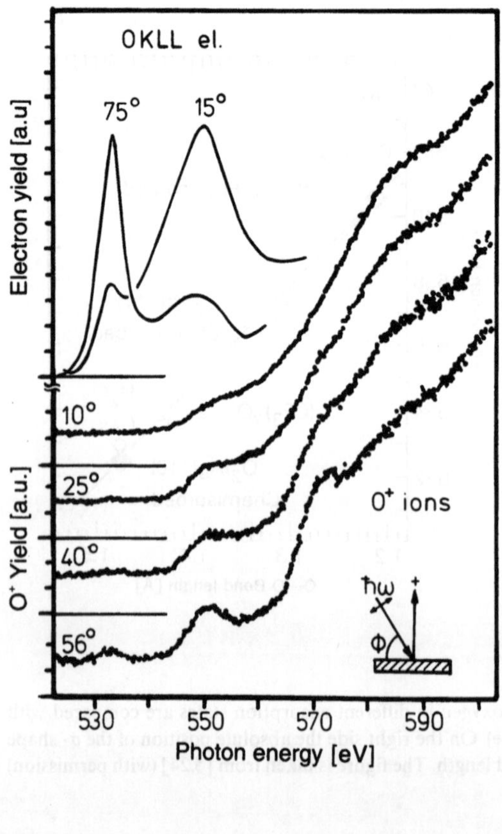

Fig. 5.10. O^+-ion spectra from CO adsorbed on Ru(001) obtained at the O–*K*-edge for different orientations of the electric field vector with respect to the surface. Electron yield spectra are shown for comparison [5.61]

particularly efficient in bond-breaking processes after core excitation [5.60–62]. We will come back to the question of hole localisation later. As a consequence multi-electron excitation resonances for chemisorbed molecules have first been detected in fragment ion desorption where again symmetry determination by polarization dependence can be used [5.60]. Figure 5.10 displays polarization-dependent O^+ desorption spectra for chemisorbed CO on Ru(0 0 1). The spectra clearly show several resonances which when compared to the partial electron yield are not correlated with dominant single electron transitions but agree very well with shake-up satellite states derived from X-ray photoemission [5.52].

5.3 Decay Dynamics

In Sect. 5.3 we discuss the evolution of resonantly excited core-hole states. As in the first part (Sect. 5.2) we will focus on small molecules consisting of low Z atoms. Hence, electronic decay is expected to proceed primarily via auto-ionization or Auger decay [5.63]. For isolated molecules these two processes can be easily separated according to their respective initial and final states, namely auto-ionization starting from a neutral initial state leading to (overall) one-hole final states vs. Auger decay starting from a one-hole initial state and leading to a two-hole final state. This is, of course, in principle also true for adsorbate molecules. However, delocalization of holes either in the initial or the final state of the decay process due to strong coupling to the substrate may lead to a virtual degeneracy of the respective processes as will be discussed below.

For isolated molecules the dynamics of the neutral (bound) core-excited states is governed by the lifetime of the core hole and the nuclear motion in the core-excited state. Recent investigations have shown that the interplay of these two timescales (decay time vs. vibrational period) is very important for the detailed understanding of the line shape of the auto-ionization lines (Chap. 6) [5.18, 23]. For resonant core-excited states above the ionisation threshold coupling of the outgoing primary electron with the secondary decay electron via post-collision interaction can affect the Auger line shape close to threshold (Chap. 6) [5.64, 65].

For adsorbed molecules the situation is even more complicated because, in addition to the intramolecular processes mentioned before, extramolecular relaxation and delocalization of holes, are competing processes which have to be taken into account [5.66–68]. Depending on the strength of the adsorbate-substrate interaction in the intermediate core-excited states extramolecular relaxation may have very different timescales relative to the core-hole decay time. Hence, the influence of adsorption on the decay dynamics of core-excited states and consequently the line shape and energetic position of decay electron spectra depends strongly on the coupling of adsorbate and substrate in the intermediate core excited state. Competition between nuclear motion and dissipation of energy to the substrate becomes especially important when tertiary bond-breaking processes such as desorption or dissociation are considered (Sect. 5.4). The yield of desorption and/or dissociation products will depend strongly on the relative timescales and hence will also be strongly influenced by the adsorbate-substrate coupling [5.14, 15].

5.3.1 Physisorbed Molecules

a) Relaxation

We will again split the discussion according to the strength of the adsorbate-substrate coupling in the ground state. First, we will review recent results for weakly coupled, physisorbed molecules and later discuss specific differences introduced by strong coupling in the case of chemisorbed molecules.

The driving force for extramolecular relaxation in the intermediate core excited state is always the existence of core-hole states lower in energy than the states occupied via the primary excitation process which are accessible by charge rearrangement either involving the substrate or (in the case of dense adsorbate layers and condensates) neighbouring molecules. The charge rearrangement involves delocalization of the excited primary electron as well as extramolecular screening of the core-hole state. For physisorbed molecules with empty bound valence levels the neutral core-excited state reached after resonant excitation to the bound valence resonance in many cases represents the energetically lowest core-hole state (Chap. 3) [5.13]. Hence, extramolecular relaxation effects are not important after resonant excitation to this state. They might, however, become important if the primary excitation energy is tuned to the Rydberg-type resonances below threshold or to states above the ionisation threshold. For Rydberg-type resonances of adsorbed rare-gas atoms delocalization of the excited electron has been shown to be important [5.13, 69]. The degree of delocalization depends on the spatial extension of the respective Rydberg states. In this case the energetically lowest core-hole state is reached by bringing the excited electron to the substrate Fermi level which also corresponds to the final state of a primary excitation high above threshold [5.13]. In the decay spectra this may even lead to different decay channels depending on the timescale of delocalization. In Fig. 5.11 Ar LMM decay spectra for an Ar monolayer after resonant excitation of an Ar $2p_{3/2}$-electron to the 4*s*-(bottom) and 3*d*, 5*s* (top)-Rydberg levels are depicted to illustrate this point. While the top spectrum represents a situation where the excited electron is fully delocalized before core-hole decay, the bottom spectrum clearly consists of two parts representing a situation where the decay starts from two different intermediate core-hole states: with and without the excited electron localized at the core-hole site. However, as mentioned above, for physisorbed molecules with large electron affinity in the core ionized state the delocalized state is not the energetically lowest core-hole state, because the bound valence resonance is

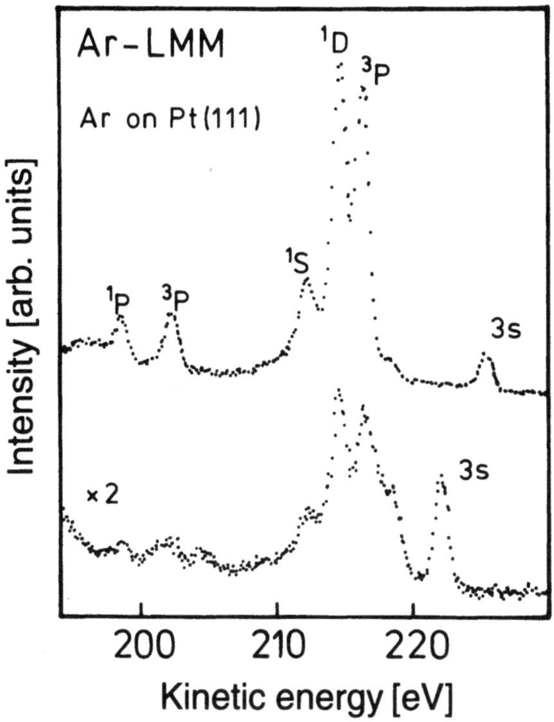

Fig. 5.11. Ar LMM decay spectra for Ar adsorbed on Pt(111) taken after resonant excitation of the Ar $2p_{3/2}$ level to the Ar $4s$ (bottom) and the $5s$, $3d$ levels (top) respectively [5.14]

even lower in energy. On the other hand, for molecules with saturated bonds a situation similar to the rare-gas atom case might easily be envisioned.

Recent investigations for molecular nitrogen adsorbed on graphite [5.70], which certainly represents a case where the bound valence resonance is the energetically lowest core-hole state, and for SF_6 adsorbed on NaCl films on Ge(100) [5.71], indicate that charge transfer screening to this particular state is important on the timescale of the decay process even if the primary excitation in the sudden limit, i.e. high above the ionisation threshold, yields negligible contributions to a screened final state. The experimental results show strong contributions of decay channels which are similar to those obtained after resonant excitation to the bound valence resonance for the respective molecules even for excitation high above the ionisation threshold. This is a strong indication for charge transfer relaxation to the energetically lowest core-hole state even in the case of weakly coupled molecules. The probability for the charge-transfer relaxed state being populated before the core-hole decay depends on the hopping time of a substrate electron relative to the core-hole lifetime. This has been nicely demonstrated in the case of SF_6 adsorbed of NaCl films on Ge(100) [5.71]. With increasing thickness of the insulating film the relative contribution of screened vs. unscreened initial states of the core-hole decay is reduced, indicating that charge transfer relaxation becomes less probable.

b) Auto-ionization and Auger Decay

After primary excitation of a bound valence resonance, physisorbed molecules show auto-ionization spectra which correspond closely to the respective spectra obtained for the isolated molecules [5.13, 72–74]. Figure 12 exhibits an example for O_2 physisorbed on Pt(111) compared to gas phase results obtained after resonant and off-resonant excitation. It is obvious that the overall line shapes after excitation to the bound π-resonance are similar (we will not discuss in detail the differences observed for the off-resonant spectra; they are not simply due to adsorption on a metallic substrate because similar spectra are obtained for condensed molecules [5.75]). A simplistic first-order approach to the interpretation of the observed decay spectra after core-to-bound excitation, which nevertheless has been quite successful in some cases, has been to divide the observed spectrum into *participant* channels, i.e., channels where the excited electron is involved in the decay process, (1 h final states) and *spectator* channels, i.e., channels where the excited electron acts as a passive screening electron, (2 h-1e final states) [5.76]. The spectator part should then be described quite well by

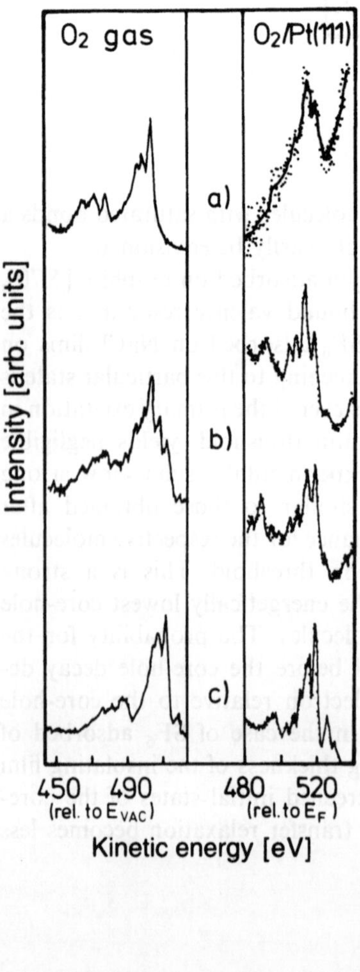

Fig. 5.12. O-KLL decay spectra after resonant (O 1s-$1\pi_g$ (c): O 1s-σ(b)) and off resonant excitation (a) for isolated O_2 (left) [5.93] compared to spectra for physisorbed molecular oxygen on Pt(111). For the physisorbed molecules two spectra are shown in case b corresponding to the two σ-type resonances (Fig. 5.5 and 9)

a shifted Auger spectrum of the respective molecule, while the participant channels should correspond to single-electron photoemission transitions. While the latter interpretation usually represents the experimental situation very well, the interpretation of the spectator contributions as a *uniformly* shifted Auger spectrum is evidently a very crude approximation which already fails for such a simple molecule like N_2 [5.76]. Configuration interaction of different two-hole one-particle (2h-1e) states makes the simple description of the excited electron in the antibonding valence level as spectator electron no longer valid and leads to strong deviations of the spectrum from a shifted normal (two-hole) Auger spectrum [5.77]. Even for the participant states a comparison with 1 h photoemission final states has only limited validity because the differences in the dynamics of nuclear motion have to be taken into account. Recent investigations for isolated molecules have shown that nuclear motion in the intermediate resonantly excited state before core-hole decay is very important for understanding the line shape of the participant channels (Chap. 6) [5.18–23]. Figure 5.13 displays a simple example of these effects for adsorbed N_2 molecules on a

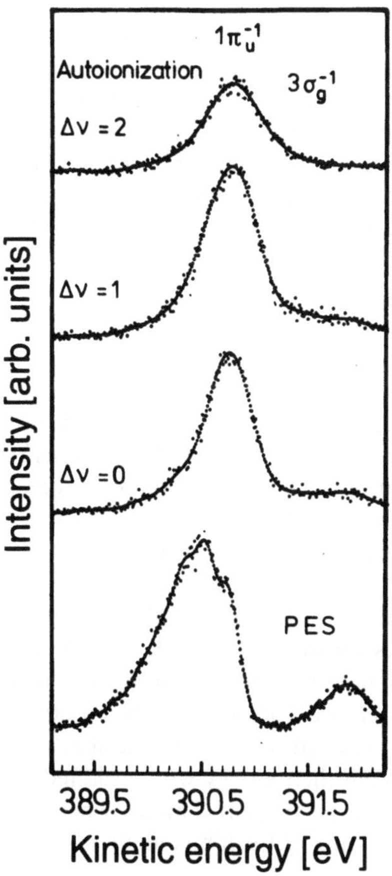

Fig. 5.13. Autoionization spectra for molecular nitrogen adsorbed on Xe/Pt(111) after vibrationally selective resonant excitation (N $1s$-$1\pi_g$; Fig. 5.3). For comparison a photoemission spectrum taken at 100 eV photon energy (shifted to align with the $\Delta v = 0$ transition energy) is shown at the bottom

Xe spacer layer on Pt(111). The figure shows the $1\pi_u^{-1}$ and $3\sigma_g^{-1}$ final states reached after vibrationally selective primary excitation (N $1s$–$1\pi_g$) in comparison to a photoemission spectrum obtained at a photon energy of 100 eV and shifted to align with the $\Delta v = 0$ excitation energy. It is evident that the dynamics in the intermediate state is important for the auto-ionization line shape and the nuclear motion in the resonantly excited state happens on a timescale comparable with core-hole decay. Similar results have been reported for N_2 on graphite [5.74]. This result is, in fact, already implicit in the resonance line shape of the X-ray absorption process which shows clearly resolved vibrational fine structure. A detailed description of the auto-ionization line shape (recent polarization dependent results for N_2 on Xe/Pt(111) indicate that there is even an orientational dependence [5.78]) requires comparison with theoretical results which explicitly take non-adiabatic effect into account [5.79, 80].

Finally, we briefly discuss the influence of the substrate on the two-hole final states resulting from core-hole decay after primary excitation into Rydberg and/or continuum states. There is a finite probability for delocalization of the excited electron before the core-hole decay takes place, as has been discussed before. Hence, the final states of the decay process will be predominantly normal Auger (two-hole) final states. However, compared to the isolated molecules, these states are, of course, shifted by extramolecular relaxation. For physisorbed molecules the dominant screening contribution will again be polarization screening which on a metal surface is dominated by image charge screening. It is important to note that this relaxation shift will lead to a decrease of the spectator shift because the extramolecular relaxation shift of the two-hole (Auger) final states will be roughly a factor of three larger than the corresponding shift of the two-hole one particle (spectator) final states [5.63]. This is, indeed, what is observed [5.13, 33, 75].

5.3.2 Chemisorbed Molecules

The situation is very different if we consider the case of strong coupling, i.e. chemisorption. In this case relaxation before core-hole decay is a very fast process which will consequently lead to the population of a charge-transfer screened core-hole ground state, independent of the primary excitation process [5.33, 66–68, 81, 82]. The importance of charge-transfer screening in the case of strong chemisorption has been demonstrated in X-ray photoemission [5.59]. There the dominant peak in the case of a strongly coupled adsorbate molecule is the adiabatic, fully relaxed core-hole state. This state is essentially neutral because the screening electron which resides on the molecule leaves a highly delocalized hole in the metal substrate which does not influence intramolecular processes. Hence, one would expect to observe a core-hole decay spectrum which is governed by two-hole one particle final states similar to the spectator states observed for isolated molecules (although their energy will be shifted by screening). The spectrum should even show some participant channels where the screening electron takes part in the decay. Figure 5.14 exhibits angle-resolved

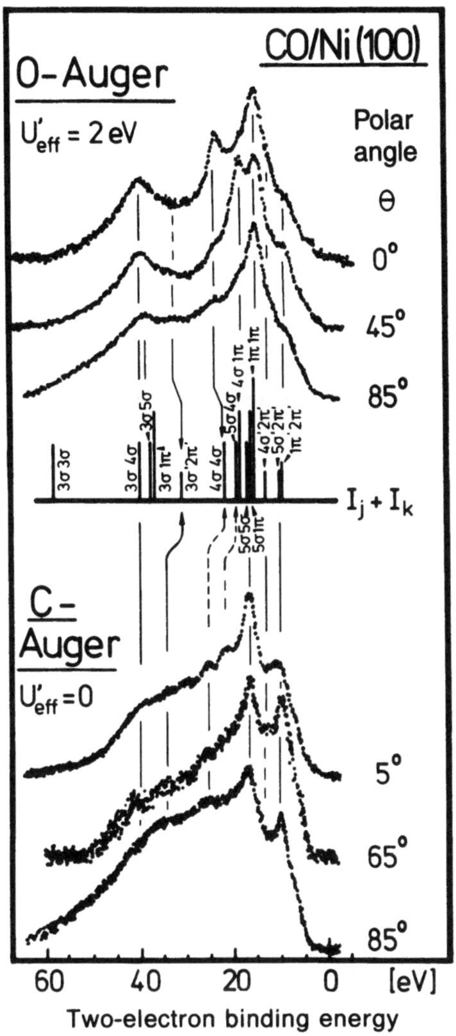

Fig. 5.14. Angle resolved C-KLL and O-KLL decay spectra for CO/Ni(100). The spectra are taken from [5.83] (with permission)

C-KLL and O-KLL spectra for CO adsorbed on Ni(100) where symmetry arguments have been used to assign the different final states [5.83, 84]. Indeed, both spectra show participant channels on the high-kinetic-energy side of the spectrum even though the primary excitation energy was chosen high above the ionisation threshold, i.e. in the sudden limit. In contrast to the results for free and/or physisorbed molecules, these participant states correspond to satellite states of the one-hole valence rather than to the pure one-hole states [5.68, 81]. The observation of these states corroborates the conclusion that charge-transfer screening is fast on the time scale of the core-hole decay process. Also, the '2h' final states will be largely screened, as seen from the decrease of U_{eff} (for CO, 18 eV in the isolated molecule [5.85], vs. 0–2 eV in the adsorbate [5.84, 86]).

We will now discuss the dependence of the core-hole decay spectra on the primary excitation process in more detail. Figure 5.15 depicts C-KLL and O-KLL decay spectra for CO adsorbed on Ni(111) obtained for different primary excitation energies. The C-KLL spectra show almost no influence of the intermediate state reached in the X-ray absorption process [5.67]. If we take into account that the π-resonantly excited state (selected by the dipole transition) is energetically higher than the adiabatic final state reached after sudden excitation (the charge transfer screened core ion), this implies that even in the case of the neutral primary excitation a fast charge rearrangement involving substrate electrons takes place [5.66–68]. It should be noted that we observe small changes in the O-KLL spectra in the energy regime of the Rydberg states [5.68, 69]. This is so far not fully understood because we would expect that in this regime the primary electron should be able to delocalize very fast. To conclude, we would like to emphasize that fast charge rearrangement under participation of substrate electrons before core-hole decay is very important for strongly coupled adsorbates.

Another interesting point is that there is no difference between the decay spectra for continuum excitation slightly above threshold, in the satellite threshold region, or far above these thresholds. Also, for CO on Cu surfaces for which final-state splitting into well-screened and poorly screened core-ionized

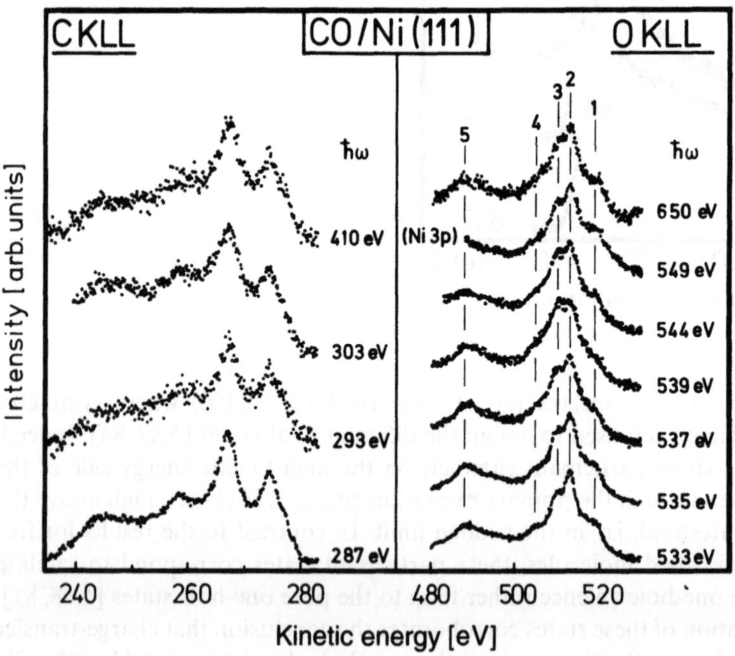

Fig. 5.15. C-KLL and O-KLL decay spectra for CO adsorbed on Ni(111) obtained after excitation with different photon energies [5.67]

states are found for continuum excitation ('giant satellites', 5.87]), no difference of the decay spectra is found for excitation below or above the threshold of these 'shake-back' resonances [5.66, 68]. All this must mean that for the overwhelming majority of the excitations all primary excitations beyond the screened ground state must have decayed by the time of core decay, even for the intermediate coupling case such as CO/Cu. We shall come back to this finding in Sect. 5.4.

5.4 Desorption and Dissociation

The core excitation induced processes lead to very efficient bond breaking. In small and particularly in diatomic molecules core ionization leads essentially to complete dissociation, because in such systems doubly ionized states are at most metastable. These processes are another case of coupling between electronic and nuclear motions which already concerned us in connection with vibrational effects. Therefore, the investigation of the corresponding processes and the influences of adsorbate-substrate coupling on them is an interesting example of core-induced dynamics. For a surface system such processes become evident as particles (fragments or entire molecules) leave the surface as ions or neutrals, i.e. by desorption. The corresponding field is therefore called Photon-Stimulated Desorption (PSD) or Desorption Induced by Electronic Transitions (DIET) [5.88] and constitutes a wide area of research [5.15, 89]. Only a small sampling of the features important in the present context can therefore be given. Again, our main focus concerns the changes induced in such processes in molecules which are brought about by adsorption. Two aspects will interest us here: the use of the desorption signals as specific monitor for certain channels, and the evidence for so called ultrafast dissociation, i.e., dissociation on a time scale which is faster than core-hole decay (here some femtoseconds).

An electronic excitation will lead to breaking of a bond if the final state of this excitation corresponds to a repulsive state of this particular interatomic coordinate. An isolated molecule will fall apart upon such an excitation unless the molecule can cross into another electronic state which is attractive for the bond concerned before dissociation can occur. If we put the molecule onto a surface, a wide range of additional such crossings will become available, because the substrate can act as a 'sponge' for excitation energy and/or charge, so that the excitation can be transferred into the substrate before it becomes effective for bond breaking. The couplings which are involved in these quenching processes are the same which operate in the electronic excitation and deexcitation processes, as discussed above. The main effect of this is that the faster the time scale of dissociation (of either an internal bond of an adsorbate, or of the adsorbate-substrate bond), the more likely the corresponding dissociation process will also operate in the adsorbed state, or the closer the coupling, the fewer dissociation channels will survive adsorption. Thus, adsorption serves as a selective means which efficiently quenches slow dissociation processes and relatively enhances fast ones. For the identical adsorbate-substrate situation, strongly localized final states will be relatively enhanced. This can be nicely seen in the dissociation of chemisorbed CO [5.90, 60, 61] or NO [5.62]. Core excitation is much more efficient for dissociation than valence excitation, with the O 1*s* hole, i.e. that on the atom furthest removed from the substrate, being much more efficient than the closely coupled C or N 1*s*, respectively. Furthermore, the highest cross sections are found for core shake-up and shake-off excitations whose decay leads to multiple-hole and multiple-hole multiple-particle states, respectively [5.60, 61]. As mentioned above (Fig. 5.10), this has been used to define satellite thresholds and symmetries of the corresponding transitions.

Some points here may appear contradictory to conclusions drawn above. This discussion has assumed that the dissociative electronic states are those reached after core-hole decay, and that these are strongly correlated and thus, strongly localized multiple-hole multiple-electron states. In the previous section on the decay spectra of chemisorbed molecules, however, we had concluded that essentially all excitations beyond the screened core ionic ground state have decayed by the time of core-hole decay. This contradiction is relieved if the latter is correct for the majority of cases which are seen in the decay spectra. A minority must retain the additional multiple excitations up to core-hole decay which then selectively shows up in fragment ion desorption. Another possibility is that at least part of the dissociation already happens before core-hole decay, i.e. as an ultrafast process. This would require a particularly repulsive C–O potential curve for the core satellite excitations (e.g., the O $1s^{-1}\,1\pi^{-1}2\pi^{+2}$ final state) which could easily be envisaged. This leads to the interesting general question of the existence of ultrafast dissociation.

As defined above, ultrafast dissociation processes are faster than core-hole decay. The core holes of interest here have lifetimes of a few femtoseconds. Therefore, the atom or ion concerned must move sufficiently far within this time frame for the bond to be broken. Obviously this is only possible if the primary core-excited state is repulsive and therefore cannot be vibrationally resolved even in the isolated molecule [5.17]; excitations exhibiting vibrational resolution (e.g., the lowest bound and continuum excitations of both CO and N_2) must, of course, possess a bonding potential curve, and the vibrational resolution of the N^+ fragment yield of both physisorbed and condensed N_2 [5.38] proves that in this case dissociation happens in repulsive core *decay* products. Dissociation on the scale of femtoseconds is more likely for more repulsive core-*excited* states and lighter fragments. An indication for such a process is a selective enhancement of dissociation for certain primary core excitations. Another indication is its persistence in the condensed and adsorbed state, since this shows it to persist against the delocalization/quenching processes thus introduced. Not surprisingly, all cases of clear ultrafast dissociation involve hydrogenic molecules such as H_2O/D_2O[5.28], H_2S[5.29], NH_3[5.91], and benzene [5.92]. For the last case, Fig. 5.16 shows an example of selective enhancement of fragment D^+ production by certain core-to-bound excitations which can be interpreted as excitations antibonding for the C–D bond. Such observations have been used to argue for valence contributions in excitations which had been disputed as valence-Rydberg mixtures [5.28, 51] since pure Rydberg excitations should not be repulsive to a specific bond. For further discussions see [5.14].

As a final point we mention the use of the angular direction of fragment ion emission as a means of selection of adsorbate orientation. In the simplest approximation, this direction is identical to the ground-state bond direction (see the discussion of this assumption, which is certainly correct for a diatomic molecule, in [5.28]). Even for randomly oriented samples, such as many condensed gases, an oriented subset can be selected by setting the direction of fragment ion detection. If the direction of primary light polarization can be set independently, the symmetry of the excitation can be uniquely determined even for random samples. An example has been given above in Fig. 5.2.

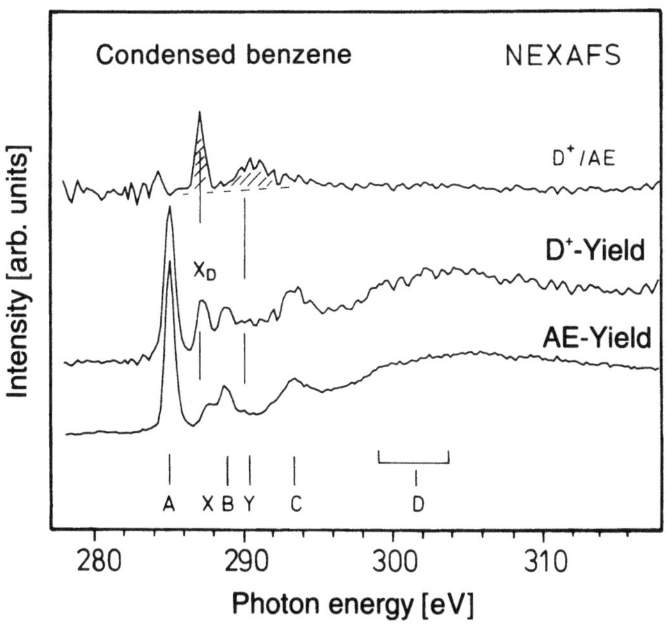

Fig. 5.16. Absorption as measured by the Auger yield (bottom), yield of desorbed D^+ (center), and their ratio (top), for multilayers of C_6D_6. The labels at the bottom give the various excitations, of which A, B, C, D are connected with the C–C bond, while *X* is a C–D excitation and *Y* a higher C–D excitation. Only *X* and *Y* show enhancement in D^+-yield [5.92]

5.5 Conclusions

This survey has attempted to show that the properties of adsorbed atoms and molecules and, in particular, the effects caused by the coupling between adsorbate and substrate can be investigated with increasing detail by well-defined and well-resolved measurements (in terms of energy and polarization of the primary light, and of energy and emission angle of secondary electrons and ions) of their core excitations, their electronic deexcitation spectra, and correlated fragmentation. The influence of graded adsorbate-substrate coupling, which can be tuned by looking at the same molecule at different substrates, e.g., CO on different metal surfaces, similar but differently coupled molecules on the same surface, e.g., CO and N_2, or the differently coupled states of O_2 on Pt and Pd, molecules on inert spacer layers and adsorbed versus condensed molecules is examined. The time scales of core-hole deexcitation, which are roughly known and for the cases investigated here are mostly between a few and 10 fs, can be used as an internal time mark against which other time scales are compared. Thus, the almost total relaxation of all excitations beyond the core ionic ground state, which are found for chemisorbed molecules on metal surfaces – both for dipole transitions to low-lying bound resonances and for core satellite excitations – from the line widths of primary excitations and from decay spectra, must mean that the time scale of energy and charge transfer in these systems is around or below 1 fs. For physisorbed atoms and molecules, on the other hand, the transfer time scale is of the same order of magnitude or even larger than core hole lifetimes, as seen again in both primary line widths and decay spectra. The time scale of nuclear motion which usually is assumed to be slower, can easily be in the fs range, as evidenced by the vibrational redistribution between primary excitation and its deexcitation (which can, of course, be resolved only in very weakly coupled systems), and by the existence of ultrafast dissociation. Thus, the relative variation of the time scales of core-hole decay, screening and charge transfer, and nuclear motion has been shown to lead to a rich variety of phenomena in this class of processes and systems.

Acknowledgements. We thank all our collegues and coworkers who have been involved in our work which has been mentioned in this survey, in particular D. Coulman, P. Feulner, G. Rocker, J. Stöhr, and E. Umbach. We are also very grateful for the support of W. Braun and H. Petersen and the staff of BESSY, Berlin, where all our mentioned results have been obtained. The financial support by the German Ministry of Research and Technology (BMFT) and of the Deutsche Forschungsgemeinschaft (DFG) is gratefully acknowledged.

References

5.1 See, e.g., Proc. 2nd Int'l. Workshop on Auger Spectroscopy and Electronic Structure. Physica Scripta T41 (1992)

5.2 C.T. Chen and F. Sette: Rev. Sci. Instrum. **60**, 1616 (1989)
5.3 M. Domke, T. Mandel, A. Puschmann, C. Xue, D.A. Shirley, G. Kaindl, H. Petersen, P. Kuske: Rev. Sci. Instrum. **63**, 80 (1992)
5.4 K. Randall, J. Feldhaus, W. Erlebach, A.M. Bradshaw, W. Eberhardt, Z. Xu, Y. Ma, P.D. Johnson: Rev. Sci. Instrum. **63**, 1367 (1992)
5.5 A. Yagashita, S. Masui, T. Toyoshima, H. Maezawa, E. Shigemasa: Rev. Sci. Instrum. **63**, 1351 (1992)
5.6 G.C. King, F.H. Read, M. Tronc: Chem. Phys. Lett. **52**, 50 (1977)
5.7 A.P. Hitchcock, C.E. Brion: J. Electron Spectrosc. **15**, 201 (1979)
5.8 D.A. Shaw, G.C. King, D. Cvejanovic, F.H. Read: J. Phys. **B 17**, 2091 (1984)
5.9 R. Scheuerer, K. Eberle, P. Feulner: to be published
5.10 D.M. Hanson, C.I. Ma, K. Lee, D. Lapiano-Smith, D.Y. Kim: J. Chem. Phys. **93**, 9200 (1990)
5.11 A. Yagashita, E. Shigemasa: Rev. Sci. Instrum. **63**, 1383 (1992)
5.12 H. Schlichting: Methoden und Mechanismen der thermischen Desorption: Adsorptions-, Desorptions-Kinetik, Epitaxie und Ordnung von Edelgasschichten auf Ru(001), Dissertation, TU Munich (1990)
5.13 W. Wurth, P. Feulner, D. Menzel: Physica Scripta **T41**, 213 (1992)
5.14 D. Menzel: In *Synchrotron Radiation and Dynamic Phenomena*, ed. by A. Beswick, AIP Conf. Proc. **258**, 387 (1992), D. Menzel: In *Desorption Induced by Electronic Transitions DIET V*, ed. by A.R. Burns, E.B. Stechel, D.R. Jennison, Springer Ser. Surf. Sci., Vol. 31 (Springer, Berlin, Heidelberg 1993)
5.15 P. Feulner, D. Menzel: In *Laser Spectroscopy and Photochemistry on Metal Surfaces*, ed. by H.-L. Dai, W. Ho (World Scientific, Singapore 1993)
5.16 Y. Ma, C.T. Chen, G. Meigs, K. Randall, F. Sette: Phys. Rev. A **44**, 1848 (1991)
5.17 J. Schirmer, A.B. Trofimov, K.J. Randall, J. Feldhaus, A.M. Bradshaw, Y. Ma, C.T. Chen, F. Sette: Phys. Rev A **47**, 1136 (1993)
5.18 T.X. Caroll, T.D. Thomas: J. Chem. Phys. **86**, 5221 (1987)
5.19 T.X. Caroll, S.E. Anderson, L. Ungier, T.D. Thomas: Phys. Rev. Lett **58**, 867 (1987)
5.20 T.X. Caroll, T.D. Thomas: J. Chem. Phys. **89**, 5983 (1988)
5.21 R. Murphy, In-Whan Lyo, W. Eberhardt: J. Chem. Phys. **88**, 6078 (1988)
5.22 W. Eberhardt, J.-E Rubensson, K.J. Randall, J. Feldhaus, A.L.D. Kilcoyne, A.M. Bradshaw, Z. Xu, P.D. Johnson, Y. Ma: Physica Scripta **T41**, 143 (1992)
5.23 M. Neeb, J.-E Rubensson, M. Biermann, W. Eberhardt, K.J. Randall, J. Feldhaus, A.M. Bradshaw, Z. Xu, P.D. Johnson, Y. Ma: Book of Abstracts, VUV 10, Tu 110, Paris (1992)
5.24 J. Stöhr: *NEXAFS Spectroscopy*, Springer Ser. Surf. Sci. Vol. 25 (Springer Berlin, Heidelberg 1992)
5.25 C.E. Brion, S. Daviel, R.N.S. Sodhi, A.P. Hitchcock: AIP Conf. Proc. **94**, 429 (1982)
C.E. Brion: Comments At. Mol. Phys. **16**, 249 (1985)
5.26 T. Tyliszcak, F. Esposto, A.P. Hitchcock: Phys. Rev. Lett **62**, 2551 (1989)
5.27 T. Tyliszcak, A.P. Hitchcock: J. Vac. Sci. Technol A **8**, 2552 (1990)
5.28 D. Coulman, A. Puschmann, W. Wurth, H.-P. Steinrück, D. Menzel: Chem. Phys. Lett. **148**, 371 (1988)
D. Coulman, A. Puschmann, U. Höfer, H.-P. Steinrück, W. Wurth, P. Feulner, D. Menzel: J. Chem. Phys. **93**, 58 (1990)
5.29 R. Scheuerer, F. Pellowski, M. Scheuer, W. Wurth, P. Feulner, D. Menzel; Book of Abstracts, VUV-10, Tu117, Paris (1992)
5.30 W. Wurth, J. Stöhr, P. Feulner, X. Pan, K.R. Bauchspiess, Y. Baba, E. Hudel, G. Rocker, D. Menzel: Phys. Rev. Lett **65**, 2426 (1990)
5.31 A. Puschmann, T. Mandel, M. Domke, C. Xue, G. Kaindl: Surf. Sci **251/252**, 354 (1991)
5.32 W. Wurth: Unpublished results (1988)
5.33 O. Björneholm, A. Sandell, A. Nilsson, N. Martensson, J.N. Andersen: Physica Scripta **T41**, 217 (1992)
5.34 R.J. Guest, A. Nilsson, O. Björneholm, B. Hernnäs, A. Sandell, R.E. Palmer, N. Martensson: Surf. Sci. **269/270**, 432 (1992)

5.35 W. Wurth, P. Feulner: To be published
5.36 D. Arvanitis, J. Singh, H. Rabus, T. Lederer, K. Baberschke: Phys. Rev. B **45**, 1518 (1992)
5.37 C.T. Chen, Y. Ma, F. Sette: Phys. Rev. A **40**, 6737 (1989)
5.38 P. Feulner, R. Scheuerer, M. Scheuer, G. Remmers, W. Wurth, D. Menzel: Appl. Phys. A **55**, 478 (1992)
5.39 W. Wurth, G. Rocker, P. Feulner, R. Scheuerer, L. Zhu, D. Menzel: Phys. Rev. B **47**, 6697 (1993)
5.40 L. Wenzel. D. Arvanitis, R. Schlögl, M. Muhler, D. Norman, K. Baberschke, G. Ertl: Phys. Rev. B **40**, 6409 (1989)
5.41 H. Rabus, D. Arvanitis, M. Domke, K. Baberschke: J. Chem. Phys. **96**, 1560 (1992)
5.42 D.A. Outka, J. Stöhr: J. Chem. Phys. **88**, 3539 (1988)
5.43 U. Ehrke and W. Wurth: To be published
5.44 F. Sette, J. Stöhr, A.P. Hitchcock: J. Chem. Phys. **81**, 4906 (1984)
5.45 J. Stöhr, K.R. Bauchspiess: Phys. Rev. Lett. **67**, 3376 (1991)
5.46 D. Arvanitis, H. Rabus, L. Wenzel, K. Baberschke: Z. Phys. D **11**, 219 (1989)
5.47 J.A. Sheehy, T.J. Gil, C.L. Winstead, R.E. Farren, P.W. Langhoff: J. Chem. Phys. **91**, 1796 (1989)
5.48 C. Winstead, P.W. Langhoff: J. Chem. Phys. **95**, 3107 (1991)
5.49 J.L. Dehmer, D. Dill, S. Wallace: Phys. Rev. Lett. **43**, 1005 (1979)
5.50 K. Müller-Dethleffs: Ann. Rev. Phys. Chem. **42**, 103 (1991)
5.51 W. Wurth: Unpublished results (1989)
5.52 A. Nilsson, H. Antonsson, N. Martensson: In *Desorption Induced by Electronic Transitions DIET IV*, ed by G. Betz, P. Varga, Springer Ser. Surf. Sci. Vol. 19, (Springer, Berlin, Heidelberg 1990) p. 105
5.53 O. Björneholm, A. Nilsson, E.O.F. Zdansky, A. Sandell, B. Hernnäs, H. Tillborg, J.N. Andersen, N. Martensson: Phys. Rev. B. **46**, 10353 (1992)
5.54 H. Antonsson, A. Nilsson, N. Martensson, I. Panas, P.E.M. Siegbahn: J. Electron Spectrosc. **54/55**, 601 (1990)
5.55 J. Singh, W. K. Walter, A. Atrei, D. A. King: Chem. Phys. Lett. **185**, 426 (1991)
5.56 O. Björneholm, A. Nilsson, E.O.F. Zdansky, A. Sandell, H. Tillborg, J.N. Andersen, N. Martensson: Phys. Rev. B in press
5.57 J. Stöhr, D.A. Outka, K. Baberschke, D. Arvanitis, J. Horsley: Phys. Rev. B **36**, 2976 (1987)
5.58 M.B. Robin: *Higher Excited States of Polyatomic Molecules*, Vols. 1 and 3 (Academic, New York 1974 and 1985)
5.59 E. Umbach: Surf. Sci. **117**, 482 (1982) and references therein
5.60 R. Treichler, W. Riedl, W. Wurth, P. Feulner, D. Menzel: Phys. Rev. Lett. **54**, 462 (1985)
5.61 R. Treichler, W. Wurth, W. Riedl, P. Feulner, D. Menzel: Chem. Phys. **153**, 259 (1991)
5.62 R. Treichler, W. Riedl, P. Feulner D. Menzel: Surf. Sci. **243**, 239 (1991)
5.63 J.C. Fuggle: *High Resolution Auger Spectroscopy of Solids and Surfaces, in: Electron Spectroscopy: Theory, Techniques, and Applications*, Vol. 4, ed. by C.R. Brundle, A.D. Baker, (Academic, London 1981) p. 85
5.64 A. Niehaus: J. Phys. B **10**, 1845 (1977)
5.65 T. Aberg: Physica Scripta **T41**, 71 (1992)
5.66 C.T. Chen, R.A. DiDio, W.K. Ford, E. W. Plummer, W. Eberhardt: Phys. Rev. B **32**, 8434 (1985)
5.67 W. Wurth, C. Schneider, R. Treichler, E. Umbach, D. Menzel: Phys. Rev. B **35**, 7741 (1987)
5.68 W. Wurth, C. Schneider, R. Treichler, E. Umbach, D. Menzel: Phys. Rev. B **37**, 8725 (1988)
5.69 A. Sandell: Autoionization studies of adsorbates. Dissertation, Uppsala University (1993)
5.70 O. Björneholm, A. Nilsson, A. Sandell, B. Hernnäs, N. Martensson: Phys. Rev. Lett **68**, 1892 (1992)
5.71 A. Klekamp, R. Reissner, E. Umbach: In *Adsorption on Ordered Surfaces of Ionic Solids and Thin Films*, ed. by H.-J. Freund, E. Umbach, Springer Ser. Surf. Sci. Vol. 33 (Springer, Berlin, Heidelberg 1993)
5.72 W. Wurth: *Proc. Int'l Workshop on Photoionization* IWP-92 Berlin (AMS, New York 1993) p. 77

5.73 T. Porwol, G. Illing, H.-J-Freund, H. Kuhlenbeck, M. Neumann, S. Bernstorff, W. Braun, W. von Niessen, C.-M. Liegener: Phys. Rev. B **41**, 10510 (1990)
5.74 O. Björneholm: Adsorbate core excitation and decay dynamics. Dissertation, Uppsala University (1992)
5.75 W. Wurth, P. Feulner: To be published
5.76 L. Ungier, T.D. Thomas: J. Chem. Phys. **82**, 346 (1985)
5.77 W. Wurth, D. Menzel: J. Electron Spectrosc. **62**, 23 (1993)
5.78 H. Tillborg: Geometrical and electronic structure of adsorbates studied by core level spectroscopies. Dissertation, Uppsala University (1993)
5.79 F. Kaspar, W. Domcke, L.S. Cederbaum: Chem. Phys. **44**, 33 (1979)
5.80 N. Correia, A. Flores-Riveros, H. Agren, K. Helenelund, L. Asplund, U. Gelius: J. Chem. Phys. **83**, 2036 (1985)
5.81 G. Illing, T. Porwol, I. Hemmerich, G. Dömötör, H. Kuhlenbeck, H.-J. Freund, C.-M. Liegener, W. von Niessen: J. Electron Spectrosc. **51**, 149 (1990)
5.82 H.-J. Freund, M. Neumann: Appl. Phys. A **47**, 3 (1988)
5.83 E. Umbach, Z. Hussain: Phys. Rev. Lett. **52**, 457 (1984)
5.84 E. Umbach: Comments Atom. Mol. Phys. **18**, 23 (1986)
5.85 W.E. Moddemann, T.A. Carlsson, M.O. Krause, B.P. Pullen, W.E. Bull, G.K. Schweitzer: J. Chem. Phys. **55**, 2317 (1971)
5.86 W. Wurth: Vacuum **40**, 3 (1990)
5.87 J.C. Fuggle, E. Umbach, D. Menzel, K. Wandelt, C.R. Brundle: Solid State Commun. **27**, 65 (1978)
O. Gunnarsson, K. Schönhammer: Phys. Rev. Lett. **41**, 1608 (1978) and **42**, 195 (1979)
5.88 See the conference series *Desorption Induced by Electronic Transitions*, DIET I to V, Springer Chem. Phys., Vol. 24 and Springer Ser. Surf. Sci. Vols. 4, 13, 19, 31 (Springer, Berlin, Heidelberg 1983 to 1993)
5.89 See e.g. R.D. Ramsier, J.T. Yates, Jr.: Surface Sci. Repts **12**, 243 (1991)
5.90 R. Franchy, D. Menzel: Phys. Rev. Lett. **43**, 865 (1979)
5.91 D. Menzel. G. Rocker, D. Coulman, P. Feulner, W. Wurth: Physica Scripta **41**, 588 (1990)
5.92 D. Menzel, G. Rocker, H.-P.Steinrück, D. Coulman, P.A. Heimann, W. Huber, P. Zebisch, D.R. Lloyd: J. Chem. Phys. **96**, 1724 (1992)
5.93 D. Lapiano-Smith, K. Lee, C.I. Ma, K. Wu, D.M. Hanson: J. Electron Spectrosc. **51**, 221 (1990)

6. Spectroscopy and Dynamics of the Electronic Decay of Core Electron Excitations in Isolated and Chemisorbed Molecules

W. Eberhardt

Over the past decade studies of the decay processes following both core-electron excitation as well as core-electron ionization have been carried out for a variety of small molecules in gas phase and in a physisorbed or chemisorbed state. This was made possible through the application of synchrotron radiation used in preparing well-defined highly-excited states. These investigations of the electronic decay of specifically-prepared electronic configurations offer unique ways to study electron correlation and dynamic charge transfer and screening processes in well characterized electronic systems. Moreover, high-resolution studies taken with a resolving power sufficient to resolve vibrational substates in both the excitation and decay channels, reveal information about the time evolution of the vibrational wave functions in the intermediate core-excited state on the time scale of a few femtoseconds. Thus, these experiments probe the limits of the quantum mechanical description of the eigenstates of the molecular system integrated over the lifetime of the core hole.

The intention of this chapter is not to summarize all the work performed up to date. Rather, the goal is to demonstrate the potential of this kind of electron spectroscopy using examples of some of the experiments selected for didactic purpose. Therefore, even though some investigations of this kind have been carried out on solid-state systems, here we will concentrate almost exclusively on studies of gas-phase molecules since, as explained in detail later, the effects are better characterized and understood for molecular systems. Additionally, we want to alert the reader to the fact that the nomenclature used in the literature to describe these processes is not unique. The process of the deexcitation of a neutral core hole excited state is correctly described by the term autoionization. Sometimes, however, it is referred to as resonant photoemission or resonant Auger decay. Additionally, in order to emphasize the fact that the object of these investigations is the electronic decay of specifically and selectively prepared highly excited (core hole) electronic states, the term Deexcitation Electron Spectroscopy (DES) was coined in a summary of the initial studies [6.1]. We will use this latter acronym throughout this review in order to stress the importance of the exact classification of both the excitation and the decay process. This also implies the expectation that this experimental technique is going to be more widely applied once the importance of these kind of studies in understanding electron correlation not only in gas phase molecules but also in solid state systems has been realized.

Springer Series in Surface Sciences, Vol. 35
Applications of Synchrotron Radiation Ed.: W. Eberhardt

6.1 Deexcitation Electron Spectroscopy

6.1.1 Historical Development

Historically, in studies of the Auger decay in diatomic and triatomic molecules it was discovered that electron-excited Auger spectra contain lines originating not only from core-ionized states but also from the decay of core-to-bound state excitations [6.2, 3]. Since the high-energy electrons generally used to excite Auger spectra can transfer any amount of energy in the collision process, up to the primary energy, a certain fraction of the core electrons is not ionized but excited into an unoccupied molecular orbital or with much smaller probability into a Rydberg state. These core-to-bound excited electronic configurations result in decay spectra which are different than the normal Auger spectra. *Moddeman* et al. [6.3] were able to give experimental proof of this hypothesis and to identify some of the lines resulting from the decay of core-to-bound state excitations by comparing Auger spectra excited with electrons and with monochromatic X-rays.

The first spectra showing specifically the electrons produced in the decay of the $1s \rightarrow \pi^*$ excitations in N_2 and CO were published in 1983 [6.4, 5]. The excited-state configurations in these experiments were prepared by an (e, 2e) coincidence technique [6.4] or by using monochromatic synchrotron radiation [6.5] tuned to the energy of the corresponding core-to-bound state transitions. In the (e, 2e) coincidence experiment only the electrons emitted in coincidence with the observation of an energy-loss event of a primary high-energy (keV) electron beam corresponding to the π^* excitation are recorded, thus ensuring the preparation of the core-excited state prior to the decay. Typically, it takes about one week to accumulate such a spectrum with reasonable statistics and resolution [6.4]. At that time the synchrotron experiments were better in terms of the counting rates. At the synchrotron it took about one day to accumulate such a spectrum. Meanwhile, however, through the development of special undulator sources and beam lines the data collection rate has improved by two to three orders of magnitude enabling investigations with much better resolution and on states that have a substantially smaller excitation cross section, e.g. the Rydberg excitations.

The dramatic development in the synchrotron radiation through the implementation of soft X-ray undulator beam lines such as X1B at the NSLS Brookhaven [6.6] is illustrated by Fig. 6.1. Shown are the Auger spectrum of N_2 and the decay spectrum of the N $1s \rightarrow 1\pi_g$ (π^*) excitation recorded with undulator radiation. For comparison, the first spectrum of the π^* decay, as it was originally recorded using a bending-magnet beam line [6.5], is also included at the bottom. All these curves were accumulated using the same electron spectrometer and gas beam experimental setup. For the bending magnet spectrum the maximum count rate was 2 counts/s for the most prominent features, whereas the undulator spectrum was recorded at a counting rate of

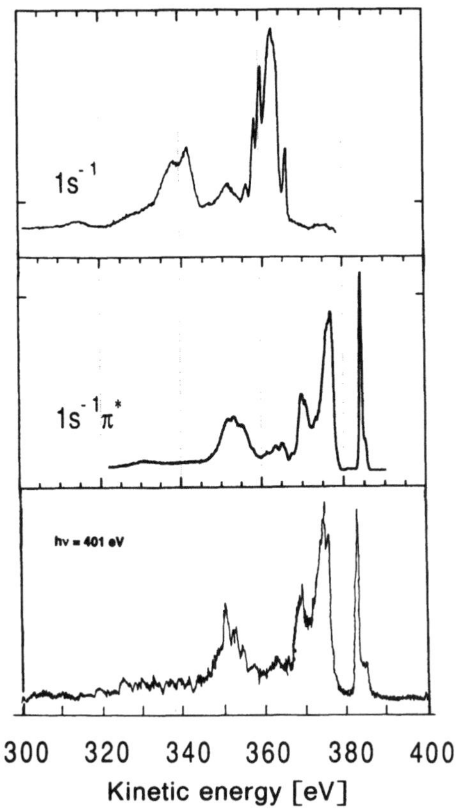

Fig. 6.1. Comparison of the Auger decay of a N 1*s* core vacancy in N_2 (top panel) with the decay of the N 1*s* → $1\pi_g$ (π^*) core-excited state (DES, center panel). Also shown at the bottom is the first deexcitation spectrum taken using a bending magnet beam line [6.5] to illustrate the impact of undulator based excitation sources for this type of spectroscopy

about 2×10^3 counts/s for these peaks. From this comparison, the dramatic improvement in the count rate, statistics and also resolution is immediately apparent.

We note that the kinetic energy of the emitted electrons is larger for the π^* decay spectrum than for the 'normal' Auger decay. This seems contradictory to the fact that the N 1*s* → π^* excited state is 9 eV lower in total energy than the N 1*s* core ionized state in N_2. We are thus faced with the paradox that the emitted electrons have larger kinetic energies for a decay from a state which requires less excitation energy. The solution to this 'puzzle', however, is readily obtained by taking into account that the π^* decay reflects a transition from a neutral core excited state into a singly ionized state, whereas the Auger decay involves a transition from a singly ionized state to a doubly ionized state of the molecule.

A first hint at the interpretation of these spectra is obtained by studying the evolution of the electron emission spectra in the photon-energy region near the onset of the core-electron excitations to above the core-ionization threshold. As the photon energy is tuned from below the onset of the C 1*s* excitations to above the C 1*s* ionization energy of CO, the observed electron emission spectra change as illustrated schematically by the curves depicted in Fig. 6.2. For reference, the absorption spectrum of CO near the C–K-edge is shown in Fig. 6.3. At photon energies well below the π^* transition energy, only the photoemission from the outer and inner valence orbitals, including all the correlation satellites is observed and the kinetic energy of the lines changes strictly in proportion to the photon energy. At a photon energy of hv = 287 eV, and only at that photon energy, the third spectrum from the bottom of this figure is measured. The

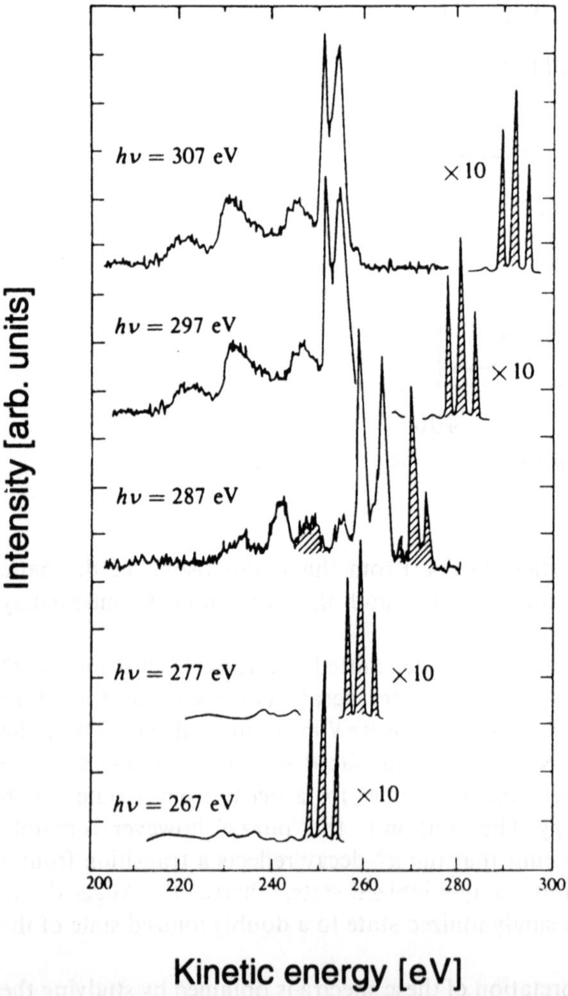

Fig. 6.2. Schematic representation of the development of the photoemission and core-hole decay spectra of CO as the excitation energy is tuned from below the C 1*s* absorption threshold through the excitation of the C 1*s* → 2π (π^*) excited state at 287 eV up to energies above the C 1*s* core electron ionisation threshold (296 eV)

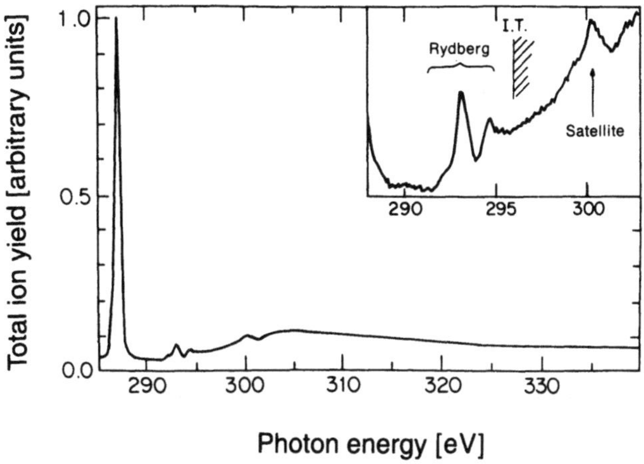

Fig. 6.3. Absorption spectrum (total ion yield) of CO showing the discrete transitions in the vicinity of the core ionisation threshold (marked IT)

intensity of the lines, where the direct photoemission of the outer valence orbitals is expected, is drastically enhanced and the relative intensity of these lines is different than in the photoemission spectra. Moreover, a whole manifold of additional emission features is observed. The strongest of these lines are located between kinetic energies of 255 and 270 eV. All these 'new' emission features disappear as soon as the photon energy is tuned off the energy of the π^* transition (Fig. 6.3). At a photon energy of 297 eV, just above the C 1s ionization threshold, the electron emission includes both the photoemission features and the conventional Auger spectrum. The dominating lines of the Auger spectrum are located between 250 and 256 eV kinetic energy. At even larger photon energies the Auger lines remain constant in kinetic energy, whereas the photoemission lines shift proportional to the excitation energy.

6.1.2 Development of an Empirical Assignment

While the electronic states populated in the Auger decay of the core ionized state have been assigned theoretically [6.7], the final electronic configurations of the deexcitation of the π^* excited state are not readily available from calculations. Therefore, we have derived an empirical scheme for this assignment [6.1], which we will introduce below. The degeneracy of some of the observed emission lines with the photoemission lines already suggests that the same final electronic states are populated. Accordingly, this process has also been referred to as a resonant photoemission process, even though the physics is incorrectly described by this nomenclature. The matrix elements leading to these final states are given by a dipole excitation of the core $\rightarrow$ bound state transition and a subsequent decay described by an Auger matrix element.

The general pathway to an assignment of the deexcitation of the decay of a core excited state in any atomic, molecular, or even solid state system is outlined in Fig. 6.4. Taking CO as example, schematically the electronic configurations involved in the excitation and decay of the C 1$s \rightarrow \pi^*$ core excited state are exhibited in Fig. 6.4. In the center the core-excited state is shown. In a single-particle notation the C 1s electron is excited into the 2π molecular orbital by the absorption of the photon. Subsequently, this state decays either via participation of the 2π electron, shown in the left hand diagram, or via a normal Auger transition while the 2π electron is present as a spectator, as displayed on the right-hand side of Fig. 6.4. The *participator process* thus leads to single-hole configurations just as the normal photoemission process does. Therefore the final electronic states are identical and the kinetic energy of the emitted electrons is degenerate with the photoemission lines.

The *spectator process*, on the other hand, is very similar to the conventional Auger decay. The presence of the 'spectator electron' will influence the matrix elements somewhat, but the largest difference comes from the interaction between the extra spectator electron and the two holes populated by the Auger transition. The dominant term of this interaction is the Coulomb screening, which lowers the final-state energy by an amount of up to 20 eV. Additional energy shifts and splittings are introduced through the coupling of the angular momenta and spins of these two-hole-one-electron final states.

This simplified (single particle) picture of the core-hole deexcitation immediately suggests plotting all spectra on a generalized binding energy scale, where

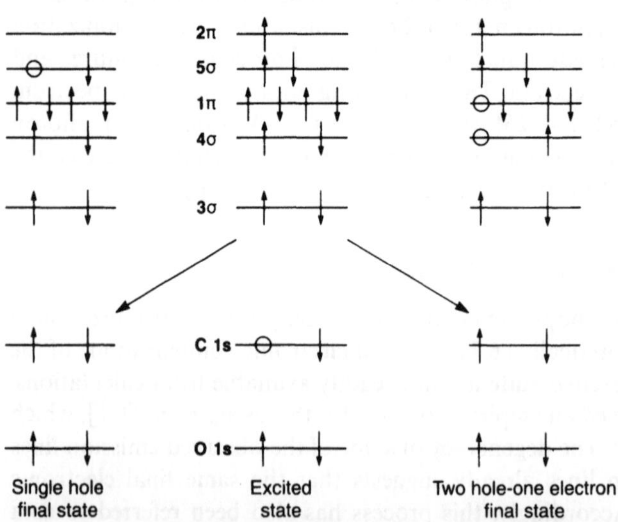

Fig. 6.4. Schematic representation of the electronic decay of the C 1$s \rightarrow 2\pi$ core excited state in CO. The core excited state configuration is shown in the center. The left (right) diagram illustrates the participator (spectator) decay channel leading to the population of single hole (two-hole one-electron) final states

the binding energy E_B is derived from the excitation energy of the core hole state E_c, and the kinetic energy of the emitted electron E_{kin} according to

$$E_B = E_c - E_{kin}. \quad (6.1)$$

For the conventional Auger spectra E_c is given by the ionization energy of the core electron, as this is the amount of energy deposited in the molecular system. The photoelectron emitted in this event carries off the excess energy. Analogously, in the case of a core-to-bound state transition, E_c is the excitation energy. E_B, thus defined, corresponds to the energy difference between the ground state of the molecular system and the energy of the final state of the decay process. Obviously, this is not only valid in the single-particle picture, but also applies to the true many-body final-state configuration.

In this scheme the participator decay processes can be identified by comparison with a photoemission spectrum (Fig. 6.5) for the C π^* deexcitation spectrum of CO. On the binding energy scale the single-hole final states in the two spectra, $5\sigma^{-1}$, $1\pi^{-1}$, and $4\sigma^{-1}$, populated either in the direct photoemission or in the participator decay process, line up, as they are supposed to, at the same final-state energy. Since the two processes are governed by different matrix elements, the relative branching ratio in photoemission and deexcitation is different. Even the $3\sigma^{-1}$ final state, which is a strongly correlated state and can only be properly described as a many-body state, has a corresponding feature in

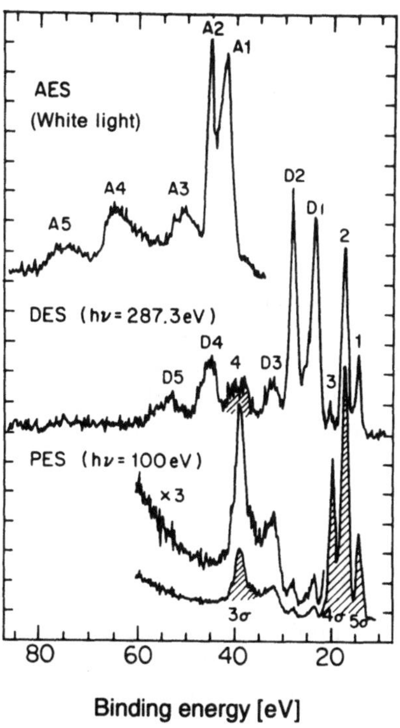

Fig. 6.5. Deexcitation of the C 1s → π^* excited state in CO (DES) compared to the Auger (AES) and the photoemission spectrum (PES) on a binding energy scale

Table 6.1. Binding energies and dominant single particle configurations for the peaks in the DES and AES spectra of Fig. 6.5.
These assignments are given in the single-particle notation, even though substantial configuration mixing might occur [6.7, 38]. Here ΔE is the energy difference between the two-hole final state and the equivalent two hole-one electron final state.

	DES spectra			AES spectra		
Label	E_B[eV]	Main Config.	Label	E_B[eV]	Main Config.	ΔE [eV]
1	14	$5\sigma^{-1}$				
2	16.8	$1\pi^{-1}$				
3	19.8	$4\sigma^{-1}$				
D1	23	$5\sigma^{-2}2\pi$	A1	42	$5\sigma^{-2}$	20
	23	$5\sigma^{-1}1\pi^{-1}2\pi$		43	$5\sigma^{-1}1\pi^{-1}$	20
D2	27.8	$4\sigma^{-1}5\sigma^{-1}2\pi$ or $5\sigma^{-2}\,2\pi$	A2	45.6	$4\sigma^{-1}5\sigma^{-1}$	18
D3	31.8	$4\sigma^{-1}1\pi^{-1}2\pi$	A3	51	$4\sigma^{-1}1\pi^{-1}$	19
4	39	$3\sigma^{-1}$				
D4	45	$3\sigma^{-1}5\sigma^{-1}2\pi$	A4	66	$3\sigma^{-1}5\sigma^{-1}$	21
D5	53	$3\sigma^{-1}1\pi^{-1}2\pi$	A5	75	$3\sigma^{-1}1\pi^{-1}$	22
		$3\sigma^{-1}4\sigma^{-1}2\pi$			$3\sigma^{-1}4\sigma$	

the deexcitation spectrum. From such a comparison a partial assignment of the deexcitation spectra for any core-hole excited state can be derived. This procedure can readily be carried out even for a molecular system not as well studied as the examples shown here.

The remaining features, labelled D1 through D5, are then assigned to spectator final states. As expected, this part of the spectrum resembles the Auger spectrum (peaks A1 through A5) and is shifted mainly by the Coulomb screening interaction in the final state. As suggested by this labelling, the average screening contribution for the spectator final states is about 20 eV. A more detailed comparison is given in Table 6.1. The differences in the lineshape and the non-homogeneous screening energy are attributed to the multiplet coupling in the spectator final state. As we will see later, the deexcitation spectra of core → Rydberg excitations resemble much more closely the Auger spectrum since the interaction with an electron in a Rydberg orbital is not as strong as with an electron in the lowest unoccupied molecular orbital, even though a sizeable Coulomb shift still remains.

6.1.3 Relationship Between Photoemission and DES

In order to emphasize the differences between photoemission and DES, one must consider the relevant matrix elements describing the two processes. The differential cross section for the photoemission process $\partial^2\sigma/\partial\omega\partial t$ is to first-order approximation described by Fermi's 'Golden Rule' according to

$$\frac{\partial^2 \sigma}{\partial \omega \partial t} \sim |\langle f|A \cdot p|i\rangle|^2 \cdot \delta(E_f - E_i - \hbar\omega) \qquad (6.2)$$

where $\langle f|$ and $|i\rangle$ are the final and initial state wave functions of the system, or the photoelectron if the less accurate single-particle description is used in the form of product wave functions, and $A \cdot p$ is the dipole operator. The δ-function insures energy conservation between the initial and final-state energies and the photon energy $\hbar\omega$.

In the "length form" of the dipole matrix element the dipole operator $A \cdot p$ is replaced by $\nabla V \cdot r$. This illustrates that the photoemission process takes place in regions where there is a strongly varying potential, i.e. ∇V is large. (A free non-relativistic electron cannot absorb a photon). However, apart from this enhancement, the photoemission process is a nonlocal process to the extent that it is not obvious at which atom within a heteronuclear molecule the photoemission process actually has taken place.

The deexcitation process on the contrary is a strongly localized process. The initial excitation is again described by a dipole transition according to (6.2) with the restriction that the final state $\langle f|$ has a bound electron rather than a free electron in the continuum. The subsequent decay of the core hole excited state is described by a Coulomb matrix element, modifying the description given by *Wentzel* [6.8] for the Auger decay

$$\langle \Psi_i^N | \frac{e^2}{r} | \Psi_f^{N-1} \phi_e \rangle \qquad (6.3)$$

where Ψ_i^N is the N – particle wave function of the core-excited system and the final state of the transition is given by the N-1 particles wave function of the molecular system Ψ_f^{N-1}, and the wave function of the emitted electron ϕ_e. Adopting the approximation of describing the total wave function as a product wave function this may be rewritten as

$$\langle \phi_1 \phi_2 | \frac{e^2}{r_{12}} | \phi_c \phi_e \rangle \langle \Psi_i^{N-2} | \Psi_f^{N-2} \rangle, \qquad (6.4)$$

where ϕ_1 and ϕ_2 are the wave functions of the two valence electrons involved in the transitions in the core excited state, and ϕ_c is the wave function of the core electron. The remaining electrons are lumped into the $(N-2)$-electron wave functions Ψ_i^{N-2} and Ψ_f^{N-2}. The major part of this interaction is thus described by the overlap of the two valence electrons participating in the transition with the wave function of the core electron in the final state and the emitted electron. Since the core-electron wave function is extremely localized, this overlap gives the deexcitation process a very strong local character. Thus to a certain extent the deexcitation intensity is a measure of the localization of the valence electron wave functions in the core excited state at the nucleus where the core hole was created. Of course, this is modified by off-diagonal terms to a similar extent as these contribute in Auger spectroscopy.

This difference in the matrix elements is also why we so strongly object to the term 'resonant photoemission' for this process. The interaction of the photon field with the electronic wave functions is fairly global, even though emphasized in the regions of a strongly variant potential. This can be easily seen in the length form of the dipole matrix element. The core electron wave function in the Auger matrix element, however, has the effect of a much stronger localization than the gradient of the potential in the dipole matrix element.

To summarize the general scheme for the assignment and interpretation of deexcitation spectra of core-hole excited states: First, the *participator final states* are assigned by comparison with a standard photoemission spectrum on the basis of the final-state energy. States that have an identical final state energy are most likely identical. The intensity distribution (branching ratio) will be different between the two spectra, reflecting the differences in the matrix elements between the two processes. Second, the *spectator final states* are assigned by comparison with the Auger spectrum. Since for the Auger and the deexcitation decay processes the matrix elements are almost identical, neglecting the modifications of the wave functions due to the presence of the spectator electron, the relative intensity of these lines should be fairly similar. The shift in energy reflects the final state interaction in the two-hole-one-electron final state of this process. The less the spectator electron couples to the two holes produced in the Auger decay, the closer the spectator spectrum resembles the Auger spectrum. As we will see later, there is actually a continuous transition at the core ionization threshold from the deexcitation spectrum of high core → Rydberg excitations into the Auger spectrum shifted by PCI (Post Collision Interaction).

Thus, the physics that can be studied by deexcitation electron spectroscopy also follows this dual aspect. The *spectator final states* offer a unique way to study electron correlation. Specific electronic configurations can be prepared and the influence of selective changes in the electronic configuration of the final states, such as the difference between a spectator electron located in a molecular orbital or Rydberg orbital or even the n-Rydberg excitation and $n + 1$ Rydberg state and the resulting changes in the $2h - 1e$ final state can be monitored very closely. The single hole *participator final states*, on the other hand, uniquely reflect upon the localization of the electronic wave functions in the core excited state. As discussed in more detail above, contrary to the 'global' coupling of the photoemission process, the single-hole final state population, created by the participator decay, is described by the Coulomb matrix element of the Auger decay process, which is very sensitive to the wave function overlap with the core hole. Since the relevant wave functions are the functions of the core excited state, this allows the study of dynamic charge rearrangement and screening processes in the molecular system occurring upon the creation of the core excited state.

Of special interest in the discussion of DES is also the question whether the excitation and decay step can be regarded as two independent processes. This is obviously related to the lifetime of the core excited state. As the lifetime decreases, the whole deexcitation process, excitation and decay, has to be treated as a coherent process. Additionally, if there are several excited states available within the energy uncertainty corresponding to the core hole lifetime, interference effects should be observed since the transition amplitudes have to be added rather than the intensities.

The various aspects of the applications of DES are going to be demonstrated and discussed in the following using deexcitation spectra of small molecules, mostly diatomics, as example.

6.2 Probing the Local Electronic Structure and Response

6.2.1 Deexcitation

Here several examples are presented highlighting the aspect of the determination of the local electronic structure and the response to the creation of the core-hole excited state. Examples of the decay of the π^* excitations of N_2, of both the terminal and central N-atom in N_2O, and of several azabenzenes will be discussed. (As information for the non-chemists: azabenzenes are benzene rings where one or several C–H units have been replaced by N-atoms).

The lowest-energy core-hole excited state, common to all of these molecules is a rather intense N $1s \to \pi^*$ transition. N_2 requires an energy of 401 eV to be excited to this state. For N_2O there are two of these transitions observed, as shown in Fig. 6.6a [6.9]. N_2O is a linear triatomic molecule with an asymmetric configuration (NNO). Due to the difference in chemical environment, the $1s$ electrons of the two N-atoms have different binding energies. The LUMO (Lowest Unoccupied Molecular Orbital) is a 3π orbital. For the terminal N-atom the transition energy N $1s \to 3\pi$ is 401.2 eV. Creating the corresponding core hole excited state for the central N-atom requires 404.9 eV.

Absorption (partial electron yield) spectra of condensed films of various azabenzenes in a photon energy range near the onset of the N $1s$ excitations in these molecules are depicted in Fig. 6.6b [6.10]. Next to each curve a diagram of the corresponding molecular structure is included. For each one of these molecular systems at least one π^* transition is observed which gives rise to the strongest features in the absorption spectra at energies near 400 eV. All these molecules have three unoccupied π^* orbitals. In the ground state of s-triazene the two lowest π^* orbitals are energetically degenerate, as they are in benzene, whereas in pyridine and pyrazine the two lowest π^* orbitals are split in energy by 0.6 and 0.9 eV, respectively [6.10]. The creation of a core hole at any of the N-atoms, however, will lower the symmetry of any one of these three molecules to C_{2v} and lift any degeneracies between the π^* orbitals that may exist in the ground state. Moreover, only transitions from the N $1s$ core levels into the lowest and highest energy π^* orbitals are dipole allowed. On the basis of these simple considerations, the assignment of the absorption spectra of these compounds is fairly straightforward [6.10].

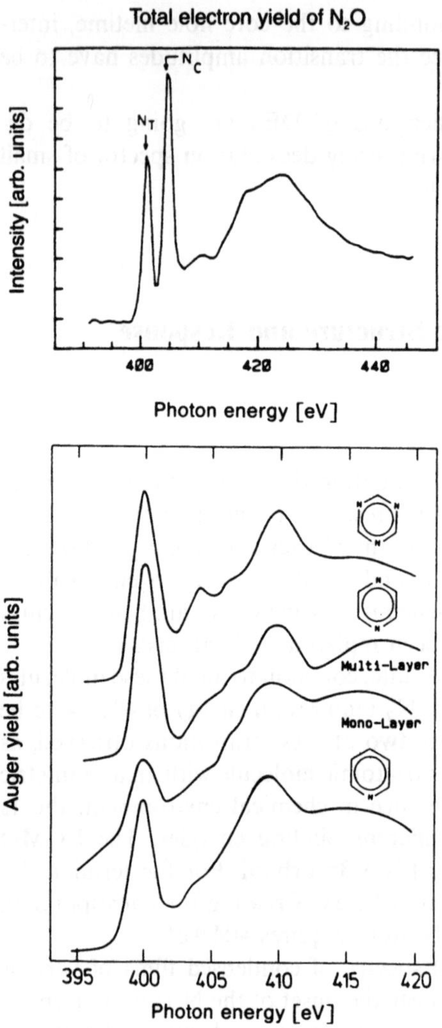

Fig. 6.6a, b. Top panel (a): Absorption (electron yield) of N_2O showing the two different π^* excitations and the onset of the N 1*s* ionization for the two inequivalent N-atoms. Bottom panel (b): Absorption (electron yield) spectra of various condensed azabenzenes

The deexcitation spectra of these molecules are compared to the conventional Auger spectra and the photoemission spectra for each species in Figs. 6.7–9. From the comparison with the photoemission spectra the assignment of the single-hole participator final states is straightforward. The only exception is the spectrum of s-triazene, where charging was observed [6.10].

In order to understand the intensities of the various single-hole final states populated in the participator decay, we have to carry out a full Auger matrix element calculation of the transition intensities. This is a considerable task. As an empirical approach, however, some intuitive, qualitative estimates for the observed transition intensities can be obtained on the basis of the valence-charge density in the intermediate core-excited state. The ground-state charge

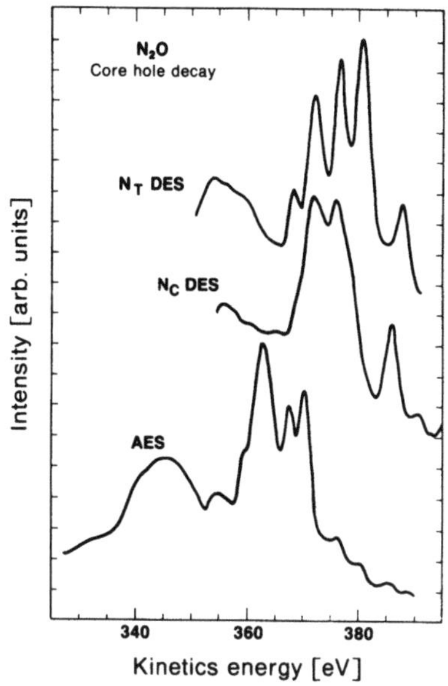

Fig. 6.7. Nitrogen Auger (AES) and deexcitation spectra (DES) of N_2O generated in the decay of the core → 3π bound state excitation at the site of the terminal N_T or the central N_C atom

density should not be taken, since the initial state of the transition is the core-hole excited state, where the charge density is modified by core-hole screening processes. The charge density in the initial state, prior to the core-hole decay, can be approximated by calculations for the corresponding molecule, where the core-excited atomic constituent is replaced by the $Z + 1$ atom. Thus for N_2 we take the calculations for NO. Some of these charge densities, taken from the book of *Jorgensen* and *Salem* [6.11], are displayed in Fig. 6.10.

On the basis of the charge-density calculations we conclude that the N 1s core hole in N_2 is predominantly screened through a charge transfer within the $1\pi_u$ orbital. The σ-orbitals also change, but since the σ-space is completely occupied, the total σ-charge at each nucleus is not dramatically changed upon the creation of the core hole. The $1\pi_u$ orbital in N_2 has a charge density, which has equal weight at either one of the two equivalent nuclei. In NO, however, the corresponding 1π orbital charge density is much higher at the O-atom then at the N-atom. According to this simple picture, the most prominent single-hole final state, showing up in the participator decay spectrum, is the $1\pi_u$ state.

Analogously in N_2O, the charge density of the 1π and 2π orbitals is distributed quite non-uniformly throughout the molecule even in the ground state (Fig. 6.10). The 1π orbital has by far largest charge density at the central N-atom within this molecule, whereas the charge of the 2π orbital is located predominantly at the O– and at the terminal N-atom. Core excitation at the central N-atom will enhance the localization of the 1π orbital at this atom and the predominant participator decay is into the $1\pi^{-1}$ single hole final state (Table 6.2). Creation of a core hole excited state at the terminal N-atom, on the other hand, results in participator processes populating the $2\pi^{-1}$ single-hole state

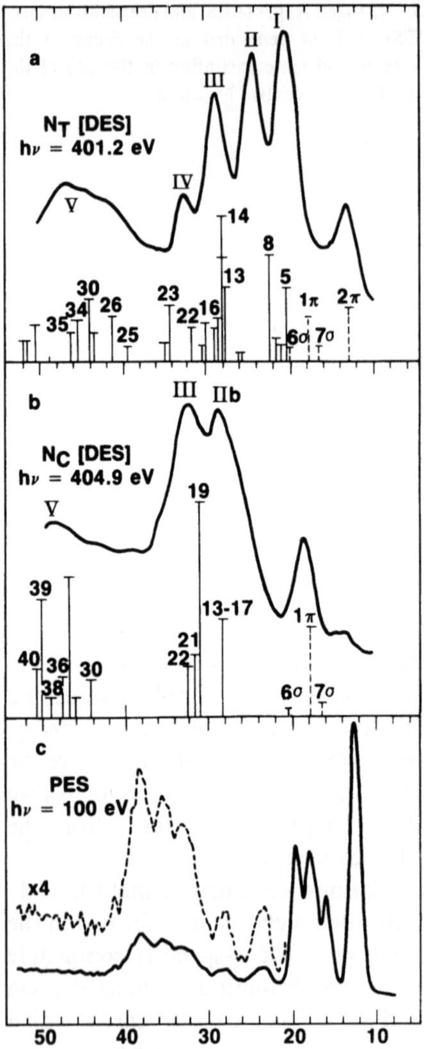

Fig. 6.8. Deexcitation spectra of the terminal N_T atom (**a**) and of the central N_C atom (**b**) in N_2O shown on the generalized binding energy scale. The photoemission spectrum, taken at a photon energy of 100 eV has been included in panel (**c**) for comparison. The bar diagrams indicate the position and intensity of the calculated spectator (solid bars) and participator (dashed bars) transitions

[6.9] favoured only slightly (4:3) over the $1\pi^{-1}$ single-hole state (Table 6.2). This indicates the importance and role of the electrons in the 1π molecular orbital in the core hole screening process. These electrons can, since the orbital exhibits no radial nodes, through a variation of charge density very easily provide the screening charge at any site within the linear triatomic molecule N_2O.

The situation is quite similar in the azabenzenes. As can be seen from the comparison of the charge density of the highest binding energy π orbital in benzene and pyridine (Fig. 6.10), a substantial charge rearrangement occurs within this orbital at the site of the $Z + 1$ nucleus. An analogous charge flow can

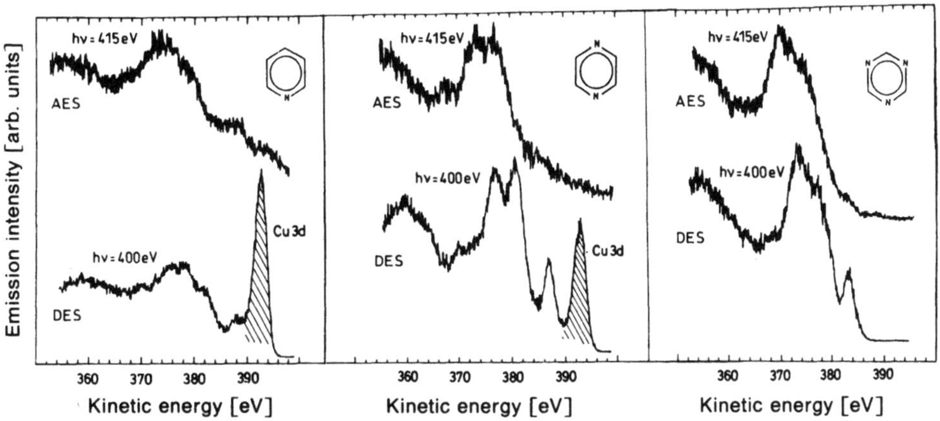

Fig. 6.9. AES and deexcitation spectra of thin solid films of pyridine (left), pyrazine (center), and s-triacine (right) condensed onto a Cu single crystal substrate. The substrate Cu 3*d* emission is shaded in the spectra

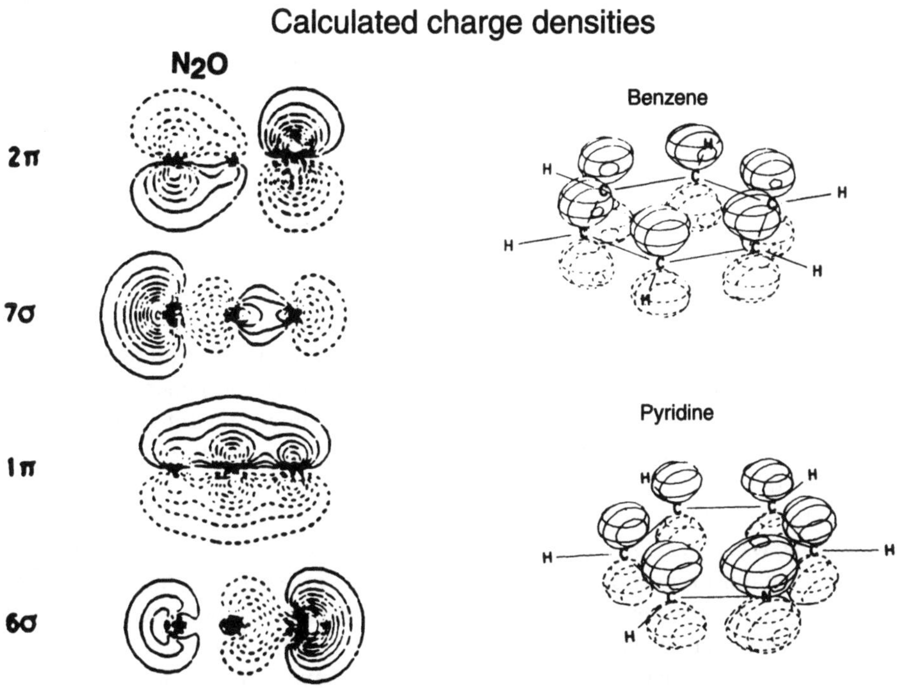

Fig. 6.10. Calculated charge densities of various orbitals in N_2O (R.P. Messmer, private commun.) in the left panel and for benzene and pyridine on the right-hand side taken from [6.11]

Table 6.2. Energies and intensities of the participator lines in the nitrogen and oxygen Auger spectra of nitrous oxide

		$N_T[3\sigma\, 2\pi]$ $E_{ex} = 401.2$ eV		$N_c[2\sigma\, 2\pi]$ $E_{ex} = 404.9$ eV		$O[1\sigma\, 2\pi]$ $E_{ex} = 534.6$ eV	
Valence Hole[v]	E_B[eV]	E_{kin}[eV]	INT[b]	E_{kin}[eV]	INT*	E_{kin}[eV]	INT[b]
2π	12.9	388.3	4	392.0	0	521.7	1
7σ	16.4	384.8	1	388.5	2	518.2	2
1π	17.7	383.0	3	386.7	18	516.4	4
6σ	20.1	381.1	1	384.8	1	514.5	2

Relative intensity estimates based upon calculations for the ONO (N_T), NOO(N_c), and NNF(O) equivalent core analogs of N_2O.

be expected upon the creation of a N 1s core hole in any one of the azabenzenes. Accordingly, the participator decay part of the deexcitation spectra is dominated by the emission from this orbital.

Another interesting trend can be observed by comparing the spectator decay processes in these molecules. The shift of the spectator part of the emission measured relative to the conventional Auger spectrum varies substantially for these molecular species. On a binding energy scale this shift reflects the screening energy of the spectator electron in the two-hole-one-electron final state of the core-hole decay process. For N_2 the additional spectator electron contributes about 18 eV in screening energy. This is the average over the various multiplet configurations in the group of lines with the highest overall intensity [6.12]. Since the exact assignment of these states is not readily determined, only an average value can be given. For N_2O the corresponding energy shift is only slightly smaller (17.5 eV) [6.9]. For the condensed azabenzenes, on the other hand, a surprisingly small value is found [6.10, 13]. For these molecules there is, on the kinetic energy scale, only an average shift of 4 eV observed between the corresponding features in the conventional Auger spectrum and the spectator part of the deexcitation of the core to π^* excited state. In addition to this 4 eV shift, one has to take into account the difference in excitation energy, which is about 5 eV lower for the deexcitation spectra. Accordingly, the screening energy contribution in the spectator final state amounts to only about 9 eV, much less than for the smaller molecules.

Intuitively this can be understood on the basis of the localisation of the electronic wave functions. The more localised the electronic states are, the larger is the Coulomb repulsion between the two holes in the Auger final state, the famous 'U'. By the same token, adding an extra electron to the system will then also provide a large screening energy in the 2h-1e final state of the spectator decay process. These Coulomb screening energies have been calculated for some atomic cases [6.9, 14]. However, as we have just seen, for the larger molecules estimates based on these atomic screening energies are off by a large amount.

Nevertheless, an estimate, guiding the interpretation of these spectra, can be obtained from some basic ionisation properties of the system.

6.2.2 Basic Parameters of the Spectator Shift

In order to derive this empirical model for the spectator shift, let us perform an 'old fashioned' 'Gedanken Experiment' using a system with continuous electronic states. First we consider the case, where the core electron is ionized right at threshold with zero excess energy. (For the moment we completely ignore PCI, which will be discussed later). In this case, the spectator spectrum will correspond to the standard Auger spectrum and the participator spectrum corresponds, with respect to energy but not intensity, to the photoemission spectrum. The highest kinetic energy of the spectator spectrum will be observed at a kinetic energy E_{KS} given by

$$E_{KS} = E_B - IP_1 - IP_2, \tag{6.5}$$

where E_B is the binding energy of the core electron, which is equivalent to the photon energy for the chosen conditions of our 'Gedanken Experiment'. IP_1 and IP_2 are the lowest (first) ionization potentials of the neutral (IP_1) and singly ionized (IP_2) system, respectively. The highest kinetic energy line of the participator decay spectrum will be observed at a kinetic energy E_{KP} of

$$E_{KP} = E_B - IP_1. \tag{6.6}$$

Both of these equations are an easy to follow, straightforward consequence of total-energy conservation. In a graphic display (Fig. 6.11), where the energy of these highest kinetic energy participator and spectator lines is shown in relationship to the binding energy of the excited core electron IP*, this marks two points at zero binding energy IP*, separated by the ionisation energy of the singly-ionized system, IP_2.

As the energy of the core-electron excited state is lowered below the core ionization threshold (IP* > 0), the kinetic energy of the participator line will be reduced strictly proportionally one-to-one. This is indicated by the solid line in Fig. 6.11. This process can be continued up to a maximum value of IP* $= IP_{HOMO*}$ in our ideal system. HOMO is the Highest Occupied Molecular Orbital. The * in this nomenclature denotes always the presence of the core hole. Note also that in our 'Gedanken Experiment', using a system with continuous electronic states, the relationship $IP_{LUMO*} = IP_{HOMO*}$ holds, since the HOMO and LUMO are not separated in energy. In this second limiting case both the spectator and participator lines coincide in energy at a kinetic energy E_{K3} given by

$$E_{K3} = E_B - IP_1 - IP_{HOMO*}. \tag{6.7}$$

In this way we have obtained a strict relationship for the participator lines as a function of the binding energy of the core electron excited state. The spectator lines have only to follow this relationship for the two limiting cases. For

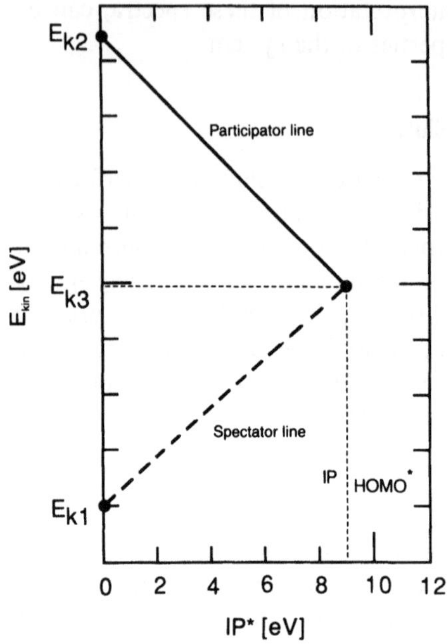

Fig. 6.11. Graphical representation of the participator and spectator line shifts as a function of the binding energy of the excited core electron. This plot is based upon the empirical parameters for benzene. For more detailed explanations see text

any other finite value of IP*, the energy of the highest spectator line has to be calculated exactly. However, a linear relationship of the shift of this line with IP*, as indicated by the dashed line in Fig. 6.11, can be taken as the first approximation. Later we will see that as long as the nature of the electronic states is not much different, i.e. in a series of Rydberg states, this linear relationship seems to hold quite well.

We have to take into account, however, that any 'real' molecular system will not have a continuous spectrum of electronic states, but rather display a HOMO-LUMO gap. In this case the 'ideal' value of IP_{HOMO*} will not be reached and the spectator and participator lines will not coincide. Nevertheless, a fairly good estimate of the line positions can be obtained for any such diagram using an intermediate value for IP_{HOMO*} near the center of the HOMO-LUMO gap.

Putting this model to a test, Fig. 6.11 has actually been drawn using the parameters for benzene. The ionization potentials can be readily taken from the literature as $IP_1 = 9.25$ eV and $IP_2 = 16.85$ eV [6.13]. The value for IP_{HOMO*} has to be estimated, since the system has a HOMO-LUMO gap. A value of 9 ± 1 eV seems quite reasonable for this parameter. Using these values we find (Fig. 6.11) that for the actual value of IP* = 5 eV, which corresponds to the case of the π^* excitations in the azabenzenes, a shift of 4 eV is predicted by this model for the shift of the spectator lines relative to the standard Auger lines on a *kinetic energy scale*. Adding to this shift the difference in excitation energy between the π^* excited state and the core ionized state, we find a predicted screening energy of 9 eV for the spectator electron in the 2h-1e final state of the deexcitation of the core $\rightarrow \pi^*$ transition. Using analogous values for the N_2 molecule, the empirical model also predicts the much larger screening energy and shift of the lines due to the spectator electron in the final state. Thus, our simple empirical model seems to cover the two most extreme cases known so far quite well.

In this day and age of modern supercomputers, which allow one to calculate all the energies of these states if so desired, one has to add some critical remarks about the usefulness of the simplistic empirical model just introduced. The justification for such an oversimplified approach comes clearly from the application in the interpretation of the spectra. This model readily predicts the large changes observed in the N deexcitation spectra for N_2 and the azabenzenes [6.13] and it is based upon a physical model, which is fairly easy to understand and follow. Moreover, even in the most sophisticated many-body configuration interaction calculations the energies calculated for these states are at best accurate within 1 to 2 eV. Thus, the empirical model does not significantly deteriorate the accuracy of the predicted assignment.

Essentially, the empirical model reduces the observed shifts to one parameter, the averaged Coulomb interaction energy of the spectator electron with the manifold of all populated Auger two-hole final states. This parameter is obtained by visually averaging the shift between corresponding Auger and 2h-1e deexcitation final states. As far as a detailed assignment and accurate multiplet splittings in these final states are concerned, the full calculations are needed. The drawback of this computational approach, however, is that sometimes the general physical picture is lost in a multitude of lines and calculated configuration interaction states all contributing to seemingly the same line observed experimentally. Therefore, we hope that the semiempirical approach will have its place even in the presence of large scale computers in uncovering the basic physical picture and guiding with the first assignment of deexcitation spectra of molecular systems of increasing complexity.

6.3 A Detailed Probe of Electron Correlation

6.3.1 The Difference Between π^* and Rydberg Excitations in N_2

In the previous section we investigated systematically the decay of the core $\rightarrow \pi^*$ excitation at the same atom (N) within increasingly larger and more complex molecules. The various deexcitation spectra revealed in the participator processes, on the one hand, the electronic states most prominently involved in the core-hole screening process and, on the other hand, directly allowed us to determine the Coulomb interaction energy between the spectator electron and the two valence holes in the 2h-1e final state of the spectator decay. Thus, deexcitation spectra offer a unique insight into the dynamics of charge transfer and screening processes, a point which we will discuss in more detail in Sect. 6.5 and also open up the possibility to directly measure the electron correlation energies for a whole variety of selectable electronic states. Along those lines in the last section we have seen that the core hole screening is largely accomplished by charge transfer within the π electron states of the molecules considered. Additionally, we obtained direct experimental data for the change of the Coulomb interaction and screening energy with the localisation of the electronic wave functions. The direct experimental determination of these quantities is a unique application for DES, which can serve as a test for the most sophisticated theoretical calculations.

Another possibility to explore is the excitation of the core electron into different electronic states within the same molecule. So far we have only considered the strongest core-hole-excited states observed in the absorption spectra of the molecules discussed. A high-resolution absorption spectrum of N_2, taken for photon energies near the onset of the core-electron excitation and ionization, is exhibited in Fig. 6.12. In particular there is the N $1s \rightarrow 1\pi_g$ (π^*) excitation at a photon energy of 401 eV, resolved with respect to the individual vibrational subbands in the excited state. Between 406 eV and the N 1s ionization threshold at 409.9 eV transitions into various Rydberg final states are observed, again with vibrational resolution. Finally, the fine structure between 414 and 415 eV, above the ionization threshold, is commonly interpreted as due

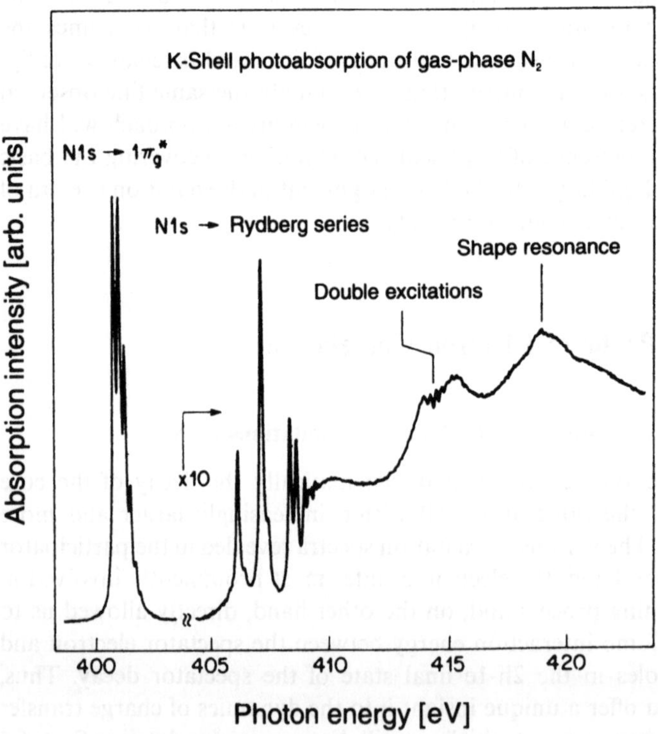

Fig. 6.12. High-resolution absorption spectrum of N_2 showing the onset of the N 1s excitations taken from [6.24]. The N 1s IP is located at 409.9 eV

to the excitation of a 2h-2e final state with the general configuration ($1s^{-1}$, $1\pi_u^{-1}$, $1\pi_g^2$). This state is superimposed on the N 1*s* continuum excitations.

The excitation cross section for the Rydberg transitions is typically one to two orders of magnitude smaller than the cross section for the excitation of the π^* transition. Thus, for intensity reasons it was not possible to study the decay processes of the Rydberg states in the molecules discussed here until very recently, when soft X-ray undulators with the appropriate high-resolution beam lines became available for this kind of spectroscopy [6.6]. In the following we will present the decay spectra of some of these core-hole excited states and discuss, at least qualitatively, the differences observed in the decay of these states.

In Fig. 6.13 deexcitation spectra of various core-hole excited states of the N_2 molecule are compared. Next to the 'standard' Auger spectrum (curve a) and the

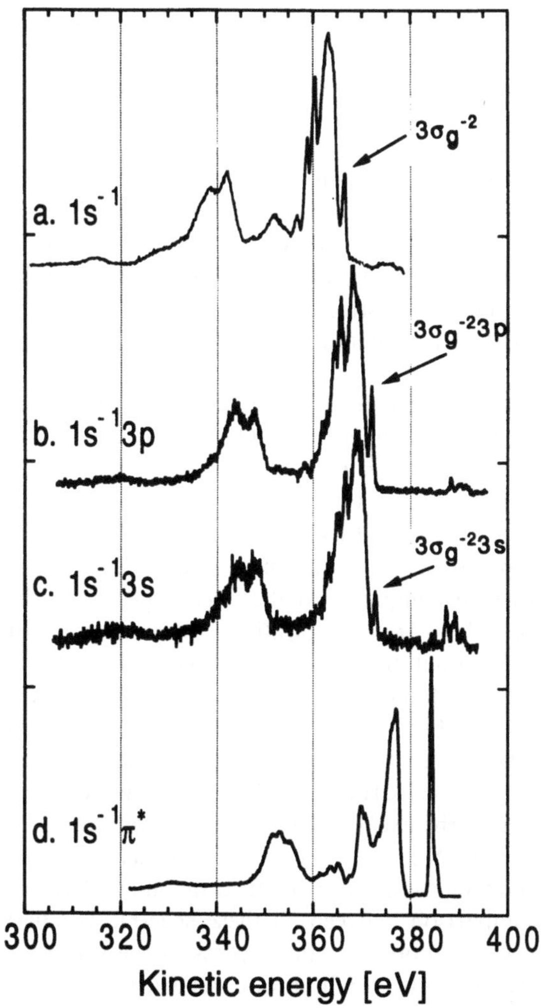

Fig. 6.13a–d. Electron emission spectra generated in the decay of various core excited states of the N_2 molecule. The top curve (**a**) shows the normal Auger decay following core electron ionization. Curves (**b**) and (**c**) show the decay of the N $1s \rightarrow 3p$ and of the N $1s \rightarrow 3s$ Rydberg state excitation respectively, and the bottom curve (**d**) shows the decay spectrum of the N $1s \rightarrow 1\pi_g$ (π^*) excitation

deexcitation spectrum of the N 1$s \to \pi^*$ transition (curve d) the deexcitation spectra of the 3p- (curve b) and the 3s-Rydberg states (curve c) are displayed. The Auger spectrum was taken using photon excitation well above the core ionization threshold. As a general impression the decay spectra of the Rydberg states resemble more closely the 'standard' Auger spectrum than the deexcitation spectrum of the π^* excitation. This similarity leads to an intuitive assignment of these spectra. In all the deexcitation spectra shown in Fig. 6.13, the most prominent group of lines is observed between 360 eV and 375 eV kinetic energy for the decay of the Rydberg states, and between 365 eV and 380 eV for the decay of the π^* resonance. These lines are due to spectator (2h-1e) final states. This assignment and observation confirms the screening model just discussed. As predicted, with increasing binding energy of the excited core electron the shift of the 2h-1e final states increases relative to the corresponding 2h final states populated in the Auger decay.

The spectator final states of the decay of the tightly bound π^* resonance are noticeably different from the spectator final states observed in the decay of the Rydberg states. The latter decay spectra resemble closely the Auger two-hole spectrum, only shifted by the Coulomb interaction with the spectator electron in the Rydberg orbital. In the decay of the π^* resonance on the other hand, the spectator electron clearly couples with its spin and angular momentum to the two holes. This produces a spectrum of lines not only shifted in energy, but also quite different in overall shape from the corresponding part of the Auger spectrum. This coincides with the intuitive notion that an electron occupying the $1\pi_g$ molecular orbital interacts much more strongly with the other electrons than an electron which is located in a rather diffuse, large radius Rydberg orbital.

From total energy considerations, as described in Sect. 6.1, the lines observed in the spectra at kinetic energies higher than 380 eV in curves (a–d) in Fig. 6.13 are all due to single-hole final states. Obviously the Auger spectrum (curve a) does not contain any of these lines. For this reason the actual curve shown in Fig. 6.13 does not even include this energy region. It is quite remarkable how much the relative intensity of the single hole states changes from the decay of the π^* resonance compared to the decay spectra of the Rydberg states. The decay of the N 1$s \to 1\pi_g$ configuration into participator or spectator states is approximately equal in probability, taking the relative population of the molecular orbitals into account. However, as the electron is excited into more diffuse Rydberg orbitals, the probability that the excited electron takes part in the decay of the core hole actually decreases rapidly. This decrease in the intensity of the single-hole (participator) final states immediately indicates a much weaker coupling of the excited electron to the core hole.

The assignment of the decay of the N 1$s \to 1\pi_g$ excitation has been discussed in detail earlier [6.12]. In the following we want to concentrate on the decay of the Rydberg states and on a comparison between all these spectra. The first point to be discussed is the question about the effectiveness of the spectator electron in screening the two-hole final state configuration. As discussed above, the screening energy can be numerically determined, by computing the apparent 'binding energy' of a given final state electronic configuration. This 'binding energy' is obtained by taking the difference between the energy deposited into the molecular system by the primary excitation through the absorption of the soft X-ray photon and the energy removed in the form of kinetic energy by the electron emitted in the decay of the core hole excited state.

In Fig. 6.14 the screening energy contribution of the spectator electron to the highest two-hole final state, the $(3\sigma_g)^{-2}$-1e state, is plotted versus the binding energy of the spectator electron in the presence of the core hole. The assignment of the $(3\sigma_g)^{-2}$-1e spectator state is straightforward for the Auger spectrum and the deexcitation spectra of the Rydberg states. In these three curves this state corresponds to the sharp peak at the high-kinetic-energy side of the dominating group of lines in the spectra. Unfortunately the relative intensity of this spectral feature decreases and in the case of the deexcitation of the $1s \rightarrow 1\pi_g$ (π^*) core excited state no sharp peak is discernable. Due to the angular momentum coupling in the spectator final state the relative energy of this 2h-1e final state has also changed in this spectrum to the extent that the feature has either moved closer to the main group of lines or even further up in kinetic energy near the single-hole states. For the sake of this evaluation, we have taken the middle of the onset of the spectator state emission as the location of the $(3\sigma_g)^{-2}$-$1\pi_g$ spectator final state.

In Fig. 6.14 the screening energy of the spectator electron in the final state of the deexcitation process is plotted versus the binding energy of the spectator electron (IP*). As discussed above, IP* is the difference of the excitation energy of the core to bound state excitation and the ionization energy of the N 1s

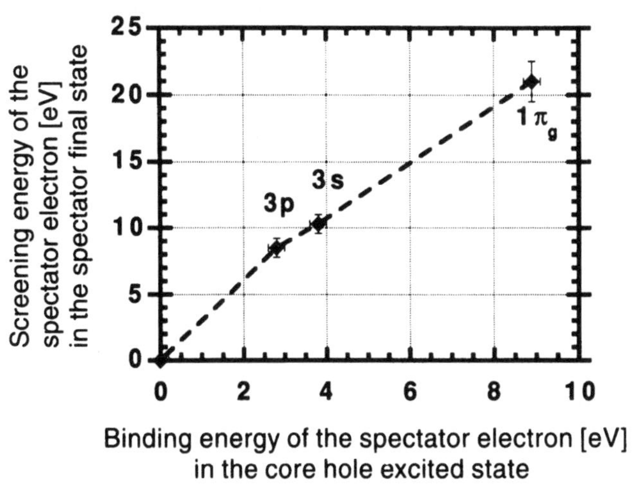

Fig. 6.14. Analysis of the screening energy contribution of the spectator electron to the final state of the core-hole decay as a function of the binding energy of the spectator electron in the core-excited state

electron (409.9 eV). The screening energy of the two-hole final state is computed from the difference between the binding energy of the (2h-1e) spectator state and the binding energy of the corresponding two-hole final state of the Auger decay. All the numerical values used to derive the plot in Fig. 6.14 are listed in Table 6.3.

From Fig. 6.14 it becomes obvious that the screening ability of the spectator electron is not a linear function of the binding energy of the spectator electron in the core-hole excited state. Such a nonlinear behaviour has also been predicted by calculations on N-atoms [6.9, 14], but these data are the first experimental verification of such a behaviour. The explanation is obviously the difference in localization of the spectator electron wave function among the $1\pi_g$, the $3s$-, and the $3p$-Rydberg orbital. This causes differences in the interaction energies in the core-hole excited state as well as in the 2h-1e spectator final state. Including higher, p-symmetry Rydberg orbitals in this compilation should probably lead to a more linear behaviour. At least this should be true for spectator electron states of the same principal symmetry.

Comparing the deexcitation spectra of the two Rydberg states, the spectator states seem to have a slightly different line shape. For example, the $3s$ deexcitation spectrum seems less well resolved and, in particular, the relative intensity of the $(3\sigma_g)^{-2}3s^1$ spectator final state is much smaller than the corresponding peak in the $3p$ Rydberg state deexcitation. This could be due to shake-up during the deexcitation process, where the spectator electron is 'promoted' into a different, higher Rydberg orbital. Shake-up in the decay of a core-hole-excited state has been observed for atoms [6.15, 16]. However, the statistics in the spectra shown in Fig. 6.13 is insufficient to unambiguously demonstrate the presence of shake-up in the Rydberg DES of N_2. In the next subsection (6.3.1) the case of Rydberg state excitation very close to the ionization threshold will be discussed using CO_2 as an example and there, shake-up will play an important role.

The triangular shaped structure in the Auger spectrum at kinetic energies between 346 and 355 eV is interpreted [6.3, 7] as due to the Auger decay of shake-up states of the primary core ionized configuration. This interpretation seems to be confirmed by the fact that a corresponding structure is missing from

Table 6.3. Screening and binding energy of the spectator electron in N_2

State	excitation energy [eV]	$(3s_g)^{-2}$-1e state kinetic energy [eV]	$(3s_g)^{-2}$-1e binding energy [eV]	screening energy [eV]
$1s \rightarrow 1\pi_g$	401	378.6	22.4	21.0
$1s \rightarrow 3s$ Rydberg	406.15	373.1	33.1	10.3
$1s \rightarrow 3p$ Rydberg	407.1	372.2	34.9	8.5
$(1s)^{-1}$ Auger	409.9	366.5	43.4	–

the deexcitation spectra of the Rydberg states. Incidentally, the fact that a related structure is seen in the deexcitation of the $1s \rightarrow 1\pi_g$ excited state is not contradictory. The main shake-up configurations involve an 'excited' electron in the $1\pi_g$ orbital and thus, from this starting configuration a scheme can be thought out that accesses similar final states as in the Auger decay of the core-hole shake-up states.

6.3.2 Threshold Behaviour in CO_2: 'There is No Threshold at the XPS-Threshold'

In the previous subsection we have dealt with the differences between the decay of neutral core-hole states where the core electron is excited into molecular states (π^* transitions) and into the more diffuse Rydberg states. Now the question arises, how the deexcitation spectra develop as the excitation energy crosses the ionization threshold, often, also called the 'XPS-threshold'. This threshold is defined as the minimum energy required to excite a core electron into continuum electronic states and thus, is a quantity determined and tabulated for most atomic or molecular species using data from XPS photo-electron spectroscopy. Since the creation of a core ionized state is the pre-requisite for the Auger decay, the XPS threshold is also commonly regarded as the onset for the Auger process.

As we have seen already, at energies slightly below the XPS-threshold, transitions into unoccupied molecular orbitals and into Rydberg states are generally observed in the absorption spectra of any molecular system. These states also involve the creation of a core hole, but the corresponding electron remains bound within the system. As discussed in detail above, the electronic decay of these states, the deexcitation process, is governed by the same matrix elements as the Auger decay. The only difference being that, in general, the transition occurs between a neutral state and a singly-ionized state.

As we have just seen for N_2, the closer the energy of the core-hole state is to the 'XPS-threshold', the closer the similarity between the deexcitation and the Auger spectrum. Moreover, from studies of atomic systems it is well known that the Auger spectrum changes as the excitation energy approaches the XPS-threshold from higher energies [6.17, 18]. These changes are attributed to PCI effects. The physics of 'PCI' entails that the emission of an Auger electron causes the Coulomb potential of the ion to change from that of a singly charged species to a doubly charged one. This essentially means a renormalization of the energy of the outgoing photoelectron, which finds itself to travel in a more attractive potential. If the kinetic energy of the outgoing photoelectron is too low, it might find itself captured in a high Rydberg state. Thus the final state of the whole process, excitation and decay, is such that the core electron, which initially was ionized into the continuum, actually remains bound within the atomic or molecular system. Total energy conservation requires that the Auger electrons then have a slightly higher kinetic energy. Additionally, a change in line shape is observed [6.19, 20]. The overall situation in PCI is more complicated than in DES since in calculating and analyzing the PCI effects one has to take into account that the experiments reflect the time-average of the capturing process taken over the lifetime of the core-hole excited state.

Based on these observations, however, the following concept can be formulated: there should be no *systematic* difference in the deexcitation spectra of core-hole states, excited with a photon energy slightly above or below the 'XPS-threshold'. The decay spectra of these states should be quite similar independent of whether the core electron actually was excited into a continuum state just above threshold or into a high Rydberg state. The reason behind this concept is that in either decay process there is a finite probability that the same final states are populated. On the one hand, this refers to the 2h-1e final states with an electron in one of the outer Rydberg orbitals. But, on the other hand, even if the initial excitation energy was below the XPS-threshold, there is also a certain probability that, due to spectator electron shake-off, the final state of the deexcitation process is a doubly ionized state and thus, two free electrons are created in continuum states. Because of this 'symmetry' of the final states of the core-hole decay for excitation below and above the 'XPS-threshold', core-hole deexcitation or auto-ionization should continuously develop into the PCI shifted Auger spectrum and then into the Auger spectrum itself, as the excitation energy is varied across the XPS-threshold [6.21]. Thus, the observation expected for these core-hole decay spectra could be phrased as: "No threshold at the XPS threshold".

We have tested this concept by studying the core hole decay of a C 1*s* vacancy in CO_2 [6.22]. The spectra were excited with synchrotron radiation, which was continuously tuned through the XPS-threshold of 297.7 eV [6.23]. Again, experiments of this kind have only recently become possible through the use of undulator radiation, which offers both the resolution and intensity required for these kinds of studies [6.6]. Figure 6.15 shows an absorption (electron yield) spectrum of CO_2, taken in the region of the onset of the C 1*s* electron excitations. This spectrum was taken at a photon-energy resolution of 210 ± 20 meV, corresponding to the photon-resolution settings used in the

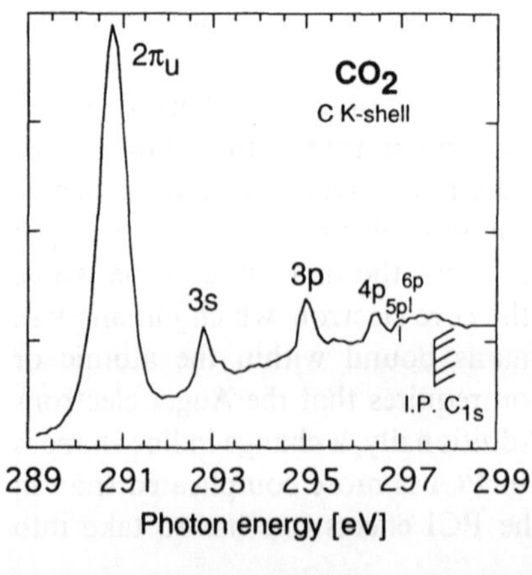

Fig. 6.15. Absorption (electron yield) spectrum of CO_2, taken near the onset of the C 1*s* core electron excitations. The ionization threshold (IP) is marked according to the value given in [6.22]

subsequent core hole decay studies. The discrete transitions observed in this spectrum are due to excitations of a C 1*s* electron into the unoccupied $2\pi_u$ molecular orbital at 290.7 eV, the 3*s* Rydberg orbital at 292.8 eV, the 3*p* Rydberg state at 295.0 eV, the transition into the 4*p* Rydberg level at 296.4 eV, and into 5*p*/6*p* at 297.1 eV. The higher Rydberg states are unresolved and overlap largely because of lifetime broadening [6.24]. The substructure observed in the lower Rydberg transitions indicates a small contribution of vibrational substates in these electronic states. The vibrational modes excited correspond to both the symmetric stretch mode as well as the bending mode of the molecule.

In the absorption spectrum the transition between excitation of the core electron into high Rydberg states or into the continuum is not distinguishable. In particular, there is no discrete structure observed at the ionization threshold. In order to test the concept of the continuous evolution of the core-hole deexcitation spectra for these core-hole excited states, we have taken deexcitation spectra for a series of excitation energies starting with the excitation of the 5*p*/6*p* C 1*s* Rydberg state through the continuum threshold at 297.7 eV up to an excitation energy of 339 eV. The latter spectrum should essentially reproduce the limit of the Auger spectrum (Fig. 6.16).

The core hole deexcitation spectra (Fig. 6.16) are characterized by three groups of lines. The lines with the highest kinetic energies around 275 to 290 eV are due to direct valence band photoemission enhanced by participator decay processes into the same single-hole states. For obvious reasons these lines are missing in the 'Auger' spectrum taken at an excitation energy of 339 eV. The next group of lines with a kinetic energy between 245 and 260 eV contain the strongest lines of the decay spectrum. A third group of lines is observed at kinetic energies between 220 eV and 240 eV.

Apart from the expected differences in the participator decay, the most striking feature of the spectra (Fig. 6.16) is the general similarity of all of these curves. The major two groups of lines, which are attributed to the Auger and spectator deexcitation decay, respectively, are nearly identical with respect to their general appearance. These lines only exhibit slight shifts in energy, as the excitation energy is varied. Especially noteworthy is the fact that these changes are continuous through the XPS threshold. Even from a careful inspection of these spectra the existence and exact location of the XPS-threshold could not be derived.

This result is emphasized by the high-resolution spectra of the central group of lines taken at a corresponding series of excitation energies (Fig. 6.17). From these spectra it becomes obvious that the shift in energy coincides with a change in lineshape of the individual lines. These effects have been predicted in the PCI framework, which was formulated for excitation into continuum states in the vicinity *above* the ionization threshold. The data in Fig. 6.17 demonstrate that the development of the spectra does not stop at the XPS-threshold. Rather the line shape and energy position of the lines evolves steadily as the photon energy is lowered through the XPS-threshold and only bound states are excited in the initial core-hole creation process.

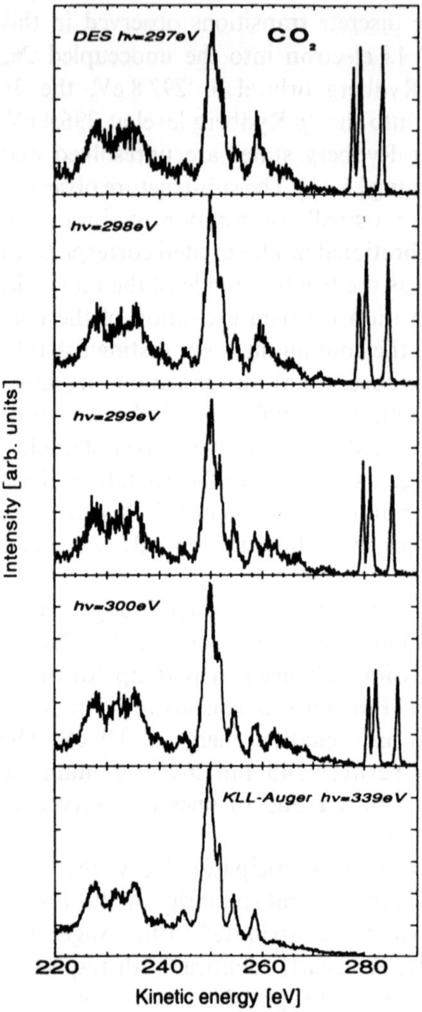

Fig. 6.16. Decay spectra observed in the decay of a C 1*s* core hole excited state at various excitation energies in the vicinity of the C 1*s* 'XPS-threshold' in CO_2. The different photon energies are given next to each curve. The 'Auger' spectrum is included in the bottom panel

Intrinsically the core-hole decay lines, in the absence of additional vibrational sidebands, have a symmetric Lorentzian line shape. Taking the PCI effect into account, in a simplistic description, the lines are predicted to develop an asymmetric tail towards larger kinetic energies. The asymmetry increases as the excitation energy is lowered towards the ionization threshold. Simultaneously, the shift in spectral weight causes the center of gravity of the line to shift also to larger kinetic energy values. A more quantitative description and various line shape formulas have been given in the literature [6.19, 20].

The change in the line shape and thus the mean energy of the lines is most clearly realized by inspecting the high-energy edge of the B4 peak near 251.6 eV kinetic energy [6.25]. It also is readily observed in the lineshape of the B2 peak at 254.5 eV. Incidentally, the B1 peak is obscured by inner valence many-body

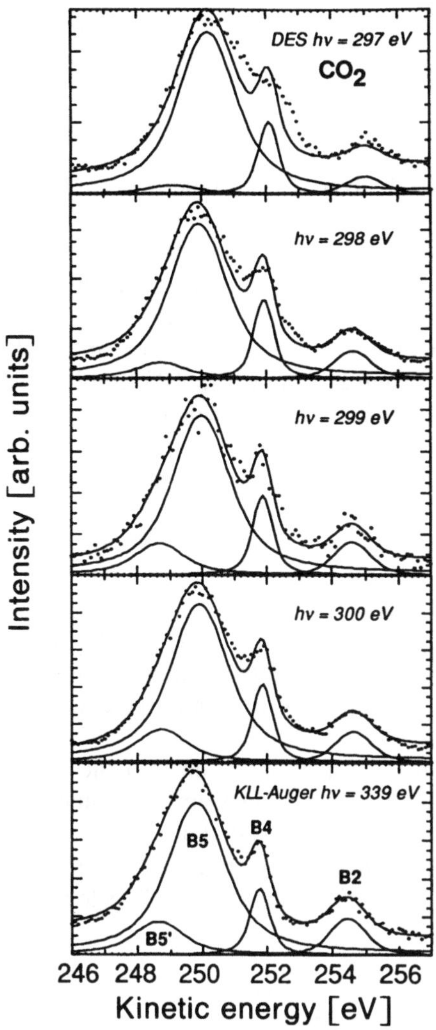

Fig. 6.17. High-resolution spectra of the major lines in the deexcitation spectra of CO_2. The original spectra are shown as points. In order to emphasize the deviations from the Auger spectrum, the components resembling shifted lines of the Auger spectrum are included in each curve. The sum of the four individual components is given by the smooth line running through the data points. (see also text and table 6.4)

final states of the participator decay at some of the photon energies chosen. Therefore the discussion here is limited to the B2, B4, and B5 peaks of the Auger spectrum in CO_2 [6.25]. A close inspection of the spectra in Fig. 6.17 clearly reveals the asymmetry of the lines to increase with lower photon energy, whereby the high-resolution deexcitation spectra of the Rydberg states are found to exhibit the broadest lines with the least distinguishable features.

One has to add here that the performance of the electron spectrometer used for these experiments was tested using the Ar 2*p* Auger lines. There only a negligible distortion was found [6.22]. Moreover, the CO_2 Auger spectrum shown in the bottom panels of Figs. 6.16 and 6.17 also exhibits perfectly symmetric line shapes. Thus the line shapes observed in the other spectra are not due to artefacts introduced by the electron spectrometer.

In an attempt to quantify these observations, we have obtained an analytical representation of the Auger spectrum taken after excitation with 339 eV photons using a fit routine [6.22]. A reasonable representation for this spectrum is given by four lines having a Voigt profile with different widths. This is in agreement with the interpretation of the Auger spectrum as given by *H. Ågren* [6.7]. The assignment of the states labelled B2 to B5′ is reproduced in Table 6.4.

Subsequently, we have taken this analytical description of the Auger spectrum and superimposed it onto the spectra taken in the near threshold region. The only variational parameters were the peak energy and intensity of the four components. The width and line shape were kept constant as obtained from the fit of the Auger spectrum. Systematic deviations in line shape and peak position are clearly recognizable from this comparison. As the XPS threshold is passed from higher energies, the peaks shift to higher kinetic energy and additional components appear as tails on the high kinetic energy side of the peaks. The additional emission increases in strength with lower excitation energy. The valley between peak B4 and B5 already starts to fill in at an excitation energy of 300 eV, but at 298 eV and 297 eV this filling is even more pronounced. For a more quantitative numerical comparison the energy values for the fit components (Fig. 6.17) are given in Table 6.4. Accordingly these spectra constitute the manifestation of PCI in molecular Auger spectra, even though this fit does not represent a rigorous PCI analysis. Such an analysis is very complicated for molecules because of the additional line shape contributions from vibrational substates.

These data show the transition from the Auger spectrum into the 2h-1e spectator deexcitation spectrum. The major observation from the numerical analysis and the spectra depicted in Fig. 6.17, however, is that all the line shape and energy changes observed in these core-hole decay spectra develop continuously from excitation energies well above threshold to well below the

Table 6.4. Peak assignments and shifts of CO_2 in the near-threshold region

Excitation Energy [eV]	Energy of final state [eV][a] $3\sigma_u^{-1}\pi_g^{-1}$ or $4\sigma_g^{-1}1\pi_g^{-1}$ (B2)	$1\pi_u^{-1}1\pi_g^{-1}$ (B4)	$1\pi_u^{-2}$ (B5)	$3\sigma_u^{-1}1\pi_u^{-1}$ or $4\sigma_g^{-1}1\pi_u^{-1}$ (B5′)
KLL[a]	254.2	251.1	249.6	
339	254.4	251.8	249.8	248.6
300	254.6	251.8	249.9	248.7
299	254.6	251.9	250.0	248.7
298	254.7	251.9	249.9	248.8
297[b]	255.0	252.1	250.2	249.1

[a] Notation and energy of KLL-Auger [6.7]
[b] 5*p*/6*p* Rydberg

XPS-threshold. Thus from the core-hole decay spectra the actual value of the XPS-threshold cannot be derived. Accordingly, a theoretical framework representing a unified picture of core hole deexcitation and Auger decay [6.21], including shake-up, shake-down, and shake-off of the spectator electron, has to be used for the proper description of the Auger and deexcitation spectra in the near threshold region.

6.3.3 Identification of the Primary Excitation Through the Decay Spectra (O_2)

In the third and last part of this section we want to discuss the application of DES in order to identify the nature of the primary core excitation. In Sect. 6.3.1 we have already seen that the deexcitation spectra are quite sensitive with respect to the energetic position of the lines and also with respect to the coupling of the spectator electron in the 2h-1e final state. The multiplet coupling in N_2 is clearly different for the decay of the Rydberg state excitations in comparison with the π^* transition. This effect can potentially be useful to help in the assignment of core excited states or to distinguish between excitations, which are energetically degenerate and/or not clearly separable in the absorption spectrum.

In general, at least for the diatomic molecules, the features in the absorption spectra are rather well understood. However, even for these quite simple molecular systems one example exists, where the assignment of the absorption spectrum is still not totally clear and very much under debate [6.26–31]. This is the location and multiplet splitting of the σ_u^* shape resonance in O_2. Whereas for CO and N_2 this resonance is degenerate with the ionization continuum, for O_2 the energy of the σ_u^* resonance, in the presence of an O 1s core hole, has dropped below the ionization threshold into a region where also Rydberg excitations are observed [6.26]. This is one of the reasons behind the rather unusual shape of the absorption spectrum of O_2 in the region between 538 and 548 eV photon energy shown enlarged in the inset of Fig. 6.18 [6.6, 29].

O_2 in the ground state ($^3\Sigma_g^-$) is paramagnetic with two unpaired electrons occupying the twofold degenerate $1\pi_g$ molecular orbital. Core ionization leads to either a doublet or a quartet state, with a difference of 1.1 eV in the ionization energy (543.1 eV for the quartet state $^4\Sigma^-$ and 544.2 eV for the doublet state $^2\Sigma^-$ [6.24, 26, 30]). Accordingly, there is a double series of Rydberg transitions converging to either one of these thresholds. Since the coupling in the outer Rydberg states is rather weak, at least in the limit of the ionization threshold, the 'spin-splitting' of the Rydberg series should be the same as the observed difference in the core hole IP. Incidentally, the π^* transition does not exhibit such a splitting, since the unoccupied $1\pi_g$ states have only one spin orientation and the excited core electron remains within the system. Therefore the π^* excited state has to be a triplet state ($^3\Pi_u$).

The strong double structure in the absorption spectrum with maxima near 541 and 543.5 eV is suggestive of a double σ_u^* shape resonance excitation upon which the two Rydberg series are superimposed. The implications of this

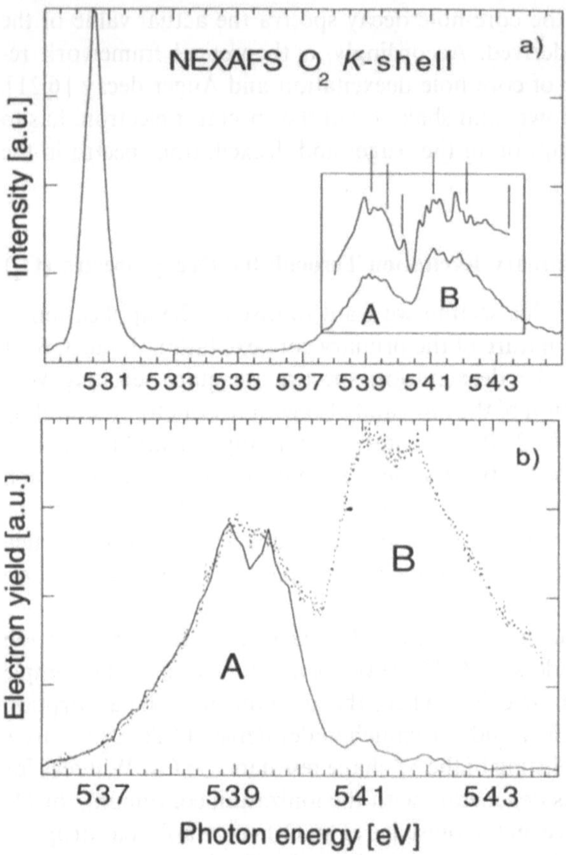

Fig. 6.18a, b. Absorption spectrum (partial electron yield) of O_2 near the O–*K*-edge. The top panel (**a**) shows the total electron-yield spectrum and in the inset a high-resolution scan across the region of the Rydberg and σ_u^* shape resonance excited states. In panel (**b**) is the total electron yield spectrum (dots) compared with a partial electron yield spectrum taken with the analyzer set to detect electrons having a kinetic energy of 510 eV only (solid curve)

intuitive assignment are that the overall structure and intensity reflects the σ_u^* excitations, whereas the finestructure details are dominated by the weaker Rydberg transitions. Since the Rydberg as well as the σ_u^* excitations are expected to exhibit vibrational fine structure, the resulting superposition becomes quite complicated and entangled. Even taking this spectrum at the currently highest possible resolution, which is substantially better than the intrinsic lifetime broadening of the structures, does not add much to the interpretation of the spectrum [6.6, 24, 30]. This situation is quite unsatisfactory, because if the intuitive assignment is correct, then another problem is created by the quite large apparent spin-splitting of the σ_u^* excitation of 2.5 eV.

In order to help with the clarification of the problem, we have taken deexcitation spectra for a whole series of excitation energies between 539 and 545 eV. These spectra are exhibited in Fig. 6.19. Contrary to what we have observed previously for CO, N_2, and CO_2, these spectra are very complex. The basic Auger spectrum of O_2 is characterized by three major features between 490 and 502 eV kinetic energy and another clear peak at 476 eV kinetic energy. For the deexcitation spectra, however, the number of the observed structures and the

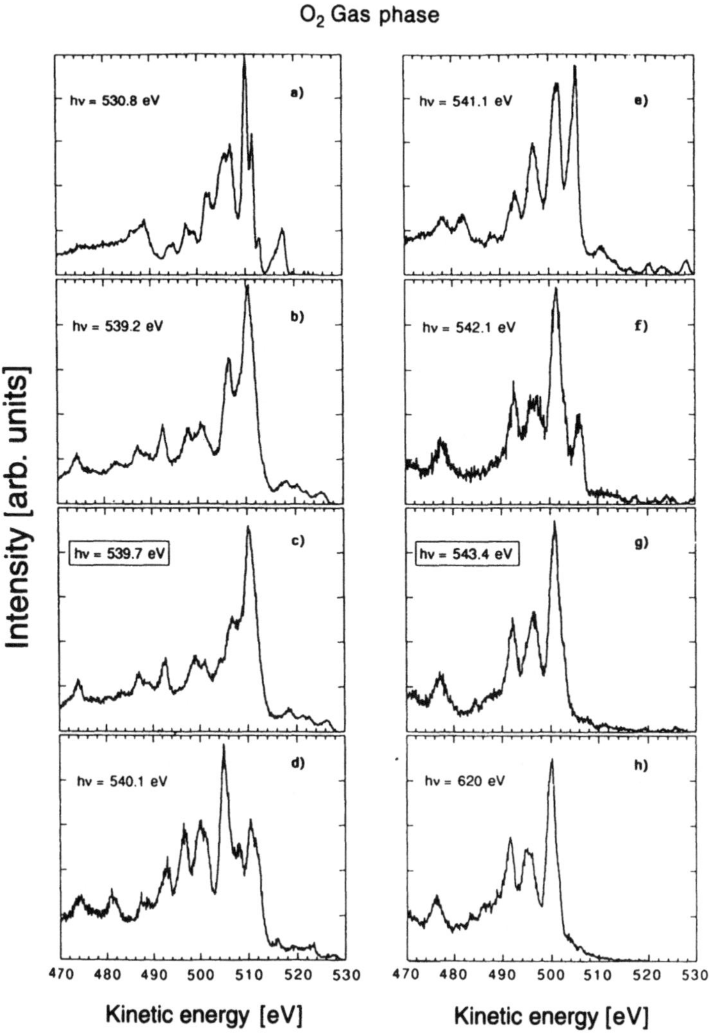

Fig. 6.19. Deexcitation spectra of O_2 taken at a series of photon energies in the range of the energetically degenerate Rydberg and σ_u^* excitations

overall shape of the spectra all vary quite remarkably upon only small changes of the excitation energy. For example, the deexcitation spectrum taken at 540.1 eV excitation energy reveals six major structures in the kinetic energy interval between 490 eV and 515 eV. This behaviour is totally unpredicted and different from the behaviour observed for the other small molecules. The N_2 spectra (Fig. 6.13) are quite representative of the deexcitation of the other molecules.

Our attempt at interpreting the DES curves of O_2 is based on the assumption that we actually observe in some of the more complex spectra the decay processes of at least two different core excited states excited simultaneously. In order to follow through with this kind of interpretation we have to identify the principal components of such a spectrum. This means we should find some spectra representative of only one type of deexcitation process. These clearly have to be the spectra, where a smaller number of structures is observed. Incidentally, the region between kinetic energies of 470 eV and 484 eV is quite indicative of the number of components present in the deexcitation spectra. The Auger spectrum exhibits only one triangular shaped structure in this region interpreted as a superposition of several configurations of 3-hole-1-electron states. Of the deexcitation spectra shown in Fig. 6.19, some exhibit one structure and some two in this energy region. Our contention now is that we take the appearance of more than one triangular shaped peak in this energy region as an indication that the corresponding deexcitation spectrum consists of a superposition of spectra resulting from the decay of different, but energetically (almost) degenerate core excited states.

According to this intuitive viewpoint, the two spectra shown in the panels (c) and (g) immediately above the Auger spectrum (panel h) in Fig. 6.19, taken at excitation energies of 539.7 and 543.4 eV, are characteristic of one dominant type of decay process. In the spectrum taken at 542.1 eV excitation energy (f) already a second minor component appears. The spectra shown in panels (d) and (e), on the other hand, clearly are indicative of at least two different types of decay processes, whereas the spectra displayed in panels (a) and (b) again are dominated by one major decay process.

In order to cut a long story short [6.29], the deexcitation spectrum (Fig. 6.19g) shows the decay of a highly-excited Rydberg state. This spectrum is very close in overall shape and appearance to the Auger spectrum, however, it has already shifted by about 1 eV to higher kinetic energy. Since the transition from the Auger spectrum to the Rydberg deexcitation spectrum is quite fluent, as just discussed, this spectrum represents the limiting case where one might still be inclined to refer to it as the decay spectrum of a Rydberg state rather than an Auger spectrum shifted by PCI effects.

The spectrum in panel (a) shows the decay of the $1s \rightarrow 1\pi_g$ (π^*) excited state and, in particular, the one in panel (c) of Fig. 6.19 exhibits the decay of the σ_u shape resonance excited state. We come to this assignment since the overall shape is quite different from the decay of a Rydberg state. This spectrum indeed resembles much more closely the decay spectrum of the π^* excitation in O_2, if one takes into account that the intensity of the participator states is much weaker in the decay spectrum of the σ_u^* shape resonance excited state. The difference in overall shape, compared to the decay of the Rydberg states, also indicates that the coupling between the valence electrons is the dominant coupling mechanism in the σ_u shape resonance excited state. Thus, the multiplet splitting of the σ_u^* shape resonance excited state has to be expected to be different than that from the Rydberg or core ionized states. As a refinement of this interpretation we have to add that the spectrum in panel (b) also has a small component from the decay of a lower Rydberg state, presumably the 3s state.

The evidence for this interpretation originates from the sharp, pointed structure at a kinetic energy of 506 eV in this spectrum and other smaller differences compared to the spectrum shown in panel (c). This leads us to assume that the spectrum in panel (c) has fewer contributions from the decay of Rydberg excited core hole states.

The assumption, introduced above, that the more complex deexcitation spectra of O_2 can be described as a superposition of at least two of the just identified building blocks is graphically tested in Fig. 6.20. The two examples chosen for this demonstration are the deexcitation spectrum taken at an excitation energy of 540.1 eV (bottom panel of Fig. 6.20 and panel d of Fig. 6.19)

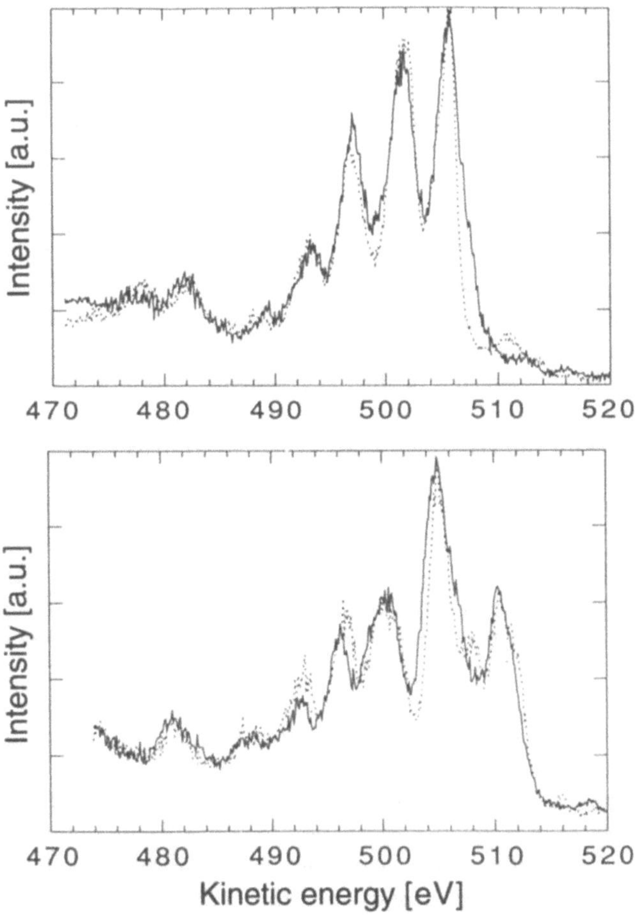

Fig. 6.20. O_2 deexcitation spectra (dots) taken at 541.1 eV (top panel) and 540.1 eV (bottom panel). The solid curves in each panel reflect the attempt of a two component reconstruction of these spectra. For a more detailed discussion see text.

and at an excitation energy of 541.1 eV (top panel of Fig. 6.20 and panel e of Fig. 6.19). The deexcitation spectrum taken at an excitation energy of 540.1 eV, is readily composed from the deexcitation spectrum of the σ_u shape resonance excitation and from a Rydberg state decay spectrum, the leading structure of which is positioned at a kinetic energy of 505 eV. The deexcitation spectrum taken at an excitation energy of 541.1 eV, on the other hand, is a composition of two Rydberg state decay spectra the strongest structures of which are located at 506 eV and 502 eV. The leading edge of this spectrum is much sharper than most of the other spectra, since it results from the decay of a 'pure' $^4\Sigma^-$ core hole state [6.29].

The question why two energetically (almost) degenerate Rydberg states produce decay spectra which are shifted by about 4 eV on the kinetic energy scale can at least qualitatively be answered by the diagram depicted in Fig. 6.14 for N_2. The two Rydberg states must belong to the two different series converging towards the $^4\Sigma^-$ and $^2\Sigma^-$ core ionized states. Thus, the binding energy IP* of the Rydberg electron differs by 1.1 eV for the two energetically almost degenerate Rydberg states. From the diagram in Fig. 6.14 for N_2 one can estimate that a difference in IP* of 1.1 eV of the Rydberg electron, results in a screening energy difference in excess of 3 eV in N_2. The situation for O_2 should be quite similar [6.14].

Having assigned the deexcitation spectra of O_2 and identified the characteristic fingerprint of the decay of the σ_u^* shape resonance excitation, we can now come back to the original question of the assignment of the absorption spectrum. The energy of the leading feature of the deexcitation spectrum of the shape resonance excitation is at 510.5 eV. None of the other deexcitation spectra, apart from the decay spectrum of the $1\pi_g$ resonance, produces significant emission intensity at this particular kinetic energy. Taking a partial electron yield spectrum with the electron spectrometer set to detect electrons emitted with a kinetic energy of 510 eV should filter out the structure in the core absorption spectrum, which is due to the σ_u^* shape resonance excitation. Such a yield spectrum is represented as the full curve in the bottom panel of Fig. 6.18. From this curve the σ_u^* shape resonance is found to contribute only to the lower energy feature of the absorption spectrum, whereas the high-energy feature of the absorption spectrum results from the superposition of the two Rydberg series.

This procedure also reveals that the shape resonance excitation thus found exhibits an energy splitting of 'only' 0.6 eV. The difference in the magnitude of this 'spin-splitting', compared to the sign slitting of the core ionized and Rydberg states (1.1 eV), results from the different coupling. As we have seen from the decay spectrum, the coupling in the σ_u^* shape resonance excited state is dominated by the coupling between the valence electrons. Since the state is created by an optical dipole transition and since the excited electron does not leave the system, both σ_u^* shape resonance excited states have overall $^3\Sigma_u$ symmetry and differ only in the spin direction of the excited core electron. The core to Rydberg transitions, on the contrary, are dominated by the doublet/quartet configurations of the core-hole coupling to the two unpaired $1\pi_g$ electrons.

The value of 0.6 eV for the 'spin-splitting' of the σ_u^* shape resonance excited state determined by our measurements [6.29] is smaller than previous estimates [6.27], but it is in much better agreement with up to date LDA calculations [6.28] and the values found for the splitting of the shape resonance in various oxides [6.31]. Moreover, absorption spectra (electron yield spectra) taken for solid condensed O_2 can be taken as a confirmation of our interpretation, since they show that the higher-energy structure of the absorption spectrum is strongly attenuated in the condensed phase. Such quenching should be expected for the Rydberg excitations but only to a much smaller extent for the excitation of the σ_u^* resonance, which is of intramolecular nature and therefore much less affected by the interactions in the solid phase. Accordingly, these authors also assigned only the lower-energy structure in the absorption spectrum to the σ_u^* shape resonance excited state and claim the same value (0.6 eV) for the exchange splitting [6.28], which is in perfect agreement with our results.

In this section we have tried to demonstrate the use of DES in the identification of different types of core hole excited states by their respective decay spectra. This in turn has led to a 'new' interpretation of the absorption spectrum of O_2. Thus the opportunity has been demonstrated to use DES as a 'fingerprint' technique in the assignment of even more complicated absorption spectra.

6.4 Dynamics of Core-Electron-Excited States

6.4.1 Coherence and Interference

The availability of high-resolution, high-brightness undulator beam lines has enabled us to perform the DES studies with vibrational resolution in both the excitation and decay channel [6.6, 32–35]. The latter kind of studies opens up the possibility to investigate the phenomenon of coherence in the excitation of the manifold of the vibrational wave functions in the intermediate, core excited state. In this framework the excitation and the decay of the core excited state are treated as a coherent process. If the lifetime of the core excited state is sufficiently short, the phase relationship of the coherently excited vibrational states is maintained. Thus interference is observed in the decay process leading to changes of the line shape of the transitions and to modulations of the population of the final-state vibrational sublevels. These experiments essentially probe the development of the wave function of the core-excited state on the time scale of the core-hole decay time, which is a few fs. The core-hole lifetime is an intrinsic atomic property, which is only slightly modified by chemical effects. However, in a study of these phenomena for core holes in different atoms such as C, N, or O, the effect of a variation of the time scale can be simulated. In this section, a systematic study of these quantum-mechanical interference effects are presented for N_2 and O_2.

Diatomic molecules are the ideal species for this kind of systematic study. These molecules have the symmetrical stretch mode as their only vibrational degree of freedom. Additionally, because of the low mass, the energy of the vibrational quanta of this stretch mode is sufficiently high to be readily resolved in photoelectron spectroscopy. Therefore these molecular systems serve as model systems where the effects of coherence and the resulting interference in electronic inner shell excitation and deexcitation can be studied.

A systematic study of this kind involves the selection of various subsets of vibronic eigenfunctions, within the same electronic state. This has previously not been possible, due to limited resolving power in the soft X-ray energy range. The vibrational substructure of these spectra is of particular interest, since it allows us to gain information on the equilibrium geometries and force constants of core-hole states or core-excited states.

Characteristic frequencies obtained for molecular vibrations of CO, N_2, and O_2 are in the range of 10^{13}–10^{14} s^{-1}. Compared to the total lifetime of inner-shell excitations in these molecules, which is determined by the lifetime of the core-hole state including the excited electron, the vibrational motions of the nuclei are slow. For bound state resonances the total lifetime is mostly determined by the Auger lifetime of the inner-shell hole and is typically in the range of 10^{-15}–10^{-14} s. This is less than the period of one vibrational cycle of the nuclei. In essence, the coherently excited nuclear wave functions develop until the core-hole decays and thus, the electronic decay process serves as a stopwatch for the nuclear motion on the fs time scale. However, since we have no control over the timing of individual decay process, the deexcitation spectra always reflect an integral over time of these processes, weighted by the actual probability in time for the decay to occur according to the lifetime of the excited state.

Nevertheless, these decay spectra allow for a stringent test of the quantum mechanical description of nuclear dynamics. Investigations of these effects are commonly performed using pump-probe arrangements with fs lasers. In our experiments, however, the inherent timescale, as given by the lifetime of a O, N, and C 1*s* core hole excited site, is less than 10 fs. This time scale is almost an order of magnitude shorter than the actual time resolution used in laser pump-probe experiments. The time resolution so far attained in laser pump-probe experiments is of the order of tens of fs, even though pulse widths shorter than 10 fs have been reported. In general, all the results which are discussed here are not only of interest in molecular physics, but are also of interest to the condensed matter community, since the work deals with some fundamental aspects of lifetime effects in relation to the electron-phonon coupling.

6.4.2 Theoretical Description of the Interference Profiles

The line shapes measured in molecular photoemission are largely determined by vibrational excitations. Within the framework of the Born-Oppenheimer approximation these vibrational progressions are described by the Franck-Condon (FC) factors determined from the overlap integral between the initial and the final vibrational states. This also explains the difference between the spectra resulting from a final electronic state which is reached in a direct process or via an intermediate state. If a long-lived intermediate state is involved, the process can be described as a decoupled two-step process. The first set of FC factors determines the population of the vibrational levels of the intermediate state. Subsequently these levels decay to the vibrational levels of the final electronic state according to the corresponding FC factors.

However, in cases where the depopulation time is on the same scale as a classical period of vibration, the classical simple two-step model is known to fail. This means that the vibrational intensity distribution can not be calculated in the same way as for stationary electronic states. This is due to the fact that the nuclear wave functions of the intermediate state are not uniquely determined on such a short time scale. Rather, this state can be described by a wave packet consisting of a superposition of coherently excited vibrational states, which evolves in time. Consequently, for any given final state the different pathways via coherently excited vibronic intermediate state wave functions give rise to significant constructive or destructive lifetime-vibrational interference, which has to be taken into account in the description of the complete process [6.36–40].

As a consequence, lifetime effects enter absorption-type spectra and emission-type spectra in different ways. Absorption type spectra are those, where the short-lived electronic state is the final state, e.g. X-ray absorption or core-electron photoemission spectra. These consist of incoherent superpositions of Lorentzian lines, whereby the Lorentzians correspond to the individual vibrational levels of the short-lived state. In contrast, emission type spectra are those, where the short-lived electronic state is the initial state such as in Auger, X-ray emission, or deexcitation spectra. Therefore, the contributions from the different, coherently populated vibrational levels of the short-lived intermediate state interfere.

The condition which must be fulfilled for the observation of nuclear-wave-function coherence in core-hole decay is that the lifetime width of the short-living intermediate state has to be comparable to the vibrational spacing. This condition is fulfilled for O $1s$ core-hole states ($\Gamma \approx 0.18$ eV) of many oxygen containing molecules, whereas for N $1s$ ($\Gamma \approx 0.13$ eV) core-hole states, the effects observed are expected to be less dramatic because of the longer lifetime.

For excitations with a large band pass we can use the formula given by conventional interference theory to describe the intensity at a given electron emission energy $I(E_{\mathrm{emi}})$ as [6.36–40]

$$I(E_{\mathrm{emi}}) \propto \sum_{\mathrm{f}} \left| \sum_{\mathrm{n}} \frac{\langle f|n\rangle \langle n|0\rangle}{E_{\mathrm{emi}} - (E_{\mathrm{n}} - E_{\mathrm{f}}) + \dfrac{\mathrm{i}\Gamma}{2}} \right|^2 \tag{6.8}$$

where $|0\rangle$, $|n\rangle$ and $|f\rangle$ refers to the vibrational eigenfunctions of the ground, intermediate and final states, respectively. E_n and E_f are the energies of the intermediate and the final states, and Γ is the lifetime width of the intermediate state.

The excitation by photons with a spectral distribution, which is narrow compared to the total width of the excited state, limits the total energy available for the process. The excitation of the intermediate core-hole states, however, is not limited, since at the time of the transition these states are not yet defined. We have tested this concept in the experiments described in the following by operating at a the photon resolution which is substantially better than the width of the vibrational progression in the absorption spectrum. Therefore, for each incoming photon the interfering pathways have to be calculated according to the overlap of the vibrational eigenfunctions of the initial, intermediate, and final state. Subsequently, these events are incoherently summed and weighted by the distribution probability of the photons described by the monochromator transmission function. Conveniently, Eq. (6.8) is rewritten so that the probability of reaching the various final states depending on the excitation energy is taken into account properly. Neglecting the width of the final state, the relation between emission and excitation energy can be written as [6.34, 35]

$$E_{emi} = E_{exc} - (E_f - E_0), \tag{6.9}$$

where E_0 is the vibrational energy of the ground state. This equation is a consequence of the fact that the lifetime width of the intermediate state is irrelevant with respect to the width of the emission line when the excitation energy is well-defined. Combining (6.8 and 9), and introducing the monochromator function $M(E_{exc})$, we get for the population probability of the final state $|f\rangle$

$$I_f(E_{exc}) \propto M(E_{exc}) \left| \sum_n \frac{\langle f|n\rangle \langle n|0\rangle}{E_{exc} - (E_n - E_0) + \frac{i\Gamma}{2}} \right|^2 . \tag{6.10}$$

For $M(E_{exc})$ we assume a Gaussian distribution of the incoming photons. We calculate the $I_f(E_{exc})$ according to Eq. (6.10) for each final state and add the spectra on the same emission energy scale, given by Eq. (6.9)

$$I(E_{emi}) \propto \sum_f I_f[E_{exc} - (E_f - E_0)] . \tag{6.11}$$

This just described FC-interference theory is based on *stationary* eigenfunctions. In general, a *time-dependent* approach to the dynamics of the intramolecular motion describes the evolution in terms of the propagation of a wave packet. The sum of the coherent, but stationary intermediate wave functions used in the previous paragraph is replaced by a non-stationary propagating wave packet. Such a propagating wave-packet model describes quite well the intensity oscillations in femtosecond multiphoton-ionization spectra. In our experiment the decay occurs even before one single oscillation is completed.

Additionally, we can vary the initial conditions for the wave packet by varying the excitation energy. This changes the speed of motion and the temporal development of the wave packet, causing the different interference patterns we observe. The time-dependent description corresponds to a wave packet which is modified by an exponential damping factor due to the finite lifetime. Therefore a decay spectrum can be seen as a weighted average of snap-shots of the wave packet at different stages of its development. This wave-packet description has been discussed more extensively in the recent work of *Tarantelli* and *Cederbaum* [6.40].

Using the X1B undulator beam line at the NSLS neutral core-hole excited states in diatomic molecules like N_2 and O_2 can be prepared with vibrational selectivity [6.6]. For these molecules we have studied the electronic participator decay as a function of the excitation energy while varying the excitation energy within the bandwidth of the core-excited intermediate state. This state is excited in a dipole transition between the $1\sigma_u$ and the empty (N_2), or in the case of O_2 doubly occupied, $1\pi_g$ orbital. The excitation curves for these transitions, as measured at X1B, are depicted in Fig. 6.21. Clearly for N_2 the vibrational substates are resolved, whereas for O_2 the lifetime broadening (180 meV) is slightly larger than the vibrational spacing (160 meV). The absorption spectrum therefore is described by a convoluted lineshape, the envelope function of which is determined by the large change in internuclear equilibrium distance between the ground state and the core-excited state.

At this time it is important to note that the absorption spectra by themselves provide an independent test for the parameters used to describe the potential curve of the intermediate core excited state. Whatever potential curve, lifetime, and vibrational parameters are used for the FC interference theory in describing the electron-emission interference profile, the same parameters have to be able to describe the absorption curves (Fig. 6.21). This consistency check is especially important, since the potential curve for the core-excited state is known to a much lesser accuracy than the other potential surfaces involved in the description of the full process. In general, the initial (ground) state potential and the final (single hole) ionic-state potentials have been quite accurately determined from various spectroscopies.

To describe the absorption spectra, we initially adopt the $Z + 1$ approximation (OF for OO*, NO for NN*) and calculate the FC profile for the transition from the lowest vibration of the ground state to the ground state levels of the $Z + 1$ molecule (adjusted for the difference in reduced mass). The FC overlaps are calculated using the semi-empirical FCFRKR (Franck-Condon Factors from Rydberg-Klein-Rees potentials) procedure [6.35, 41]. In a second refinement step the $Z + 1$ parameters are modified slightly to get a better match with the experimental absorption data. The calculated curves are shown as thick solid lines superimposed on the experimental data points (triangles) in Fig. 6.21. From this comparison it is obvious, that the experimental absorption spectra of both molecules are quite well reproduced using this procedure. Since the details of the absorption curves are quite sensitive to the exact choice of the parameters,

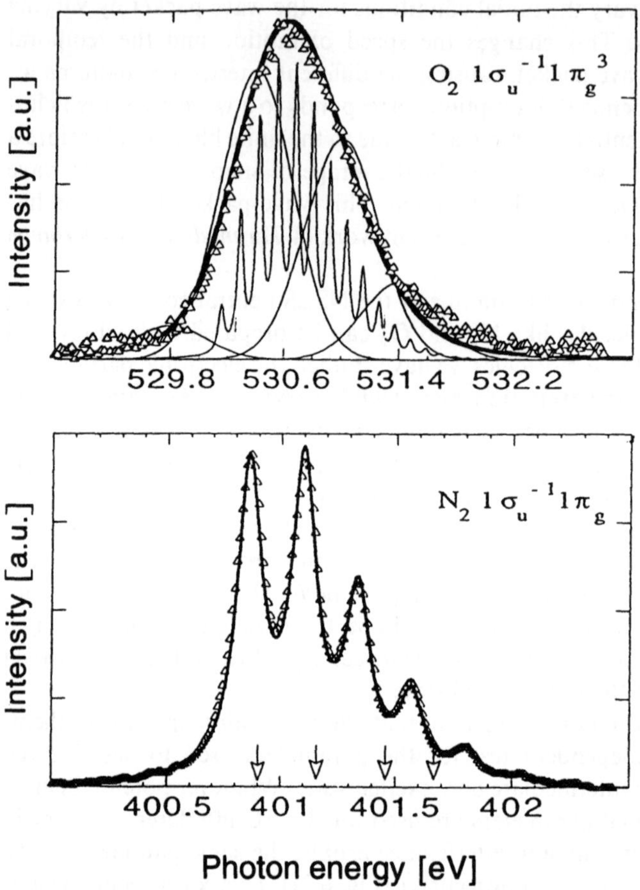

Fig. 6.21. Absorption spectra showing the creation of the core → π^* excited states of N_2 (bottom) and O_2 (top) taken with vibrational resolution. The experimental data points are shown as triangles

this is a fairly tight constraint for the description of the potential energy curves. The parameters of the potential curves for the various electronic configurations of the molecules thus determined are listed in Table 6.5 [6.35].

In the study of the deexcitation of these core-excited states we have specifically concentrated on the participator transitions to the ionic ground state and the first excited ionic states in these molecules. For O_2 this is the $X^2\Pi_g$ state of the molecular ion. For N_2 the two lowest final states of the ionized molecule ($X^2\Sigma_g^+$, $A^2\Pi_u$) have sufficient spectral weight to be included in our vibrationally resolved analysis.

We are able to selectively emphasize various vibrational pathways for N_2 with a resolution comparable to the lifetime width of the intermediate state [6.33, 35]. For O_2 different subsets of the vibrational states were excited as indicated by the four Gaussian curves in Fig. 6.21 [6.34, 35]. Furthermore, the

Table 6.5. Spectroscopic constants for O_2 and N_2 used for the description of the potential energy surfaces of the various electronic states involved in the participator decay of the core → π^* excited states

	D_e	ω_e	$\omega_e X_e$	r_e	α_e
$O_2\,(X^3\Sigma_g^-)$	41259.76	1580.19	11.98	1.2074	0.016
$O^*O\,(^3\Pi_u)$	20915.35	1097.18	10.75	1.3540	0.014
$O_2^+\,(X^2\Pi_g)$	53740.29	1904.77	16.26	1.1162	0.020
$N_2\,(X^1\Sigma_g^+)$	59466.93	2359.61	14.46	1.0943	0.020
$N^*N\,(^1\Pi_u)$	59452.21	1966.24	14.90	1.1641	0.018
$N_2^+\,(X^2\Sigma_g^+)$	70272.91	2207.00	16.10	1.1163	0.019
$N_2^+\,(A^2\Pi_u)$	60276.45	1903.7	15.02	1.1747	0.019

high photon intensity at the undulator beam line [6.6] allowed us to record the deexcitation spectra with a good statistics at a resolution comparable to the resolution in the excitation channel. This enabled us to perform a detailed comparison with lifetime-vibrational interference theory.

Now let us concentrate first on the results for N_2, where the individual vibronic states are fairly well resolved. The decay of the broadband excited $1s^{-1}\,1\pi_g\,(^1\Pi_u)$ state in N_2 has earlier been measured with vibrational resolution in the electron channel and discussed by *Murphy* et al. [6.42]. Here, we present results with vibrational selectivity in excitation and decay and show that the extended theory as formulated above [6.33–35] also accurately describes the excitation-energy dependence of the deexcitation spectra.

6.4.3 Vibrational Resolution in Excitation and Decay (N_2)

Ideally, one would like to separately monitor the decay of each vibrational excitation. In cases where the spacing between the vibrational levels is comparable to the lifetime width, as in O_2, this becomes impossible due to lifetime broadening in the intermediate core-excited state. The $1s^{-1}\,1\pi_g$ state in N_2, on the other hand, has vibrational spacings, that exceed the lifetime-width by a factor of 2. With sufficient resolution it is here possible to separately monitor the decay of selected vibronic eigenfunctions, thus taking vibrationally-resolved decay spectra.

In Fig. 6.22 we present the observed (triangles) and calculated (solid line) spectra corresponding to the decay of the $1s^{-1}\,1\pi_g$ core excited state of N_2 into the $1\pi_u^{-1}$ (384–384.5 eV) and $3\sigma_g^{-1}$ (384.5–386.5 eV) ionic final states as a function of excitation energy. Within the uncertainty given by the statistics of the experiment, the agreement between theory and experiment is excellent. The theoretical curves are all calculated using a lifetime of $\Gamma = 0.132$ eV. The resolution of the monochromator for these studies was set to 0.1 eV, resulting in the excitation (absorption) spectrum shown in Fig. 6.21. The excitation energies chosen are indicated by arrows in the absorption spectrum of Fig. 6.21b. The

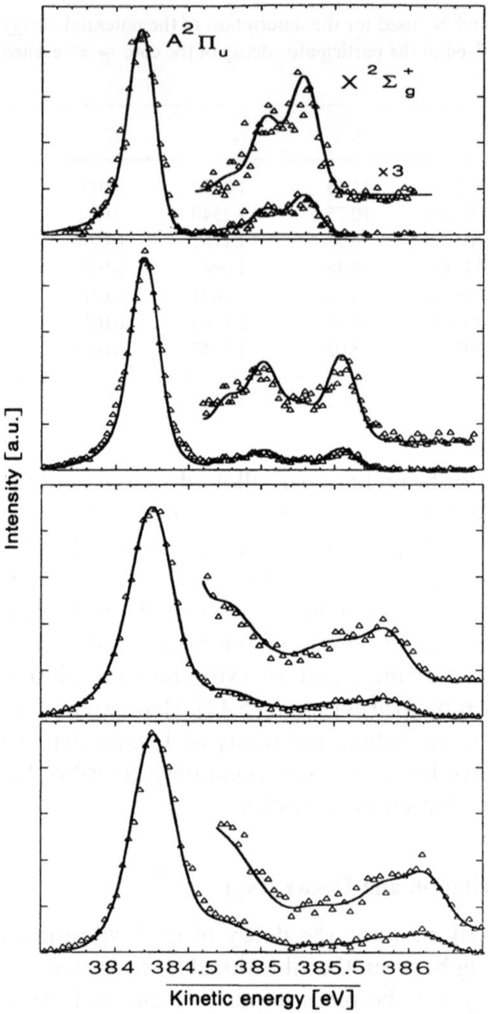

Fig. 6.22. Experimental (triangles) and calculated (solid line) participator deexcitation spectra of the $1\sigma_u^{-1}\,1\pi_g^1$ state of N_2. The excitation energies are from top to bottom 400.90 eV, 401.16 eV, 401.46 eV, and 401.68 eV, as indicated by the arrows in Fig. 6.21. The resolution of the electron spectrometer of 0.25 eV (0.35 eV) for the top (bottom) two curves has been included in calculating the spectra

relative weight of the two electronic transitions $I(3\sigma_g^{-1})/I(1\pi_u^{-1})$ is determined from the comparison between theory and experiment to be 0.4. This value is independent of excitation energy within the accuracy of our measurements. The energy spacing between the $v = 0$ final electronic states is 1.11 eV.

When the $v_i = 0$ vibration dominates in the intermediate state, the band corresponding to the ionic ground state shows distinct vibrational fine structure that can readily be assigned to $0 \leq V_f \leq 2$ final states. A clear asymmetry of the $1\pi_u^{-1}$ line also indicates vibrational excitations. Going to higher excitation energies, the shape of the $1\pi_u^{-1}$ transition remains fairly constant, whereas the $3\sigma_g^{-1}$ emission displays remarkable changes. For the two higher excitation energies we worked at slightly lower electron spectrometer resolution, resulting in additional broadening (FWHM = 0.35 eV instead of 0.25 for the two lower excitation energies). The different behaviour of the two final electronic states can be correlated with the differences in the potential surfaces. The π final state has a very similar potential surface as the core excited state ($\Delta r_e = 0.011$ Å), favouring transitions between vibrational levels of identical quantum number ($\Delta v = 0$). All the transitions therefore have roughly the same energy, and one narrow (< 0.3 eV) peak is seen in the spectrum. Hence, even though the spectral shape is constant, the dominating final state is $v_f = 0, 1, 2$ and 3 as $v_i = 0, 1, 2$ and 3 dominate the intermediate state, respectively.

Whereas the $1\pi_u^{-1}$ participator final state does not change much in line shape, we find that the spectral shape of the $3\sigma_g^{-1}$ final state changes dramatically as the excitation energy is tuned within the band width of the $1s^{-1}\,1\pi_g$ core-excited state. This is mainly caused by the large differences in the equilibrium internuclear distance for the various states involved. The ionic ground state has a much shorter bond length than the core excited state ($\Delta r_e = -0.048$ Å), which gives rise to vibrational progressions from each of the vibronic wave functions of the core-excited state. The result is a broad (> 1.5 eV) feature with some distinct structures, that can be related to the vibrational excitations. Peaks can readily be assigned to specific vibrational final states.

In the following we will, therefore, concentrate our discussion on the transition to the electronic ground state $3\sigma_g^{-1}$ of the ionized molecule. A high-kinetic-energy peak moves with the excitation energy. This peak can be assigned to the $v_f = 0$ final state. At $v_i > 0$ energy is transferred from vibrational energy in the intermediate state into kinetic energy of the emitted electron (phonon gain, $\Delta v_{fi} < 0$). In the decay of the $v_i = 1$ excitation we see two sharp peaks corresponding to $v_f = 0$ and $v_f = 2$. There is also an intensity minimum at 385.2 eV, roughly the position of the main peak in the spectrum excited at $v_i = 0$, and a shoulder at the position of $v_f = 3$. At higher excitation energies the intensity minimum at the $\Delta v_{fi} = 0$ position divides the spectra into two groups: a high kinetic energy phonon gain group and a low-kinetic-energy group, where $\Delta v_{fi} > 0$. At the highest-excitation-energy vibrations up to $v = 6$ can be discerned before the vibrational progression disappears under the $1\pi_u^{-1}$ final-state peak.

We can carry out the theoretical analysis of these spectra in various stages of refinement. The various stages of this analysis are exhibited in Fig. 6.23. In the simplest treatment, corresponding to the one-vibration-at-a-time picture, we neglect interference effects and assume infinite monochromator resolution. The probability to populate the various final states $|f\rangle$ in the decay of the intermediate states $|n\rangle$ is then determined by

$$I_f(|n\rangle) \propto |\langle f|n\rangle|^2 . \tag{6.12}$$

The calculated $I_f(|n\rangle)$ intensities are added using the correct energy scale and broadened with an electron spectrometer Gaussian function with FWHM = 200 meV to give the theoretically predicted results (dotted line in Fig. 6.23). As our second level of approximation we use (6.11), still assuming infinite

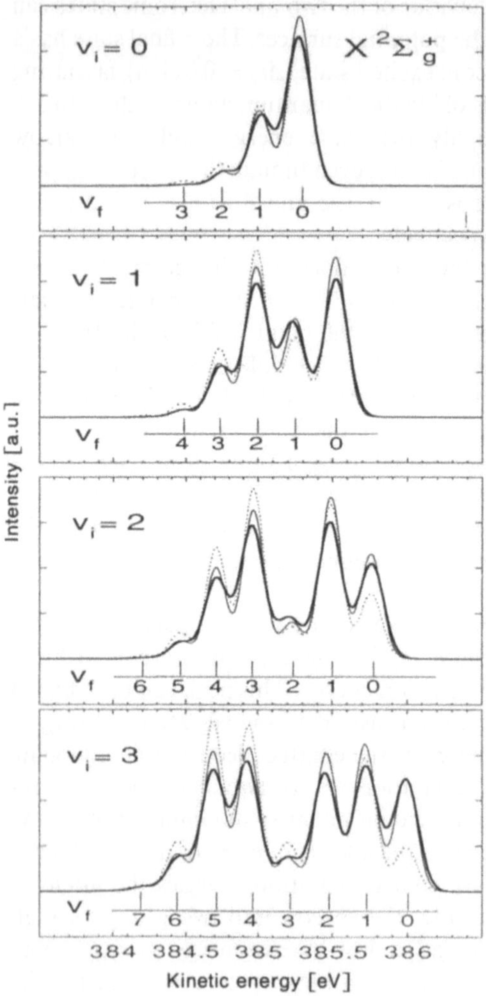

Fig. 6.23. Theoretical analysis of the $3\sigma_g^{-1} \leftarrow 1\sigma_u^{-1}1\pi_g^1$ transition of N_2 calculated at three different levels of approximation. The dotted line is the result of treating the decay of the various vibronic substates separately, the thin solid line includes inerference effects ($\Gamma = 0.132$ eV), and the thick line additionally includes the effect of the limited monochromator resolution (100 meV)

monochromator resolution. This result is represented as thin lines in Fig. 6.23. We notice that the interference effects tend to shift intensity from lower to higher kinetic energy, but it is obvious that these shifts are small, and they are probably not observable within the present experimental accuracy. As a third level of approximation we introduce a realistic monochromator function, assuming a Gaussian with FWHM = 100 meV, situated at the absorption peak maxima. The result is shown as thick lines in Fig. 6.23. We see that the limited resolution in the excitation step only has a small impact on the shape of the spectra. It contributes only slightly to the broadening of the features.

The overall agreement between the three levels of approximation we take as a justification for the simple one-vibration-at-a-time picture. The assignment of the spectral structures is straightforward, and we accordingly mark the final vibrational states in Fig. 6.23. The difference between the full theory curves in Fig. 6.23 and 22 is due to the fact that the excitation energy in Fig. 6.23 is assumed to be centered on the various maxima of the absorption spectrum, whereas in Fig. 6.22 the excitation energy is set according to the real experimental values as indicated by the arrows in Fig. 6.21b in addition, the slightly narrower electron analyzer function causes the structures in Fig. 6.23 to appear sharper.

6.4.4 Time Development of the Eigenstates on the FS Time Scale (O_2)

As we just have seen, for N_2 the intensity shifts due to interference are, as expected, quite small and the analysis of the spectra can be carried out in great detail, leading to specific assignments of all the vibrational substates involved in the transitions. For O_2, however, the situation is remarkably different. Vibrationally non-resolved spectra of the participator decay in O_2 have been taken by *Thomas* and *Carroll* earlier using the e-2e coincidence technique [6.43]. However, the interference effects become really dramatic when studied with vibrational resolution and high counting statistics, as shown in the following [6.34, 35].

In Fig. 6.24 we present the observed (triangles) and calculated (solid line) electron emission spectra produced in the decay of the $1\sigma_u^{-1}\,1\pi_g^3$ core-excited state into the $1\pi_g^1$ $(X^2\Pi_g)$ ground state of the molecular ion. The four panels represent the selection of different subsets of vibrational states in the excitation using photon energies, as indicated by the four Gaussian curves represented as thin solid lines in Fig. 6.21a. To demonstrate the importance of the interference effects, we have also included in Fig. 6.24 the theoretical curves calculated neglecting the interference effects (dotted line). As demonstrated in the four panels of Fig. 6.24, we observe a dramatic change in line shape in the emission feature upon only slight changes of the excitation energy. At low energies the main emission peak is almost symmetric with a width of about 1 eV. With increasing excitation energy a distinct low-energy tail develops, and simulataneously the kinetic energy of the leading edge increases. The full width of the structure at the highest excitation energy is roughly 5 eV. The peak at a kinetic energy near 512.5 eV corresponds to the $1\pi_u^3\,1\pi_g^2$ final-state configuration, which we will not discuss here. The calculations are performed using the parameters listed in Table 6.5. The lifetime width of the core-excited state was taken as $\Gamma = 0.18$ eV and a Gaussian monochromator function with a FWHM of 0.5 eV was assumed for the excitation spectrum (Fig. 6.21a).

In general, we find that the theoretical curves are in excellent agreement with experimental observations for O_2, as they were for N_2. Specifically in the case of the O_2 molecule, where the more dramatic lifetime-interference effects are observed, even the large changes in the spectra are quite well predicted by theory. Moreover, in order to understand the experimental observations it is not even needed to carry out these extensive calculations. The basic physics and the gross features of the spectra can be understood in a simple two-step scheme,

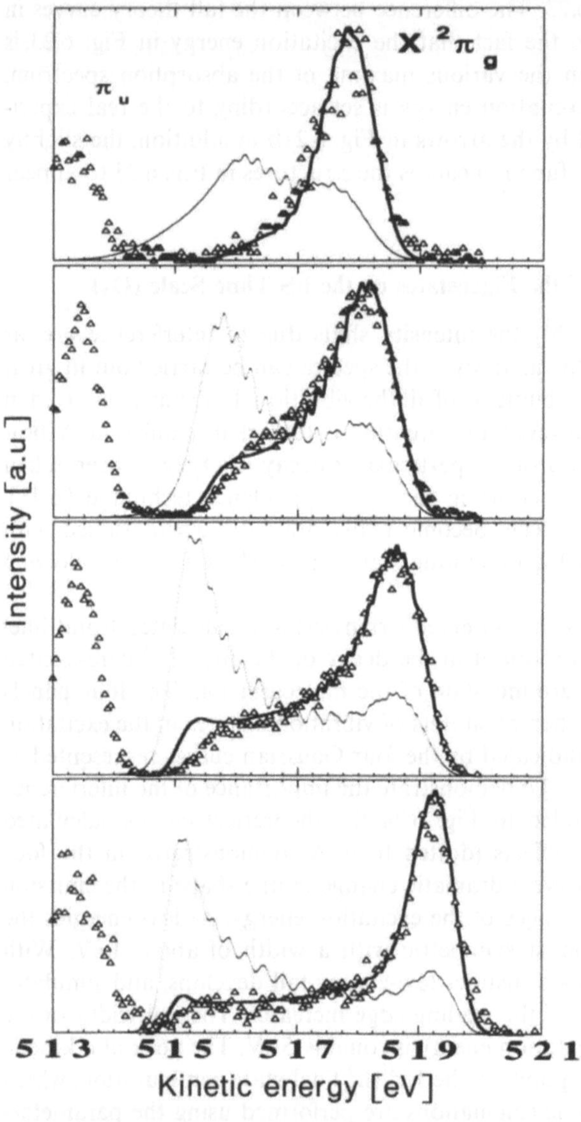

Fig. 6.24. Experimental (triangles) and calculated (solid line) participator decay spectra of the $1\sigma_u^{-1}\,1\pi_g^3$ state into the ground state of the molecular ion $X^2\Pi_g$. The excitation energies were varied from top to bottom as indicated in Fig. 6.21 from 598.8 eV, 530.4 eV, 531.0 eV, to 531.4 eV. The dotted lines show the theoretical result when the lifetime vibrational interference is neglected. The curves are normalized arbitarily

using only an intuitive wave packet concept. Due to the large differences in equilibrium bond lengths (Fig. 6.25), very high vibrational excitations can be reached in the intermediate ($v \geq 15$) and final states ($v \geq 35$). The Franck-Condon overlap integrals between wave functions of such high quantum

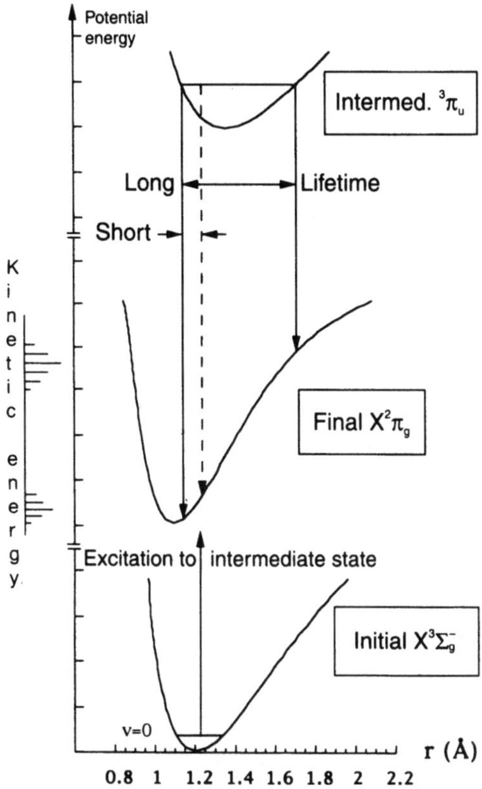

Fig. 6.25. Calculated RKR potential curves for the states of O_2 involved in the decay processes shown in Fig. 6.24. The parameters for these curves are listed in Table 6.5

numbers are dominated by the contributions corresponding to the classical turning points. Consequently, in the long-lifetime limit a two-peak structure corresponding to the two turning points of the intermediate state potential is expected. That is also what the two-step FC theory without interference (dotted line in Fig. 6.24) predicts.

However, since the core-excited state is strongly antibonding, as is illustrated in Fig. 6.25, and because of the short lifetime only transitions near the inner turning point of the core-excited state are contributing predominantly. Therefore, the high-kinetic-energy peak, corresponding to a fast decay near the inner turning point of the core-excited state, is observed prominently in the experimental spectra. The kinetic energy of this feature increases proportional to the excitation energy because the repulsive branch of the potential curve of the intermediate state is located over the minimum of the final-state potential curve. Accordingly, transitions down to the $v_f = 0$ level of the final state are observed, which cause this part of the emission to shift exactly proportional with the photon energy.

At the energy position corresponding to the outer turning point of the intermediate state potential only a weak tailing of the spectrum is observed. This can be qualitatively understood by considering the time evolution of the nuclear distance probability distribution. It suggests that the evolution of a wave packet cannot reach the outer turning point before the decay occurs. Moreover, at this turning point only highly excited vibrational states of the final state can be reached, since the final-state potential is steeper in the attractive region and since the equilibrium distance of the final state is significantly shorter than the corresponding value for the core-excited state (Table 6.5). Accordingly, much more electronic energy is transferred into vibrational energy for transitions occurring when the wave packet has evolved into that region. As a consequence, the electron emission is observed at significantly lower kinetic energy, since part of the energy is transferred into vibronic excitation, whereby the total width of the emission feature approaches 5 eV in energy.

In order to carry out the full quantum mechanical interference calculation with the FC-interference theory, the potential energy curves of the different electronic states involved have to be known quite accurately. For the initial and final electronic states, these curves are well known [6.44]. The potential of the core-exicted state is less readily available, but can be independently obtained from the absorption spectrum, as outlined above.

At all excitation energies it is obvious that for O_2 the lifetime-vibrational interference is of utmost importance for the understanding of the spectra. Neglecting the cross terms of (6.10) always leads to a two-peak structure with maxima corresponding to the two turning points, i.e., neglecting the coherence between the excitation and the decay leads to spectra that have very little resemblance with the experimental results as pointed out above. In detail we notice in the theoretical decay spectra, which are excited by the two highest excitation energies, a small low-energy peak that is much weaker than predicted in the experimental spectrum. This low-energy peak predicted too strong by theory might be a consequence of the wave functions used in the theory. The evolution of the nuclear wave functions is described in the basis of the bound vibronic eigenfunctions. This basis is, however, not complete since the *continuum* functions are not included. The extra peak corresponding to the outer turning point of the oscillation might be overemphasized in the theory since only the bound functions are included.

In summarizing this section, we presented and analyzed various electronic participator decay processes of the *K*-shell excited short-lived π^*-state with *vibrational selectivity* for the *N* and *O*-edge of N_2 and O_2 molecules. Due to the short lifetime of the core-excited state of O_2, the most dramatic vibrational lifetime interference effects are observed for this molecule. These effects can be intuitively explained by a semiclassical wavepacket concept, which is close to the description in a time-dependent framework of the lifetime interference effects.

For N_2 we have demonstrated that it is possible to measure the decay of vibrationally selected core excited states as it is postulated in the one-vibration-at-a-time picture. An unequivocal assignment of the vibrational structure in the deexcitation of the $1\sigma_u^{-1}\,1\pi_g$ state to the ionic ground state could be given. Altogether these studies with vibrational resolution in the excitation and decay give insight into the fsec dynamics of electronic-vibrational coupling. For the proper description of these phenomena the excitation and decay have to be treated as a one-step coherent process in a time-dependent quantum-mechanical framework. An intuitive wave-packet picture, however, serves quite well to illustrate the basic phenomena.

6.5 Direct Observation of Charge Transfer and Screening Processes

6.5.1 Deexcitation of Transition Metal Carbonyls and CO Chemisorbed on Surfaces

So far we have seen that the deexcitation spectra reveal the dynamics of the response to the creation of the core-hole excited state. The question arises how the response is modified when the particular molecule is chemisorbed onto a surface or bound in a transition-metal complex. In an isolated molecule it is straightforward to keep track of the electron count and occupation numbers within the molecular entity. In a chemisorbed molecule, however, charge transfer and screening processes are not limited to the molecular complex but have to include the substrate electronic states. On the other hand, a comparison of these processes in the isolated molecule and in the chemisorption system will reflect upon the modification of the electronic structure due to the formation of the chemical bond.

One has to keep in mind that all of the electronic states, and also the vibronic states, which are observed in high-resolution studies, are changed by the interaction with the substrate. In absorption or photoemission studies this refers to the changes of the (initial) ground state and also to the differences observed in the final state of the process, which are due to modifications of the screening process. For DES this implies that the changes of the three different electronic configurations, the ground state, the core-excited state and the final valence-hole state have to be taken into account in the interpretation of the spectra.

The physics of the core-hole excitation and screening processes in molecular adsorbates has been addressed previously by core-level photoemission (XPS) studies [6.45–49] and by Auger spectroscopy [6.50–52]. Some of the chemisorption induced changes of the electronic structure are also reflected in the differences in the absorption spectrum. This has been more extensively discussed by *W. Wurth* and *D. Menzel* in Chap. 5. One general result of these investigations is, that core $\rightarrow \pi^*$ transitions are still observed, albeit modified, for essentially all chemisorption systems whenever this kind of excitation exists for the isolated molecules. Rydberg excitations, however, change their character to a much larger extent due to the fact that the wave function of the Rydberg electron is more delocalized and accordingly hybridization with the substrate states plays a much larger role. Therefore, in the following I will concentrate on the decay of core → π^* excited states and also restrict myself in this discussion to the electronic excitations. Some more data concerning the vibronic coupling are presented by *N. Mårtensson* and *A. Nielsson* in Chap. 3.

The first discussion within the framework of DES of the chemisorption-induced changes in the core-hole decay spectra was orginally presented in the context of one of the DIET (Desorption Induced by Electronic Transitions) meltings [6.53]. These data, a comparison of the Auger spectra and the decay spectra of the π^* excited states for CO, transition metal carbonyls, and CO on Cu(110), are shown in Fig. 2.26. The most important observation to be made from this compilation of data is the increase in similarity of the Auger spectra and the deexcitation spectra of the π^* resonances for the coordinated compounds. As already extensively discussed above, the Auger and the π^*-deexcitation spectrum of the isolated CO molecule are distinctly different. However, when CO is interacting with a metal atom as in the carbonyls or in a situation

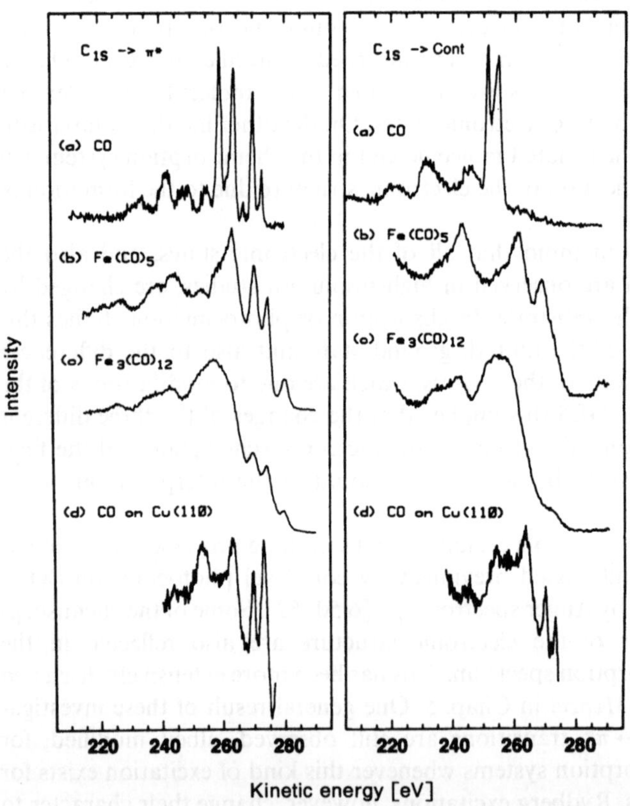

Fig. 6.26. Comparison between the deexcitation of the C 1*s* → π^* excited states of CO, FE(CO)$_5$, Fe$_3$(CO)$_{12}$, and CO chemisorbed on Cu(110) in the left panel with the photon excited Auger spectra of the same compounds shown in the right-hand panel

where CO is bound to a metallic surface, then the Auger spectrum starts to resemble more and more the decay spectrum of the π^* resonance. It is important to note here that indeed the Auger spectrum becomes closer in appearance to the deexcitation spectrum rather than that the deexcitation spectrum develops a closer similarity to the Auger spectrum. The physics behind this is: the core ionized state of the chemisorbed molecule is not a true ionic state, but rather corresponds to an electronic configuration, where the C 1s core hole is screened by a charge transfer process from the metal into the 2π orbital of the chemisorption complex. Accordingly, the state that leads to the production of the decay spectra (Fig. 6.26) has an electron occupying the 2π orbital of the chemisorption complex. For the coordinated CO molecules the 2π occupation thus is either a consequence of the direct excitation process or a result of the screening process of the core hole following C 1s ionization. One immediate conclusion to be drawn from these spectra is also that the charge-transfer screening process is much faster than the lifetime of the core hole excited state. In Chap. 3 systems have been discussed where the charge-transfer process is substantially slower because of the choice of graphite as the substrate [6.54]. In this case core ionization results in the simultaneous observation of two superimposed but distinctly different decay spectra. One of these contributions corresponds to the decay of the screened core hole state and the other one results from the decay of the 'unscreened' core-hole state.

The Auger spectrum of CO on Cu(110) depicted in Fig. 6.26 is also quite remarkable from another point of view. From XPS investigations of CO on Cu surfaces [6.46, 48, 49] it is known that the core lines exhibit very intense shake-up satellites. An example is shown in Fig. 6.27 taken from [6.48]. This means, that there are several core-hole states created in the photoemission process, which have a total spread in energy of more than 7 eV in the case of the C 1s photoemission. Nevertheless, the lines observed in the Auger spectrum are quite sharp, as shown in Fig. 6.28, and correspond in a first-order approximation only to the decay of the fully screened state [6.55]. Thus, all of the excited core-hole configurations decay into the fully screened core-hole state with a lifetime, which has to be much shorter than the lifetime of the core hole itself. A careful inspection of the spectra displayed in Fig. 6.28 reveals, that the lines of the Auger spectrum are somewhat broader than the corresponding lines in the DES data. This can be taken as an indication of contributions from the decay of higher energy core-hole states other than the fully screened state.

The exact nature of the final states observed in these spectra has been the subject of an intense discussion still ongoing in the literature [6.54–63]. Initially, it was observed that, while the peaks and structures in the deexcitation and the Auger spectrum coincide on the kinetic-energy scale, there is a discrepancy in energy when the deexcitation spectrum is plotted on a binding-energy scale and compared to the direct photoemission results (Fig. 6.28). On this binding energy scale the DES peaks seem to correspond to larger binding energies than the corresponding photoemission peaks. If the final states of these two spectroscopies are indeed equivalent, then the question arises about the origin of the

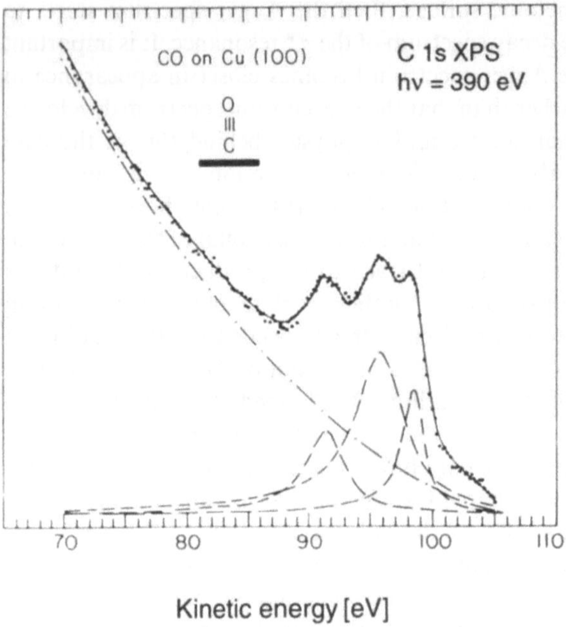

Fig. 6.27. C 1*s* XPS of CO on Cu(100) showing the simultaneous creation of several core-hole states in the XPS process. An attempt at deconvoluting the shake-up structures is indicated by the dashed curves

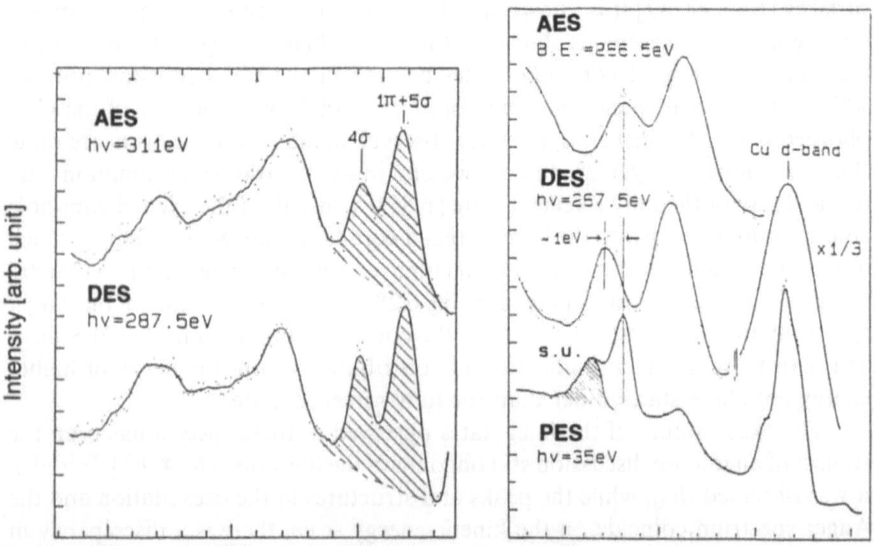

Fig. 6.28. AES and DES of CO on Cu(110) compared on a kinetic energy as well as on a binding energy scale. The direct photoemission spectrum is included in the right-hand panel (PES)

nature of the mechanism responsible for the apparent 'energy loss' of 1 eV in DES. This leads to the interpretation that not only the core ionized state relaxes prior to the core-hole decay, but that there is also a relaxation mechanism active for the core $\rightarrow \pi^*$ excited state, which is much faster in time than the core-hole lifetime [6.55, 57]. This is actually equivalent to the observation, that for most adsorbate systems the energy of the core $\rightarrow \pi^*$ resonance excitation is slightly larger than the energy of the fully-screened core-hole state measured in XPS. One possibility of a relaxation mechanism originates from the fact that the triplet configuration of the core $\rightarrow \pi^*$ excitation in CO and N_2 has an energy about 1 eV lower than the corresponding singulet state, which is the only one that may be excited in a dipole transition [6.64]. The triplet configuration, however, is energetically favoured for an electron scattering from the substrate into an adsorbate molecular orbital in the screening process of the core hole.

Another possible interpretation is to propose that the final states in photoemission and in DES are not equivalent in this case [6.57, 63]. This is a difficult question to answer, since the photoemission final states are also most likely not 'pure' single hole states, but are screened by charge transfered from the substrate [6.56, 58, 59]. Experimentally the nature of these states might be resolved by measuring the angular emission characteristics of these states and comparing the results to detailed calculations [6.57, 59]. However, even this at present does not give unambiguous results [6.57, 59, 62, 63].

The difference in nature of the deexcitation process and photoemission definitely leads to the population of final states different from the ones populated in photoemission [6.58, 65]. This is emphasized by the deexcitation spectra shown for CO chemisorbed on Co(0001) in Fig. 6.29. There is a well pronounced satellite visible at a binding energy of 12.6 eV. This satellite line is only observed in DES but not in photoemission. The observation of this type of satellite structure indicates the existence of a second less-well-screened valence-hole state rather than the fully-screened valence-hole state, where one electron has been transferred into the '2π' molecular orbital. Actually this orbital hybridizes with substrate states and in the simplest picture forms a bonding-antibonding orbital combination. The valence-hole final state may be screened either by charge transfer into the 2π bonding orbital, which has most of its weight at the CO molecular entity or by an electron occupying the antibonding part of the 2π orbital, which is mostly located in the metal. Even though this intuitive explanation has also been used to explain the core-hole satellites, there is always the possibility that the peak called a *satellite* here actually belongs to a completely different electronic configuration.

The existence of these kinds of satellites is also proposed for CO on Pt(111) [6.58] and CO on Ni(110) [6.59]. The major open question is why both these states are populated in DES, whereas one clearly dominates in photoemission. This is definitely one of the still open questions in DES of adsorbates. Especially, since there almost seems to be an anti-correlation between the appearance of these satellites in photoemission and in DES. For CO on Cu there are large satellites observed in photoemission, but not in DES, whereas for CO on Pt and

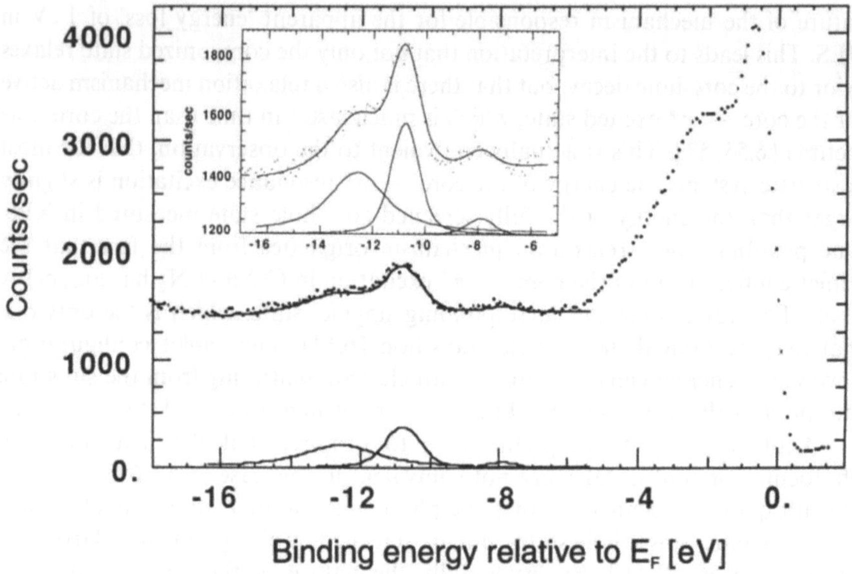

Fig. 6.29. Deexcitation spectrum of the C 1s → π^* excited state of CO on Co(0001). The inset shows an enlarged view of the intrinsic CO emission and a deconvolution of the emission features

on Co there are no satellites observed in photoemission, but the satellite features seen in DES are quite intense. This is definitely an area of research, where future angle-dependent studies with high resolution and excellent signal to noise, taken on undulator sources, will shed some light on the interpretation when the experimental results are compared with detailed, high quality, theoretical calculations.

6.6 Summary and Outlook

In this chapter I have tried to review the status of the experimental situation in the study of the decay of neutral core-hole-excited states in molecular systems. In contrast to the conventional Auger spectroscopy this has been named DES (Deexcitation Electron Spectroscopy), even though technically it is an autoionization process. This chapter is not and was never meant to be a review of all the work that has been done in this field. Rather, I have tried to illustrate the physics addressed in these studies from different viewpoints. The major progress in this field comes from the possibility to prepare highly excited neutral core-hole states in various molecular systems with a previously unprecedented precision. The study of the decay of these states in great detail gives new insight into the nature of the electronic and vibronic interactions in these systems. Specifically there are four different areas of interest and possible future studies using this type of spectroscopy.

First, the study of the process of the deexcitation of neutral core-excited states in series of chemically slightly different molecules. Here one can make use of the atomic specificity and the local nature of the deexcitation process. These studies should reveal the variations of the charge densities of the individual molecular wave functions at the various inequivalent atomic sites within any given molecule. For these specific experiments it would be extremely valuable to have calculations at hand for this type of core-hole decay, since the intensities are hardly understood beyond the 'handwaving' picture presented in some of the discussion above. For example, even for a simple molecule like CO the extremely low intensity of the 4σ single-hole participator state in the O-π^* DES data is one of the still unresolved puzzles.

The second major avenue is the systematic study of the decay of different classes of core-hole states excited within the same molecule. For example, taking a series of deexcitation spectra for different core $\rightarrow$ Rydberg state excitations, reveals detailed information about the coupling of the excited electron to various molecular electronic configurations. This difference in coupling manifests itself in the change of the relative intensity of the participator lines and in the change of the screening energy contribution of the Rydberg electron as a spectator in the two-hole, one-electron final states. Furthermore, the differences in coupling between Rydberg states and excitation into a molecular orbital like the π^* or σ^* orbitals needs to be investigated and understood in detail. Conversely, as a fingerprint the decay spectra can be used to unambiguously identify the nature of the core excited states, as demonstrated above for O_2. These studies allow in a unique fashion the determination of electronic interactions at a previously unprecedented precision. Establishing a solid data base of this kind should lead to improved theoretical understanding of the electronic interaction in molecules and solids.

The third direction of these studies, and possibly the most exciting one at present, is the area of interference phenomena in the coherent excitation and decay of various vibrational sublevels of the electronic states. This sheds light onto the dynamics of the electronic-vibrational coupling on the fs time scale and serves as a testing ground for quantum-mechanical models. On this time scale the excitation and decay have to be treated as coherent processes, which leads to interference phenomena, resulting in modifications of the line shapes and transition intensities. At present there is no other experimental technique available that allows testing the models of quantum mechanics, such as the wave-packet description, on such a short time scale. Even though these concepts are quite sophisticated, one special advantage of this area is the use of fairly simple computational schemes that allow one to calculate the complex line shapes observed in the experimental data.

The fourth major direction is the study of chemisorption systems. Here DES is extremely well suited to study the dynamics of the core-hole relaxation and screening processes and the modifications introduced by the chemical bond with a substrate. Mapping out the angular characteristics of the emission features might help to derive a definitive assignment of the final-state electronic configurations.

Altogether this is electron spectroscopy at an enhanced level, where the reaction to the probe, the photon, plays a major role in the process itself. Without the development of undulator sources some of these experiments might not have been possible, since high resolution in the excitation and the decay channel is required, which at the limited density of a gas phase molecular beam also is a matter of counting statistics and thus flux of the incident photons.

I would like to thank my collaborators over time for the many stimulating discussions and all their contributions. Among these I have to mention especially for the earlier results C.T. Chen, W.K. Ford, In-Whan Lyo, R. Murphy, and E.W. Plummer from the University of Pennsylvania and for the more recent experiments M. Neeb and J.E. Rubensson. The most recent results also would not have been possible without the efforts of the team that made X1B undulator beam line come into existence.

References

6.1 W. Eberhardt, E.W. Plummer, C.T. Chen, W.K. Ford: Austr. J. Phys. **39**, 853 (1986)
6.2 K. Siegbahn, C. Nordling, G. Johansson, J. Hedman, P.F. Heden, K. Hamrin, U. Gelius, T. Bergmark, L.O. Werme, R. Manne, Y. Baer: In *ESCA Applied to Free Molecules* (North Holland, Amsterdam. 1969)
6.3 W.E. Moddeman, T.A. Carlson, M.O. Krause, B.P. Pullen, W.E. Bull, G.K. Schweizer: J. Chem. Phys. **55**, 2317 (1971)
6.4 L. Ungier, T.D. Thomas: Chem. Phys. Lett. **96**, 247 (1983) and J. Chem. Phys. **82**, 3146 (1985)
6.5 W. Eberhardt, J. Stöhr, J. Feldhaus, E.W. Plummer, F. Sette: Phys. Rev. Lett. **51**, 2370 (1983)
6.6 K. Randall, J. Feldhaus, W. Erlebach, A.M. Bradshaw, W. Eberhardt, X. Zu, Y. Ma, P.D. Johnson: Rev. Sci. Instrum. **63**, 1367 (1992)
6.7 H. Ågren: J. Chem. Phys. **75**, 1267 (1981)
6.8 G. Wentzel: Z. Phys. **43**, 524 (1927)
6.9 F.P. Larkins, W. Eberhardt, In-Whan Lyo, R. Murphy, E.W. Plummer: J. Chem. Phys. **88**, 2948 (1988)
6.10 R. Dudde, M.L.M. Rocco, E.E. Koch, S. Bernstorff, W. Eberhardt: J. Chem. Phys. **91**, 20 (1989)
6.11 W.L. Jorgensen and L. Salem: *The Organic Chemist's Book of Orbitals*, (Academic, New York 1973)
6.12 W. Eberhardt, E.W. Plummer, In-Whan Lyo, R. Murphy, R. Carr, W.K. Ford: J. Physique **C9**, 679 (1987)
6.13 W. Eberhardt, R. Dudde, M.L.M. Rocco, E.E. Koch, S. Bernstorff: J. Electr. Spectr. Rel. Phen. **51**, 373 (1990)
6.14 F.P. Larkins: J. Electr. Spectr. el. Phen. **51**, 115 (1990)
6.15 M. Meyer, E.v. Raven, M. Richter, B. Sonntag, R.D. Cowan, J.E. Hansen: Phys. Rev. A **39**, 4319 (1989)
6.16 H. Aksela, S. Aksela, A Mäntykenttä, J. Tulkki, E. Shigamasa, Y. Yagishita, Y. Furusawa: Physica Scripta T **41**, 122 (1992)
6.17 V. Schmidt: J. Phys. C **9**, 401 (1987) and references therein
6.18 B. Crasemann: Comm. At. Mol. Phys. **22**, 163 (1989) and references therein

6.19 A. Niehaus: J. Phys. B **10**, 1845 (1977)
6.20 A. Russek, W. Mehlhorn: J. Phys. B **19**, 911 (1986)
6.21 T. Åberg: Physica Scripta T **41**, 71 (1992)
6.22 M. Neeb, M. Biermann, W. Eberhardt: J. Electr. Spectrosc. (1994)
6.23 T.D. Thomas, R.W. Shaw: J. Electr. Spectr. **5**, 1081 (1974)
6.24 Y. Ma, C.T. Chen, G. Meigs, K. Randall, F. Sette: Phys. Rev. A **44**, 1848 (1991)
6.25 We assume the notation and assignment of the peaks as given in [6]. The existence of the B5′ state is derived from these calculations but previously was not labelled as such in the spectra.
6.26 A.P. Hitchcock, C.E. Brion: J. Electr. Spectrosc. **18**, 1 (1980)
6.27 W. Wurth, J. Stöhr, P. Feulner, X. Pan, K.R. Bauchspiess, Y. Baba, E. Hudel, G. Rocker, D. Menzel: Phys. Rev. Lett. **65**, 2426 (1990)
6.28 P. Kuiper, M.W. Ruckman, B.I. Dunlap, L.H. Tjeng: **71**, 3091
6.29 M. Neeb, J.E. Rubensson, M. Biermann, W. Eberhardt: Phys. Rev. Lett. (1993)
6.30 N. Kosugi, E. Shigemasa, A. Yagishita: Chem. Phys. Lett. **190**, 481 (1992)
6.31 M.W. Ruckman, Jie Chen, S.L. Qiu, P. Kuiper, M. Strongin, B.I. Dunlap: Phys. Rev. Lett. **67**, 2533 (1991)
6.32 W. Eberhardt, J.E. Rubensson, K.J. Randall, J. Feldhaus, A.L.D. Kilcoyne, A.M. Bradshaw, Z. Xu, P.D. Johnson, Y. Ma: Physica Scripta T **41**, 143 (1992)
6.33 J.E. Rubensson, M. Neeb, M. Biermann, Z. Xu, W. Eberhardt: J. Chem. Phys. **99**, 1633 (1993)
6.34 M. Neeb, J.E. Rubensson, M. Biermann, W. Eberhardt, K.J. Randall, J. Feldhaus, A.L.D. Kilcoyne, A.M. Bradshaw, Z. Xu, P.D. Johnson, Y. Ma: Chem. Phys. Lett. **212**, 205 (1993)
6.35 M. Neeb, J.E. Rubensson, M. Biermann, W. Eberhardt: J. Electr. Spectr. (1994)
6.36 F.K. Gel'mulkhanov, L. N. Mazalov, A.V. Nikolaev, A.V. Kondratenko: Chem. Phys. Lett. **46**, 133 (1977)
6.37 F. Kaspar, W. Domcke and L.S. Cederbaum: Chem. Phys. **44**, 33 (1979)
6.38 N. Correia, A. Flores-Riveros, H. Ågren, K. Helenelund, L. Asplund and U. Gelius: J. Chem. Phys., **83**, 2035 (1985)
6.39 A. Cesar, H. Ågren, V. Carravetta: Phys Rev. A **40**, 187 (1989)
6.40 L.S. Cederbaum, F. Tarantelli: J. Chem. Phys. **98**, 9691 (1993)
6.41 H. Telle, U. Telle: Comp. Phys. Commun. **28**, 1 (1982)
6.42 R. Murphy, In-Whan Lyo, W. Eberhardt: J. Chem. Phys. **88** 6078 (1988)
6.43 T.X. Carroll, T.D. Thomas: J. Chem. Phys. **92**, 7171 (1990)
T.X. Carroll, T.D. Thomas: Chem. Phys. Lett., **185**, 31 (1991)
6.44 K.P. Huber, G. Herzberg: *Molecular Spectra and Molecular Structure IV, Constants of Diatomic Molecules* (Van Nostrand Reinhold, New York 1979).
6.45 J.C. Fuggle, E. Umbach, D. Menzel, K. Wandelt, C.R. Brundle: Sol. State Commun. **27**, 65 (1978)
6.46 P.R. Norton, R.L. Tapping, J.W. Goodale: Surf. Sci. **72**, 33 (1978)
6.47 E.W. Plummer, W.R. Salaneck, J.S. Miller: Phys. Rev. B **18**, 1673 (1978)
6.48 H.J. Freund, E.W. Plummer: Phys. Rev. B **23**, 4859 (1981)
6.49 W. Eberhardt, J. Stöhr, D. Outka, R.J. Madix, Sol. State Commun. **54**, 493 (1985)
6.50 D.R. Jennison, G.D. Stuccky, R.R. Rye, J.A. Kelber: Phys. Rev. Lett. **46**, 911 (1981)
J.A. Kebler, D.R. Jennison, R.R Rye: J. Chem. Phys. **75**, 652 (1981)
6.51 E. Umbach, Z. Hussain: Phys. Rev. Lett. **52**, 457 (1984)
6.52 M.D. Baker, N.D.S. Canning, M.A. Chesters: Surf. Sci. **111**, 452 (1981)
6.53 W. Eberhardt, C.T. Chen, W.K. Ford, E.W. Plummer, H.R. Moser: In *DIET II*, ed. by W. Brenig, D. Menzel, Springer Ser. Surf. Sci. Vol. 4, (Springer, Berlin, Heidelberg 1985). p. 50
6.54 O. Björneholm, A. Nilsson, H. Sandell, B. Hernhäs, N. Mårtensson: Phys. Rev. Lett. **68**, 1892 (1992)
6.55 C.T. Chen, R.A. DiDio, W.K. Ford, E.W. Plummer, W. Eberhardt: Phys. Rev. B **32**, 8434 (1985)
6.56 E.W Plummer, C.T. Chen, W.K. Ford, W. Eberhardt, R.P. Messmer, H.J. Freund: Surf. Sci. **158**, 58 (1985)
6.57 W. Wurth, C. Schneider, R. Treichler, E. Umbach, D. Menzel: Phys. Rev. B **35**, 7741 (1987) and Phys. Rev. B **37**, 8725 (1988)

6.58 R. Murphy, E.W. Plummer, C.T. Chen, W. Eberhardt, R. Carr: Phys. Rev. B **39**, 7517 (1989)
6.59 G. Illing, T. Porwol, I. Hemmerich, G. Dömötör, H. Kuhlenbeck, H.J. Freund, C.M. Liegener, W. von Niessen: J. Electr. Spectrosc. **51**, 149 (1990)
6.60 W. Wurth, P. Feulner, D. Menzel: Physica Scripta T **41**, 213 (1992)
6.61 O. Björneholm, A. Sandell, A. Nilsson, N. Mårtensson, J.N. Andersen: Physica Scripta T **41**, 217 (1992)
6.62 M. Ohno: Phys. Rev. **B45**, 3865 (1992)
6.63 W. Wurth, D. Menzel, E. Umbach: Phys. Rev. **B45**, 3868 (1992)
6.64 D.A. Shaw, G.C. King, F.H. Read, D. Cvejanovic: J. Phys. B **15**, 1785 (1982)
6.65 D. Sondericker, Z. Fu, W. Eberhardt: unpublished

7. Synchrotron Radiation in the Far Infrared: Infrared Reflection Absorption Spectroscopy

F.M. Hoffmann and *G.P. Williams*

Vibrational spectroscopy at surfaces has developed in recent years into a powerful surface science tool for studying the adsorption of molecules on well-defined metal surfaces. Two techniques have emerged as the most widely used vibrational probes – Infrared Reflection Absorption Spectroscopy (IRAS) [7.1–3] and high-resolution Electron Energy Loss Spectroscopy (EELS) [7.4]. Both techniques are shown schematically in Fig. 7.1. EELS is at present the most widely used technique due to its high sensitivity (detection limit $< 1/1000$ of a CO monolayer) and its large spectral range (4000–50 cm^{-1}). However, the limited spectral resolution of most instruments (10–40 cm^{-1}) has prevented so far its use in vibrational line-shape analysis and other applications requiring high spectral resolution. In the past few years IRAS has gained wide acceptance as a high-resolution vibrational probe due to drastic improvements in sensitivity by the use of Fourier Transform Spectrometers (detection limit $< 1/1000$ of a CO monolayer). Rapid scanning Fourier Transform InfraRed Spectrometers (FTIR) also permit time-resolved spectroscopy on a millisecond time scale (10–100 ms). Further advantages of this technique are the inherent high spectral resolution (< 0.1 cm^{-1}) and the ability to perform measurements both at low pressure (UHV) and at atmospheric pressures. The latter is of importance in the study of catalytically relevant surface reactions which in many cases require high gas pressures. One of the limitations of infrared spectroscopy has been its restricted spectral range (4000–600 cm^{-1}). The conventionally used thermal emitter sources for infrared radiation have a drastically reduced intensity in the Far Infrared (FIR) region below 600 cm^{-1}. In addition, photoconductive detectors cease to operate much below 400 cm^{-1} and less sensitive bolometers have to be used, further reducing the signal-to-noise. This is unfortunate because this region is of particular importance since it contains most of the molecule-substrate vibrations and hence, information about the chemical bond between the adsorbed molecule and the metal substrate. The recent development of synchrotron light sources in the FIR [7.5–8] offers an unique opportunity in overcoming the inherent limitations of thermal emitter sources to access this important class of vibrations in the FIR region.

In this chapter we present examples which demonstrate the breakdown of the surface selection rule in the *far infrared* as a result of resonant adsorbate-substrate interactions, and the application of IRAS to the observation of phonons at surfaces.

Springer Series in Surface Sciences, Vol. 35
Applications of Synchrotron Radiation Ed.: W. Eberhardt

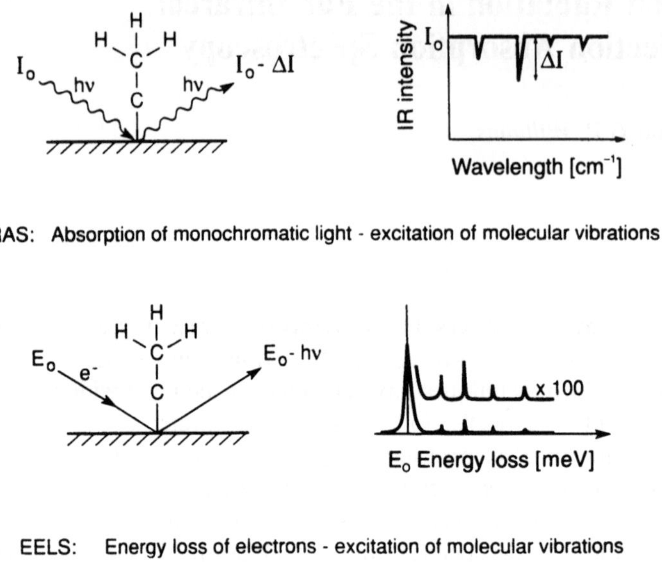

Fig. 7.1. Schematic representation of Infrared Reflection Absorption Spectroscopy (IRAS) and high resolution Electron Energy Loss Spectroscopy (EELS). IRAS: The absorption of monochromatic light leads to the excitation of molecular vibrations which are recorded in an IR absorption spectrum. EELS: Scattering of low energy electrons results in an energy loss as molecular vibrations are excited. The recorded energy loss spectrum is the equivalent of the IR absorption spectrum

7.1 IR Spectroscopy at Surfaces

The absorption of IR radiation by a molecule adsorbed on a surface is dominated by the dielectric behavior of the metal surface, since the electric field of the incident light interacts with the metal electrons. The reflection of IR light from a *clean* metal substrate can be treated theoretically by using Fresnel's equations [7.9]. With the complex index of refraction $n' = n + ik$ and assuming $n^2 + k^2 \gg 1$ (valid for the infrared), one obtains for the reflected light incident on a *clean metal surface*

$$R_s = \frac{(n - \sec\phi)^2 + k^2}{(n + \sec\phi)^2 + k^2} \tag{7.1}$$

$$R_p = \frac{(n - \cos\phi)^2 + k^2}{(n + \cos\phi)^2 + k^2} \tag{7.2}$$

$$\tan\Delta = \tan(d_p - d_s) = \frac{2k\tan\phi\sin\phi}{\tan^2\phi - (n^2 + k^2)} \tag{7.3}$$

where R_s and R_p are reflected intensities of *s*- and *p*-polarized light, d_s and d_p are the phase shifts upon reflection and ϕ is the angle of incidence as defined in Fig.

7.2. E_p and E_s define the electric-field components parallel and perpendicular to the plane of incidence. For the sake of simplicity and because we deal mainly with reflection at grazing angle of incidence, we neglect in E_p the component parallel to the surface ($E_{p\parallel}$). Calculations for a typical metal (Cu) have been performed by *Pritchard* [7.10]. Figure 7.3a shows that the *s*-component of the electric field at the surface suffers a uniform phase shift of 180° for all angles of incidence. This means that the electric component E_s is almost completely suppressed at the surface due to destructive interference of the incoming and outgoing waves. This effect is the reason for the *surface dipole selection rule in* IRAS, which states that *only vibrations with a dipole moment perpendicular to the metal surface can be observed* with IRAS. However, since there is a small but finite component of the electric field parallel to the surface there are exceptions to this rule in the FIR, as discussed further below [7.11].

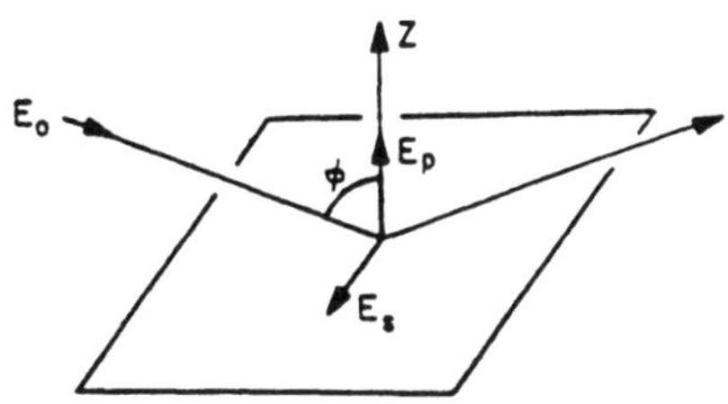

Fig 7.2. Reflection of light at a surface. The electric field components are denoted E_p and E_s

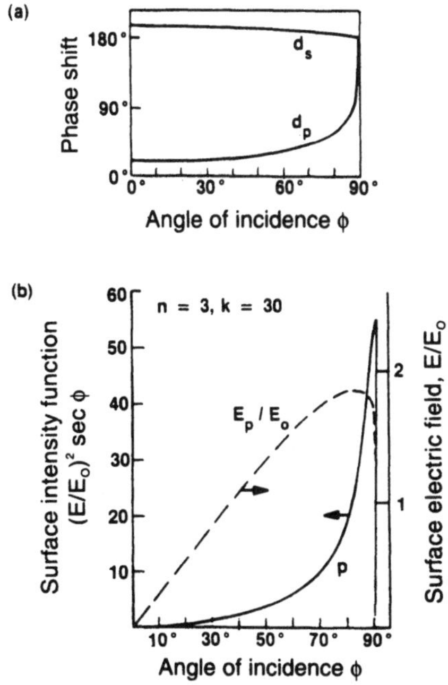

Fig. 7.3a, b. Reflection of IR light at a *clean* metal surface (Cu). (**a**) Phase shifts d_s and d_p vs. angle of incidence ϕ (**b**) Electric field and intensity of *p*-light as function of the angle of incidence ϕ. From Ref. 7.10, with permission

In contrast to E_s which is close to zero at all angles, the electric field E_p at the surface exhibits a pronounced angular dependence. E_p increases with the angle of incidence and reaches a maximum close to grazing incidence. Since the intensity of absorption is proportional to E_p^2 (and to the number of molecules over which the field is exerted), it is useful to obtain the surface intensity function $E_p^2 \sec\phi$. The plot of this function in Fig. 7.3b shows that the surface intensity is *strongly peaked at very high angles of incidence.* In Table 7.1, the optimum angle of reflectance ϕ_{opt} has been calculated by *Greenler* [7.12] for several typical metals in the mid-IR region at 2100 cm^{-1} (e.g., for the C–O stretching vibration of adsorbed CO) and the FIR (Far InfraRed) region at 500 cm^{-1} (e.g., for the *M*–CO stretch of adsorbed CO). The table demonstrates that *the optimum angle of incidence in the FIR region is close to grazing incidence.* This means that the ideal FIR source should have a small angular spread or high brilliance in order to achieve a high angle of incidence.

For a more exact description including the presence of adsorbates, one has to take into account the optical properties of the adsorbate. This can be done with a three-phase model, as illustrated in Fig. 7.4, using a classical macroscopic description with isotropic complex dielectric constants $\varepsilon' = n + ik$ for the different regions of interest: the metal substrate (3), the adsorbate film (2) and the vacuum above the film (1). The theoretical treatment of this problem by *McIntyre* and *Aspnes* [7.13] results in expressions for the reflectivity change ΔR at an *adsorbate covered* surface

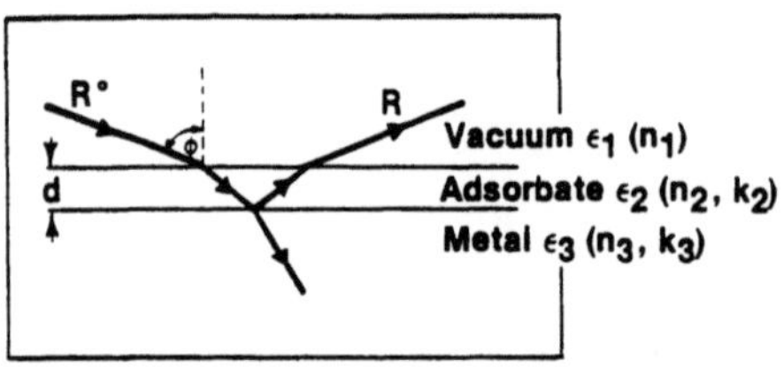

Fig. 7.4. Three phase model for reflection of light at a surface covered with an adsorbate

Table 7.1. Optical constants and optimum angle of incidence in IRAS for metals in the Mid-Infrared and the Far-Infrared [7.12]

	2100 cm^{-1}			500 cm^{-1}		
Metal	n	k	ϕ_{opt}	n	k	ϕ_{opt}
Ag	2.9	33.7	87.2	36.8	127.9	89.6
Au	4.2	27.6	87.8	43.0	93.0	89.0
Cu	3.5	30.0	88.0	36.2	104.9	89.3
Ni	5.4	18.6	86.2	18.6	64.4	88.9
Pd	6.3	15.3	85.0	24.9	70.2	89.0
Pt	3.0	20.0	86.7	31.4	67.0	88.5

$$\frac{\Delta R}{R_s} = (8\pi d/\lambda)\cos\phi \,\mathrm{Im}\left\{\frac{\varepsilon_2 - \varepsilon_3}{1 - \varepsilon_3}\right\} \tag{7.4}$$

$$\frac{\Delta R_p}{R_p} = (8\pi d/\lambda)\cos\phi \,\mathrm{Im}\left\{\left(\frac{\varepsilon_2 - \varepsilon_3}{1 - \varepsilon_3}\right)\frac{1 - (\sin^2\phi/\varepsilon_2 + \sin^2\phi/\varepsilon_3)}{1 - (\sin^2\phi + \sin^2\phi/\varepsilon_3)}\right\} \tag{7.5}$$

λ is the wavelength of the light and s, p refer to s- and p-light, as previously defined in Fig. 7.2. Taking into account that for IR frequencies $|\varepsilon_3| \gg |\varepsilon_2| \approx 1$ and $d/\lambda \ll 1$, one immediately realizes that the reflectivity change of s-light is negligibly small. This demonstrates again the *surface selection rule* according to which dipole-exited modes parallel to the surface are not observed [7.11]. Equation (7.5) can be further simplified according to *Ibach* [7.4], assuming again $|\varepsilon_3| \gg |\varepsilon_2|$ and $\cos^2\phi > \varepsilon_3^{-1}$ (a condition satisfied in most experimental cases)

$$\frac{\Delta R_p}{R_p} = \left(\frac{8\pi d \sin^2\phi}{\lambda\cos\phi}\right)\mathrm{Im}\left\{\frac{1}{(1 - \varepsilon_2)}\right\} \tag{7.6}$$

This demonstrates again the necessity of working at high angles of incidence in an IRAS experiment. We also see from equation (7.6) that the absorption signal $\Delta R_p/R_p$ decreases with the wavelength as $1/\lambda$. This means that *molecule–substrate vibrations in the Far-Infrared* ($< 500\,\mathrm{cm}^{-1}$) *region are weaker by approximately one order of magnitude* in comparison to the internal vibrations of a molecule in the Mid-Infrared.

In conclusion, the theoretical treatment described here gives us a guidance for experiments in IRAS:

1) The *surface selection rule*, according to which only dipole moments perpendicular to the surface can be observed, makes it *necessary to use p-light*, since s-light does not give an appreciable contribution to the absorption band.
2) The requirement to work with *reflections at grazing angle of incidence*, in particular in the FIR makes it advantageous to use radiation sources with high brilliance.
3) The *decrease of the absorption intensity with increasing wavelength* presents a serious sensitivity problem in the Far Infrared spectral region mandating the use of a radiation source of high intensity and stability.

7.2 Synchrotron Radiation in the Far Infrared

The conventional radiation sources used for vibrational spectroscopy are of the *thermal emitter* type such as SiC globars or Nernst glowers. The spectral distribution of a black-body source is described by Planck's radiation law:

$$P_{\mathrm{bb}}(\lambda) = A\frac{2\pi hc^2}{\lambda^5}\frac{d\lambda}{(e^{hc/\lambda kT} - 1)} \tag{7.7}$$

where P is the power, λ the wavelength, T the temperature and A the total area of the emitter. In the limit of long wavelength we obtain the Raleigh-Jeans law

$$P_{bb}(\lambda) \approx \text{const } \lambda^{-4} \tag{7.8}$$

This equation shows that the power of a black body drastically falls off at long wavelengths. In fact, it changes by more than an order of magnitude each time λ doubles!

The spectral distribution of a *synchrotron radiation source* in the limit of long wavelength is given by [7.14, 15]

$$P_{syn}(\lambda) \approx \text{const } I\,\theta\,\rho^{1/3}\,\lambda^{-7/3} \tag{7.9}$$

where I is the electron current in the storage ring, θ the horizontal collection angle and ρ the radius of the bending magnets in the storage ring. The ratio of (7.8 and 7.9)

$$\frac{P_{syn}}{P_{bb}} = \text{const } \lambda^{5/3} \tag{7.10}$$

indicates that the power advantage of the synchrotron increases as λ increases! The critical issue for grazing incidence reflection spectroscopy, however, is not only the power. The relatively small size of single metal crystals (typ. 10×10 mm) and the necessity for grazing incidence, mean that this power has to be concentrated into a relatively small "phase space". Thus, if the angle of incidence is $87 \pm 2.5°$, giving an opening angle of 5°, (f 10) or ≈ 0.1 radian, and if we allow the angular spread in the other direction (out of the plane of incidence) to be the same, then the throughput (or etendue – solid-angle × area) is $10 \times 10 \times (\cos 87°) \times 0.1^2$ or $\approx 0.05\ \text{mm}^2$ steradian. Now, since a 10×10 mm black body source emitting into 2π steradian has an emittance of $10 \times 10 \times 2\pi$ or $\approx 600\ \text{mm}^2$ steradian, then only one photon in 10000 or so can be utilized by the IRAS experiment. In contrast, for the synchrotron, the emittance is a strong function of wavelength but in this region is typically $\approx 0.1\ \text{mm}^2$ steradian, thus matching the experiment perfectly.

In Table 7.2, we reproduce the results of a theoretical calculation for the opening angle, source size, emittance and brightness of synchrotron radiation in the IR and FIR at the NSLS at Brookhaven with a stored beam of 1 A [7.16]. Table 7.2 shows that both the radiation opening angle and the source size increase with lower wave numbers. This limits the useful spectral range to $10\ \text{cm}^{-1}$ at the lower limit, since the radiation opening angles are limited both horizontally (H) and vertically (V) to 100 mrad at the NSLS by the mechanical design the finite aperture, and diffraction causes some additional losses at the long wavelength end ($10\ \text{cm}^{-1}$). Table 7.2 also gives a comparison between the brightness of the synchrotron and that of a black-body source at 2000 K. For the black body we have assumed that only $10\ \text{mm}^2$ of the area could be used into an angle of 100×100 mrad. These are typical values for an IRAS experiment using a 15 mm × 15 mm single crystal at 87° angle of incidence. The comparison demonstrates a brightness advantage of synchrotron radiation of about three orders of magnitude in the FIR.

Table 7.2. Theoretical comparison of synchrotron radiation and black-body source brightness in the IR (see text for details) [7.16]

Wavenumber [cm^{-1}]	NSLS radiation Opening angle V × H [mrad]	NSLS source size V × H [mm]	NSLS emittance [mm^2 sr]	NSLS brightness [flux/emitt.]	Black-body brightness [flux/emitt.]
10,000	9 × 100	0.3 × 3	8×10^{-4}	7.3×10^{19}	2.2×10^{16}
1,000	20 × 100	0.5 × 3	3×10^{-3}	9.7×10^{18}	2.9×10^{16}
100	75 × 100	1.3 × 3	3×10^{-2}	4.0×10^{17}	4.0×10^{14}
10	100 × 100	10 × 10	1	5.6×10^{15}	4.1×10^{12}

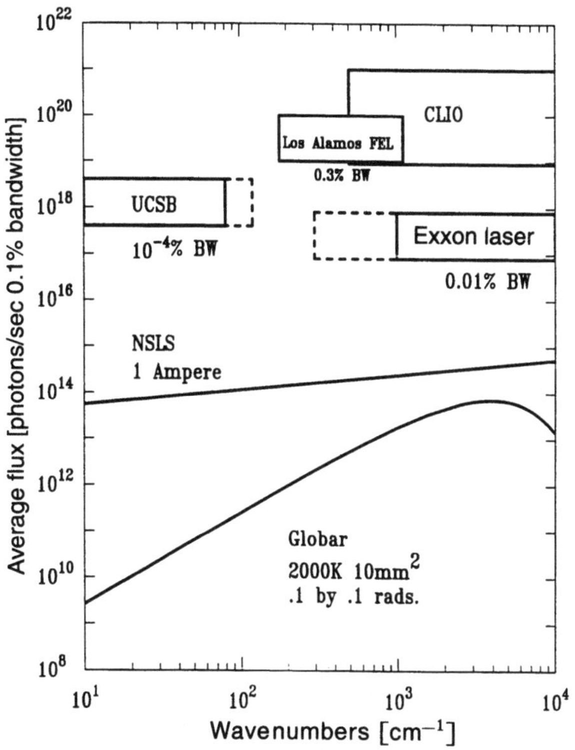

Fig. 7.5. Comparison of radiation sources in the infrared: Average photon flux vs. wavenumber for black body, synchrotron, free electron lasers (FEL) and a tunable IR laser. Calculations after *Duncan* and *Williams* [7.14, 16]

In Fig. 7.5 the average photon flux for synchrotron radiation at the NSLS and for the black body are shown as a function of the wavelength, and compared to various IR lasers. The plot demonstrates that over the infrared region, the synchrotron offers an increase in detected light of 10–10,000 over a conventional globar for experiments with limited throughput, with particular gains at longer wavelength. However, we note that for the IRAS experiment, out-of-plane opening angles can be, in principal, higher than the present 100 mrad limit imposed by the present design of the beam line. We also show for comparison in Fig. 7.5 the output of 3 free-electron lasers: at the University of California Santa Barbara, at CLIO (LURE, Orsay) and at Los Alamos [7.17], and the output from the tunable IR Raman laser at Exxon [7.18]. It is evident that lasers are far superior in delivering high photon fluxes. In fact, the comparison in Fig. 7.5 represents average photon fluxes integrated over time. The peak photon fluxes of lasers can be many orders of magnitude higher due to their pulsed time structure, which has led to their use in time-resolved spectroscopy and nonlinear optical Sum Frequency Generation (SFG). However, for surface IRAS, where access to a large spectral range and high source stability are usually required, tunable IR lasers have found only limited use due to their limited tunability and their inherent instability, which can lead to large pulse-to-pulse intensity fluctuations [7.19, 20]. Therefore, broadband sources such as thermal emitters and synchrotron radiation, which permit the use of rapid scanning Fourier transform infrared spectrometers, have proven to deliver superior signal-to-noise performance and spectra range for surface IRAS.

Measurements which compare the spectral distribution of radiation emitted from the synchrotron with that of a globar source are presented in Fig. 7.6 [7.21]. The solid line shows the spectral distribution of synchrotron radiation emitted from the NSLS Infrared Beam line at Brookhaven (100 mrad, 750 mA

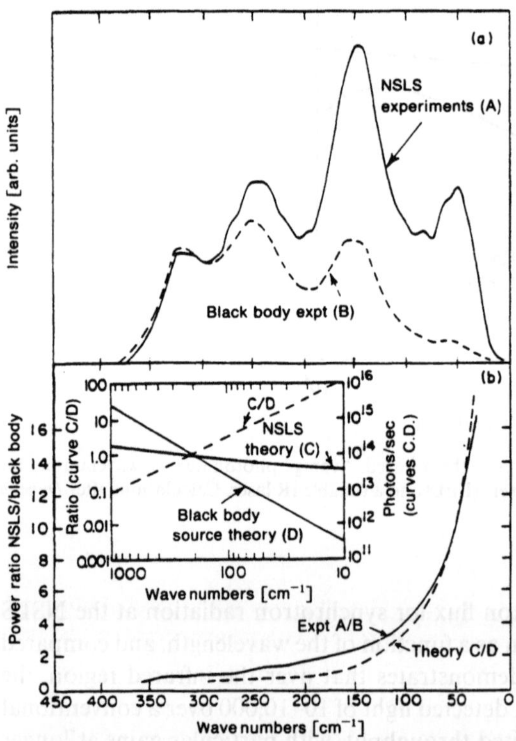

Fig. 7.6a, b. Experimental comparison between globar and synchrotron radiation intensities. (**a**) Experimentally determined intensity from globar (dashed curve) and synchrotron radiation from the NSLS IR beamline with a stored current of 750 mA (solid line). (**b**) Experimental (solid line) versus calculated (dashed line) power ratio of NSLS/globar. The inset shows the calculated intensities for the NSLS (C) and for the blackbody (D), from which the ratio C/D is derived [7.21]

beam current), while the broken line represents black-body radiation emitted from a globar (285 mrad or $f/3.5$). The experimental arrangement used for this purpose is described further below (Fig. 7.8). From Fig. 7.6a we note that the total photon flux from the synchrotron dominates in the FIR below 200 cm^{-1}. This is in agreement with the theoretical predictions indicated in the inset of Fig. 7.6b. Note that the decrease in intensity at higher wave numbers in both cases is due to the cut-off at high frequency of the filter used with the liquid-He-cooled bolometer. The comparison between synchrotron and globar improves drastically in favor of synchrotron radiation if we take the higher brightness of the latter source into account. Figure 7.7 shows the intensity ratio between the synchrotron and globar for radiation transmitted through small apertures of 1, 2, 5 and 10 mm. This comparison shows that for experiments with small samples the intensity advantage of the synchrotron is about two orders of magnitude. If we further take the smaller emission angle of the synchrotron radiation into account – 100 mrad ($f/10$) vs. 285 mrad ($f/3.5$) – the ratio improves another order of magnitude. Thus, the experimental data agree well with the theoretical predictions made earlier.

The optical layout of a FIR beam line is depicted in Fig. 7.8, which represents the U4IR-line at the UV ring at the National Synchrotron Light Source at Brookhaven [7.7]. The storage ring is operated at an energy of 750 MeV with stored electron currents of up to 850 mA. The IR beam line uses a specially constructed extraction port with a 6 cm × 6 cm aperature, which allows radiation to be extracted with a 100 × 100 mrad collection angle. This radiation is reflected vertically upwards to an experimental platform 3 m above the experimental floor and then focussed horizontally onto a diamond window (250 μm

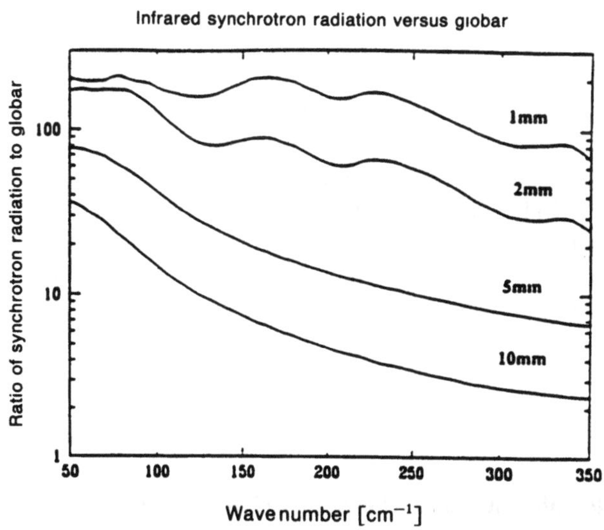

Fig. 7.7. Comparison of Synchrotron radiation versus globar intensity transmitted through small apertures of 1, 2, 5 and 10 mm

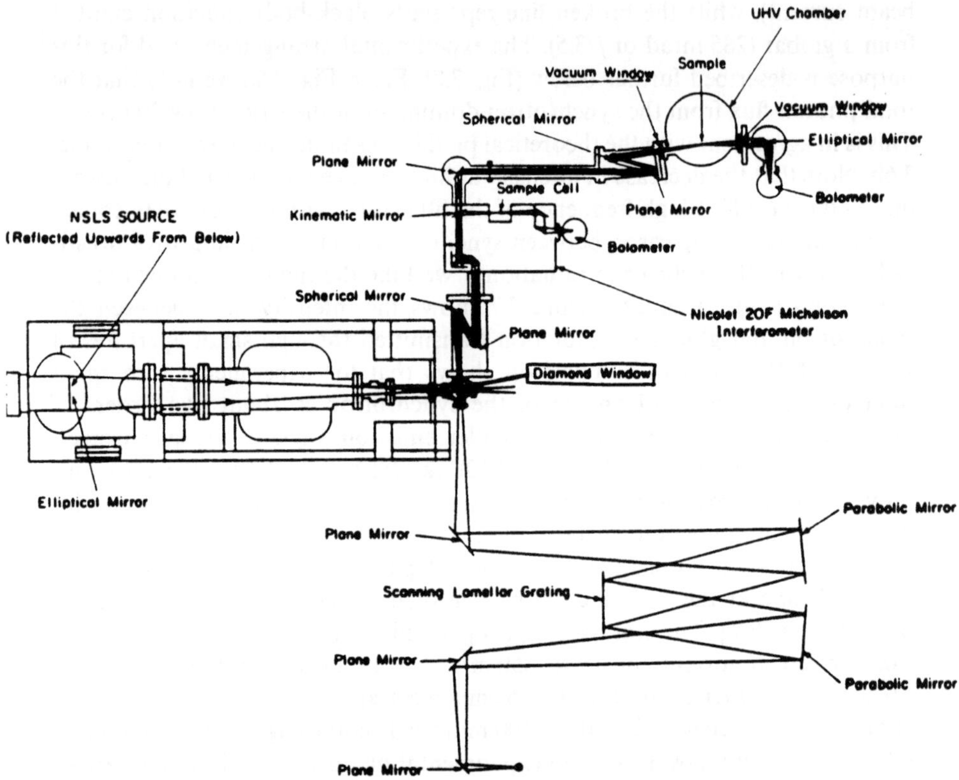

Fig. 7.8. Optical Layout of Infrared Beamline at the NSLS Brookhaven [7.7]

thick, 9 mm diameter). This window separates the Ultra High Vacuum (UHV) of the storage ring (10^{-10} Torr) from the high vacuum of the infrared optics (10^{-2} Torr). Following the diamond window the IR beam can then be directed either to a scanning lamellar grating which operates in the lower end of the FIR region from 1–100 cm^{-1} or to an evacuated Interferometer operating in the FIR from 50–800 cm^{-1} (Nicolet 20 F). The exit beam from the interferometer is then brought into a UHV chamber where it is focussed at glancing angle of incidence (87°–89°) onto a single crystal sample, and then focussed onto a liquid-He-cooled boron-doped silicon bolometer.

7.3 Examples

7.3.1 Resonant Adsorbate-Substrate Interactions for CO Adsorbed on Cu(100)

The intensity and brightness advantage of synchrotron radiation in the FIR gives us access to the spectral region which contains the adsorbate-substrate vibrations. These modes are of particular importance since they contain information about the chemisorption bond that forms when a molecule or atom is adsorbed on a surface. One of the classical model systems is the adsorption of CO on single crystal Cu(100). This system has been characterized extensively and is thought to be well understood. For CO adsorbed on Cu(100) we expect four vibrational modes as schematically represented in Fig. 7.9 [7.2]. There is one *internal* mode, the C–O stretching vibration ν_1, and three *external* modes, the frustrated translation perpendicular to the surface ν_2 (M–CO stretch), the frustrated rotation ν_3 (M–CO bend) and the frustrated translation parallel to the surface ν_4. The external modes correspond to the rotational and translation modes of the molecule in the gas-phase. They are sometimes also called *frustrated or hindered* modes because the bonding to the surface restricts the motion of the molecule. If the molecules are adsorbed in the highest symmetry configuration, i.e., in a linear on-top site with four-fold rotational symmetry, then according to the *surface selection rule* only the modes with a dipole moment change perpendicular to the surface or modes with A_1 symmetry should be observed. This means that ν_1 and ν_2 are allowed and both ν_3 and ν_4 are forbidden. That the molecule is in fact adsorbed in such a high symmetry site can be experimentally determined by combining infrared spectroscopy with LEED (Low Energy Electron Diffraction). Figure 7.10 shows vibrational spectra of the

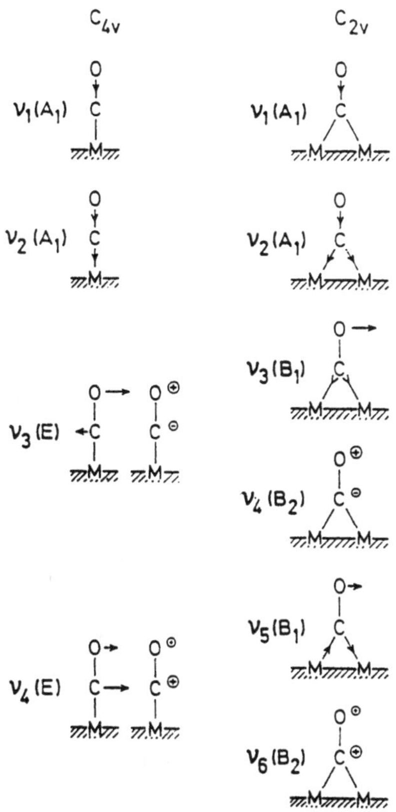

Fig. 7.9. The normal mode vibrations for CO adsorbed on a metal in an on-top site with four-fold rotational symmetry (left) and in a bridge site with two-fold rotational symmetry (right)

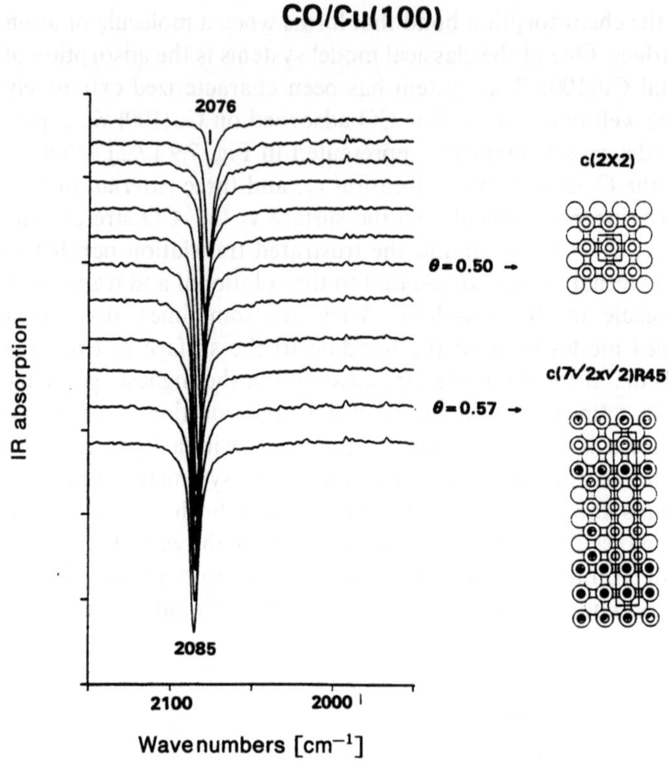

Fig. 7.10. Vibrational spectra of the C–O stretch region for CO adsorbed on Cu(100) as a function of increasing coverage show the adsorption in the on-top site. LEED indicates the formation of ordered structures, the $c(2\times2)$ at $\theta = 0.50$ and the $c(7\sqrt{2}x\sqrt{2})$-$R45°$ at $\theta = 0.57$ (LEED structures reproduced with permission from [7.23])

C–O stretching region as a function of increasing coverage with CO. At all coverages only one band is observed, which shifts slightly with increasing coverage from 2076 to 2085 cm^{-1} in agreement with earlier work by *Pritchard* et al. [7.22]. The frequency of this band indicates that the CO is adsorbed in a *linear* (on-top) site. A more detailed discussion of the relationship between adsorption sites and C–O stretch frequency can be found in [7.1]. Since we observe only *one* C–O stretch band, we expect that *all molecules are adsorbed in on-top sites.* The use of LEED allows us to relate the vibrational spectra to the structure of the CO adlayer. At a CO coverage of $\theta = 0.5$ an ordered $c(2\times2)$ overlayer is observed. This structure accommodates all of the molecules in identical on-top adsorption sites with C_{4v} symmetry (Fig. 7.10). At higher coverage $\theta > 0.50$ a compression structure is observed, which results in the occupation of low symmetry sites where CO molecules are located on adjacent copper sites [7.23].

Vibrational spectra of the region containing the external CO/Cu modes at three different coverages are shown in Fig. 7.11 [7.24]. There, vibrational spectra of the low-frequency modes are depicted in the left-hand panel. For comparison the C–O stretch band in the high-frequency region is displayed in the right-hand panel. Two bands at 283 cm^{-1} and 343 cm^{-1} are observed in the low-frequency region Fig. 7.11a. We notice immediately a drastic difference in absorption intensity between the internal C–O stretch of $\approx 6\%$ and the external modes $\approx 0.15\%$. The weakness of the internal modes confirms the theoretically predicted dependence of the absorption intensity on the wavelength in (7.6). As a consequence, the signal-to-noise ratio for the external modes is correspondingly lower than for the internal mode. A similarly weak *M*–CO stretching band has also been observed for CO adsorbed on Pt(111) by several groups [7.25] confirming for other metals that external modes result in absorption bands that are typically one to two order of magnitudes smaller in intensity than the internal C–O stretch. Based on the frequency, it is clear that the two bands are due to the *M*–CO stretch ν_2 and the frustrated rotation ν_3 [7.23]. The frustrated translation parallel to the surface, ν_4, is expected to have a frequency below 100 cm^{-1}. In a recent Raman study of CO adsorbed on disordered copper films *Akemann* and *Otto* report a frequency of 24 cm^{-1} for this mode [7.26].

In order to assign the two low frequency bands to the ν_2 and ν_3 modes, respectively, we utilize the isotopic substitution method, where the C- and O-atoms in CO are selectively replaced by their heavier isotopes ^{13}C and ^{18}O. The results presented in Fig. 7.12 reveal that the mode at 345 cm^{-1} (a) shifts to 334 cm^{-1} for $^{13}C^{18}O$ (b), and 337 cm^{-1} for $^{12}C^{18}O$ (c). These isotopic shifts are in agreement with the assignment of this band to the Cu–CO stretch mode

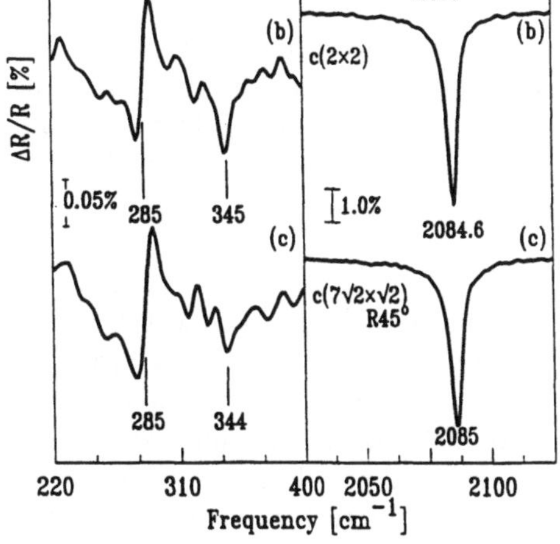

Fig. 7.11a–c. Vibrational spectra for CO adsorbed on Cu(100) with increasing coverage in the *M*–C stretching region and in the C–O stretch region; (**a**) $\theta = 0.34$; (**b**) $\theta = 0.50$; $c(2\times 2)$; (**c**) $\theta = 0.57$, $c(7\sqrt{2}\times\sqrt{2})$-*R45°*. The data in the *M*–C region were obtained with synchrotron radiation at 8 cm^{-1} resolution. The data for the C–O region were obtained at 2 cm^{-1} resolution using a conventional globar source [7.24]

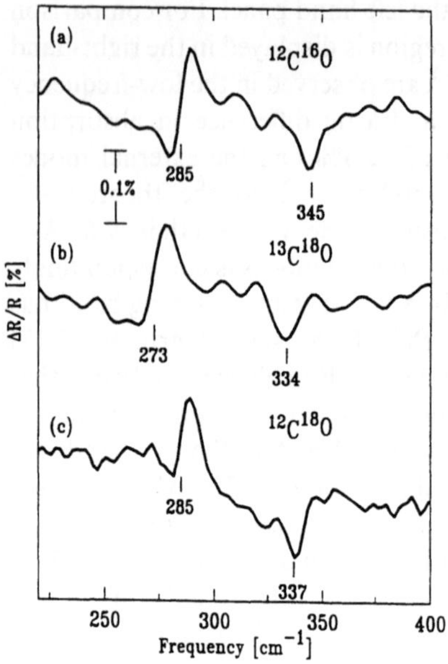

Fig. 7.12a–c. Vibrational spectra of the *M*–CO stretch region for CO adsorbed in the *c*(2×2) structure on Cu(100) ($\theta = 0.5$) at 100 K. **a)** $^{12}C^{18}O$; **b)** $^{13}C^{18}O$; **c)** $^{12}C^{18}O$ [7.24]

[7.27]. The second absorption band observed at 285 cm^{-1} (a) is remarkable due to its unusual line shape, which will be discussed later. This band shifts with $^{13}C^{18}O$ substitution to 273 cm^{-1}. However, it does *not* shift for $^{12}C^{18}O$ substitution (c). The absence of the ^{18}O shift is characteristic for the CO bending mode (frustrated rotation ν_3). The frequency of the frustrated rotation ν_3 of a linear molecule (Cu–C–O) is given by [7.27]:

$$(2\pi\nu_3)^2 = \frac{k\delta}{l_1 l_2}\left(\frac{l_1^2}{m_O} + \frac{l_2^2}{m_{Cu}} + \frac{(l_1 + l_2)^2}{m_C}\right) \tag{7.11}$$

where k is the force constant and $l_1 = d_{Cu\ C} \approx 1.92$ Å, $l_2 = d_{C\ O} \approx 1.18$ Å, $m_{Cu} = 63.5$, $m_O = 16$ amu and $m_C = 12$ amu. This can be approximated to $\nu_3 \sim 1/\sqrt{m_C}$. This means that the frequency for the frustrated rotation will shift with ^{13}C substitution, but *does not shift with ^{18}O substitution.* In other words, most of the vibrational motion is from the C-atom while the O-atom essentially does not move. Isotopic substitution thus allows us to assign the band at 285 cm^{-1} to the frustrated rotation and the band at 345 cm^{-1} to the Cu–CO stretching vibration.

Having assigned the bands at 345 and 285 cm^{-1}, we have to explain both the infrared activity and the line shape of the latter band. We recall from our previous discussion of the surface selection rule that in a high symmetry configuration only the C–O stretch ν_1 and the *M*–CO stretch ν_2 (modes with a dipole moment change perpendicular to the surface) are allowed and the frustrated rotation ν_3 and the frustrated translation parallel to the surface ν_4 are forbidden. We further recall that for the $c(2\times2)$ structure at $\theta = 0.5$ in Fig.

7.11b, the C_{4v} symmetry of the adsorbed CO molecule has been established experimentally with LEED and IR. It is thus clear that *the observation of the frustrated rotation at 285 cm^{-1} constitutes a breakdown of the surface selection rule.*

The explanation for the observation of a forbidden mode at 285 cm^{-1} in Figs. 7.11 and 7.12 can be found in its particular line shape, which is often referred to as a Fano line shape. Such Fano line shapes can arise from the coupling of a sharp vibrational mode to a broad absorption band. The interference of the sharp vibrational mode with such a 'continuum' band then results in a modulation of the latter as schematically depicted in Fig. 7.13. Such a broad IR absorption band is observed experimentally (Fig. 7.14). There, we observe a broad band IR absorption spanning the accessible spectral range from 180 to 500 cm^{-1}. Since the broad-band absorption increases with CO coverage,

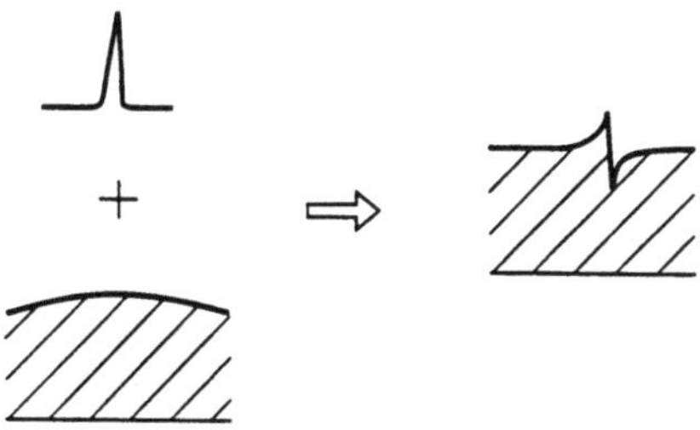

Fig. 7.13. Fano line shapes originate from the coupling of sharp vibrational modes with a broad continuum. This resonant interaction results in the modulation of the continuum and hence the Faco line shape of the mode

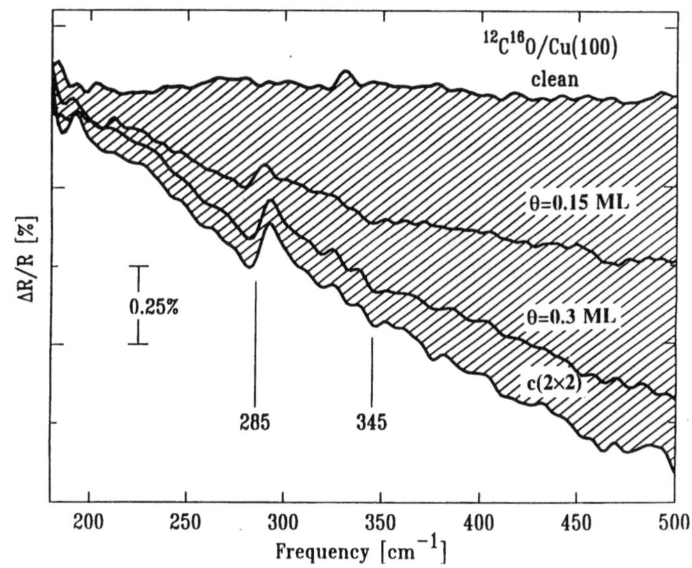

Fig. 7.14. Broad band absorption resulting from the adsorption of CO on Cu(100) as a function of increasing coverage [7.24]

it must be related to the adsorption of the CO. It is clear that the origin of this background must be due to the *parallel electric field E_s inside the metal surface.* We have previously shown in Sec. 7.1 that *outside* of the metal surface the electric field is dominated by the electric field perpendicular to the surface E_p [7.28]. As discussed in detail by *Persson* [7.29], the ratio of the electric field components *outside* of the surface is given by $|E_s/E_p| \sim \omega/\omega_p$, where ω is the frequency range of external vibrations $|E_s/E_p| \sim \omega/\omega_p \approx 10^{-2}$. *Inside* of the metal surface, however, the *parallel field E_s is the dominating field* since the perpendicular field E_p is even smaller, as schematically sketched in Fig. 7.15, now the ratio $|E_s/E_p| \sim (\omega/\omega_p)^{-1} \approx 10^2$ and the corresponding intensity ratio is $I_s/I_p = |E_s/E_p|^2 \approx 10^4$.

Two models have been suggested to explain the origin of this broad band IR absorption:

I) Excitation of an electronic interband transition between the metal and the adsorbed CO [7.24]. Due to symmetry requirements this transition must be parallel to the surface in order to couple to a vibrational mode parallel to the surface. Strong coupling then leads to a modulation of the broad band absorption and the observed Fano line shape.

II) The 'surface resistivity' model proposed by *Persson* [7.30] considers a thin metallic film with a layer of adsorbed molecules. An oscillating electric field acting on the electrons in the films induces a collective drift motion of the electrons. Inelastic scattering of the electrons leads to the excitation of electron-hole pairs, i.e., the electrons experience a friction or resistance in their motion. The presence of a layer of adsorbed molecules results in an increase of the friction which the electrons experience and hence in an uniform background absorption which is proportional to the number of adsorbed molecules. In addition, the oscillating electric field can also excite vibrational motions of the molecules parallel to the surface. The molecule and the drifting electrons can therefore move together in resonance at the frequency of the molecular vibration and the additional friction due to the adsorbate disappears. The IR reflectivity returns at this frequency to the value it had in the absence of the CO adsorbate thus resulting in a sharp anti-absorption band. In reality, damping results in a finite lifetime of the electron-hole pairs and the sharp resonance is broadened to exhibit the line shape observed in Fig. 7.14.

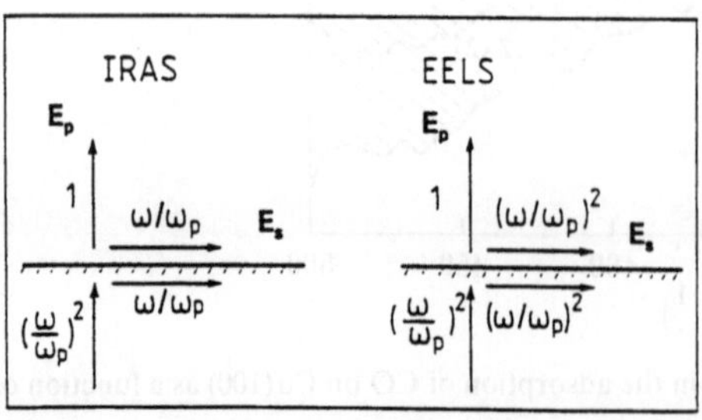

Fig. 7.15. The relative strength of the normal and parallel electric field vector components at a metal surface in IRAS and EELS [7.29]

At present neither of these models have been verified with independent experiments. There is, however, an important difference between these two models, which could be addressed in future experiments. Model (I) is specific to the adsorbed molecule and the substrate metal since it assumes an electronic excitation from the metal into a molecular orbital of the adsorbed molecule. Therefore, the existence of a broad band absorption depends on the electronic band structure of the adsorbate-substrate complex. The latter should be verifiable with photoemission experiments. In contrast, model (II) will lead to a broad-band absorption in most cases, since the scattering of the electrons on the adsorbed molecules is similar for most chemisorbed molecules. This suggests that such anti-absorption bands should be observed for a variety of molecules and metal surfaces. In fact, recent work has shown a similar anti-absorption band for the frustrated rotation of NO adsorbed on Cu(111) [7.31] in support of the scattering model (II) proposed by *Persson*.

We finally note that such a breakdown of the surface selection rule has only been observed, so far, in IRAS. It has not been observed in EELS, where for dipole scattering the surface selection rule still holds [7.23]. The reason for this discrepancy comes from an important difference between IRAS and EELS which is illustrated in Fig. 7.15. In EELS the electric field inside the metal surface is $|E_s| \sim (\omega/\omega_p)^2$, i.e. the field is about two orders of magnitudes weaker than in IRAS resulting in absorption intensities that are weaker by four orders of magnitude. This explains why for CO/Cu(100) the forbidden frustrated rotation is observed in IRAS but does not result in a corresponding dipole scattering loss in EELS.

7.3.2 Potassium–Oxygen Bond Formation

The bond formation between alkali metals and oxygen is of interest in a variety of catalytic reactions where alkalis are used as promoters [7.32]. Alkali metals are very reactive towards oxygen or molecules that contain oxygen such as water, methanol or carbon dioxide. Since these molecules are often present in the reactant gas, the active state of the alkali promoter is usually in the form of an alkali compound such as an alkali oxide, hydroxide, methoxide or carbonate. Vibrational spectroscopy is an important probe in the characterization of the K–O bonds formed in these compounds. Previous EELS experiments have established K–O bond formation for a variety of alkali compounds [7.33], but the limited resolution of EELS in these experiments (typ. 30 cm^{-1}) has prevented a detailed vibrational analysis. The use of high resolution IRAS with a conventional thermal emitter source has not been possible due to the low frequency of the K–O stretching vibrations of these compounds (200–400 cm^{-1}). The utilization of synchrotron radiation, therefore allows alkali compound formation in thin films on metal surfaces to be studied for the first time with high resolution vibrational spectroscopy.

Time-evolved vibrational spectra obtained during oxidation of a thin film of potassium at room temperature are displayed in Fig. 7.16 [7.34]. The potassium

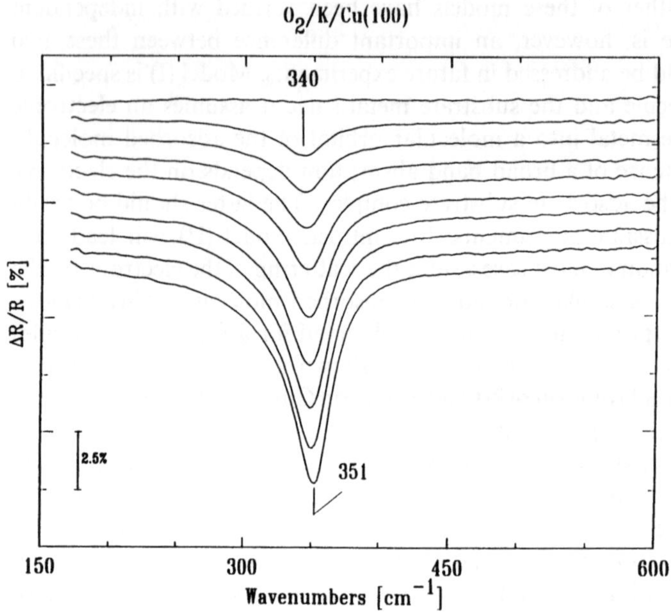

Fig. 7.16. Time-evolved vibrational spectra demonstrating K–O bond formation during exposure of oxygen to a potassium multilayer adsorbed on Cu(100) at room temperature. The spectra have been collected at time intervals of 100 seconds and correspond to oxygen exposures of 1 L each [7.34]

film was prepared by evaporation of approximately 10 monolayers of potassium onto a Cu(100) surface. The vibrational spectra were taken during exposure of the potassium film to oxygen. The time interval between the spectra is 100 s, and since the oxygen pressure was kept at 10^{-8} Torr, each spectrum corresponds to an exposure of 1 L (1 L = 10^{-6} Torr s). The spectra exhibit a single band, which appears at 340 cm^{-1} and shifts with increasing oxygen exposure to higher frequency until it saturates at 351 cm^{-1} after an oxygen exposure of 15 L. The band at 351 cm^{-1} can be assigned to a K–O stretching vibration and therefore characterizes K–O bond formation. Similar K–O bond stretch frequencies (260–340 cm^{-1}) have been observed previously with EELS for a K–O_2 layer adsorbed on Ru(001) [7.35], and for matrix isolated potassium superoxide (307 cm^{-1}) [7.36]. Auger spectra obtained from these layers indicate a K:O stoichiometry of 1:1.7 in agreement with potassium superoxide KO_2 [7.34]. This assignment is further supported by recent model cluster calculations by *Shao* and *Paul* [7.37], who find for a superoxide O_2^- species in a potassium cluster a dominant vibrational band at 330–350 cm^{-1}.

The high reactivity of potassium towards oxygen is demonstrated by the fact that oxidation of the potassium film is observed already at low temperature (Fig. 7.17). Oxidation of a potassium multilayer at 100 K results in a vibrational spectrum characterized by an asymmetric band at 347 cm^{-1}. Annealing of the layer to 350 K results in an increase in intensity and a change towards a more symmetrical line shape. The intensity increase and the change in line shape suggest ordering of the film as a result of the annealing. Further annealing of the film to 650 K results in the decomposition of the potassium super oxide as evidenced by the disappearance of the K–O vibration in Fig. 7.17. Simultaneously we observe the appearance of a band at 621 cm^{-1} which indicates the formation of cuprous oxide Cu_2O in agreement with literature data reported for bulk CuO_2 [7.38, 39]. This assignment is further confirmed with isotopic substitution spectra obtained with $^{18}O_2$. Figure 7.18 shows vibrational spectra obtained after oxidizing potassium layers with O_2 (a) and $^{18}O_2$ (b), respectively, and subsequent annealing to 650 K. The spectra, displayed over an extended spectral range from 50–650 cm^{-1}, display an isotopic shift for the high-frequency band from 629 to 605 cm^{-1}, while the low-frequency bands remain essentially unaffected. Table 7.3 presents a summary of the vibrational assignments of the observed modes together with the results of calculations for a (perfect) Cu_2O lattice, as discussed elsewhere in detail [7.34]. It is evident from Table 7.3 that the band at 629 cm^{-1} and the low-frequency band at 159 cm^{-1} can be assigned to the two infrared active transverse optical phonon bands F_{1u} of the ideal Cu_2O crystal (O_h symmetry). A second low-frequency band at 124 cm^{-1} is assigned to a E_u mode which is inactive in the perfect lattice, but could be activated by bulk or surface defects. The calculations further show that only the high-frequency band at 629 cm^{-1} shifts with isotopic substitution,

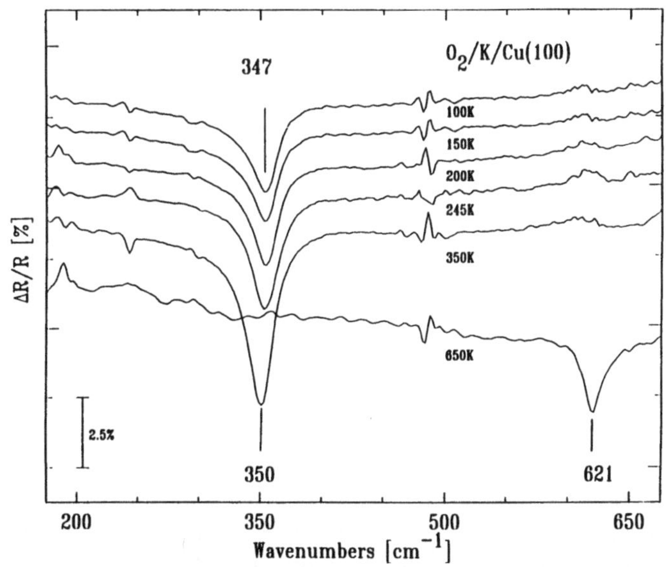

Fig. 7.17. Oxidation of a potassium layer at low temperature (100 K) results in a vibrational line shape characteristic of a disordered layer. Annealing to 350 K results in ordering of the layer. Further annealing to 650 K leads to decomposition of the potassium superoxide and the formation of cuprous oxide Cu_2O [7.34]

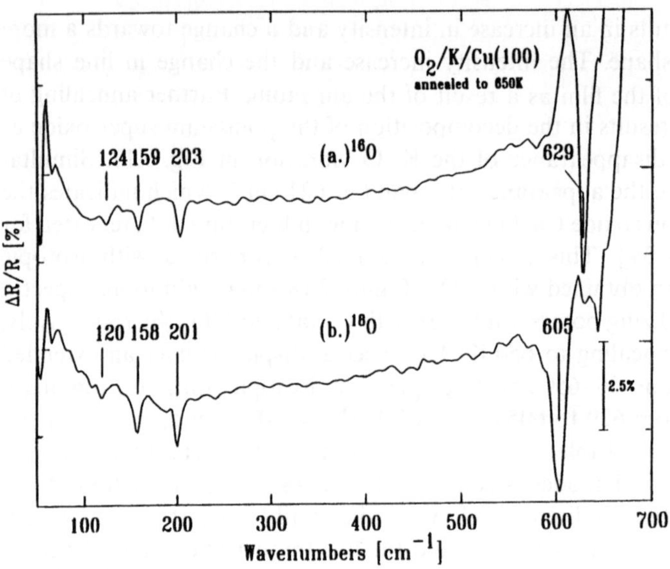

Fig. 7.18. Effect of isotopic substitution with ^{18}O on the vibrational spectra obtained after annealing of a potassium superoxide layer to 650 K [7.34]

Table 7.3. Calculated phonon frequencies for the ideal Cu_2O crystal [cm^{-1}] with force constants adjusted to match the experimentally observed transverse F_{1u} frequencies for ^{16}O ($\alpha = 0.767$ dyne/cm, $\beta = 0.603$ dyne/cm, $e^* = 0.35e$) [7.34]

Isotope	Mode F_{2u}	E_u	A_{2u}	F_{2g}	F_{1u} (long.)		F_{1u} (trans.)	
Calculated								
^{16}O	110	122	316	570	660	173	629	159
^{18}O	110	122	316	538	630	172	601	158
Experimental								
^{16}O	–	124	–	–	–	–	629	159
^{18}O	–	120	–	–	–	–	605	158

while the low-frequency modes do not exhibit such a shift. This is a result of the nature of the modes, since the high-frequency mode involves motion of the copper and the oxygen ion in opposite direction, while both low frequency modes involve largely motion of the copper lattice [7.39]. Finally, we note that the third low-frequency band at 203 cm^{-1} cannot be assigned to a Cu_2O phonon since no fundamentals are found in this region and is probably due to a vibration involving the potassium remaining at the surface.

The comparison of the observed IR intensity of the Cu_2O band at 621 cm^{-1} in Fig. 7.17 with calculations performed by *Greenler* [7.38] can be used to estimate the thickness of the Cu_2O film to ≈ 20 Å or approximately six layers of cuprous oxide [7.34]. Since the total exposure of the surface to oxygen has been ≈ 20 L, one can estimate a sticking coefficient in the order of 0.1 to 1 for the dissociative adsorption of oxygen. In the absence of potassium, a comparable oxide layer on Cu(111) requires oxygen exposures of 10^6 to 10^9 L [7.40]. This means that the presence of potassium promotes the rate of oxidation by more than six orders of magnitude. The reason for this promotional effect of the potassium is obviously due to the bond formation between oxygen and potassium which increases the sticking coefficient for molecular oxygen on the surface. The decomposition of the potassium super oxide at higher temperatures then results in the release of atomic oxygen which reacts with the underlying copper substrate to form Cu_2O. On the clean copper surface the dissociation of oxygen is found to be the rate-limiting step in the oxidation process, since with the formation of the first layer of chemisorbed atomic oxygen the adsorption and dissociation of additional oxygen is inhibited.

7.4 Conclusion – Outlook

Synchrotron radiation in the FIR offers a broad-band radiation source of high intensity and brightness. The comparison with thermal emitters shows that for synchrotron radiation in the FIR the intensity advantage can be 2–3 orders of magnitude in cases where small sample size and small emittance angles are required. These properties make this radiation source particularly suitable for Infrared Reflection Absorption Spectroscopy providing access to the important spectral range of adsorbate-substrate vibrations below 500 cm^{-1}. The example of the CO/Cu(100) system demonstrates that resonant adsorbate-substrate interactions can lead to the breakdown of the surface selection rule in the Far Infrared. This results in the observation of vibrational modes parallel to the surface due to non-vanishing parallel electric fields inside the metal surface. The example of K–O bond formation on copper demonstrates that the full spectral range down to 50 cm^{-1} is now accessible with IRAS. The observation of phonons and their assignments based on isotopic shifts illustrates the future potential of this radiation source to study low-frequency vibrations with high spectral resolution. The observation of anti-absorption bands demonstrates that access to the FIR region opens up a new "frontier" in surface vibrational spectroscopy which holds many surprises [7.44] and presents new opportunities and challenges in surface vibrational spectroscopy. Although the present work constitutes an important first step in extending surface vibrational spectroscopy into the Far Infrared, it is clear that the inherent weakness of absorption bands in the FIR demands further improvements in sensitivity. Progress in this direction will be made at several fronts by improvements in synchrotron-source stability [7.41], by beam-line design optimized for stability and large radiation collection angles, and by developments of new IR detectors with better sensitivity. Finally, the use of the time structure of the synchrotron in single-bunch mode opens new possibilities for time-resolved experiments on a nanosecond time scale with an unparalleled tuning range from the UV to the FIR region [7.43].

Acknowledgement. The authors are very grateful to *C.J. Hirschmugl, Y.J. Chabal, P. Dumas, K.C. Lin,* and *R.G. Tobin* who have contributed greatly to this work, and also to D. Carlson, E. Sullivan, M. Weisel, P. Hollins and D. Slater for their assistance with some of the experiments reported here. The NSLS is supported under the auspices of the U.S. Department of Energy, under contract DE-AC02-76CH00016.

References

7.1 F.M. Hoffmann: Infrared Reflection Absorption Spectroscopy, Surf. Sci. Rep. **3**, 107 (1983)

7.2 A.M. Bradshaw, E. Schweitzer: "Infrared reflection adsorption spectroscopy of adsorbed molecules". In *Advances in Spectroscopy*, ed. by R.J.H. Clark and R.E. Hester (Wiley, New York 1988)

7.3 Y.J. Chabal: Surface Infrared Spectroscopy. Surf. Sci. Rep. **8**, 211 (1988)

7.4 H. Ibach, D.L. Mills: *Electron Energy Loss Spectroscopy and Surface Vibrations* (Academic, New York 1982) H. Ibach, Electron Energy Loss Spectrometers, Springer Series in Optical Science **63**, Springer, Berlin, 1991

7.5 J. Yarwood, T. Shuttleworth, J.B. Hasted, T. Namba: Nature **317**, 743 (1984) T. Namba: Rev. Sci. Instr. **60**, 1689 (1989)

7.6 E. Schweitzer, J. Nagel, W. Braun, E. Lippert, A.M. Bradshaw: Nucl. Instrum. Methods Phys. Res. Sect. A **239**, 630 (1985)

7.7 G.P. Williams, C.J. Hirschmugl, E.M. Kneedler, E.A. Sullivan, D.P. Siddons, Y.J. Chabal, F.M. Hoffmann, K.D. Moeller: Rev. Sci. Instr. **60**, 2176 (1989)

7.8 D.A. Slater, P. Hollins, M.A. Chesters, J. Pritchard, D.H. Martin, M. Surman, D.A. Shaw, I.H. Munro: Rev. Sci. Instr. **63**, 1547 (1992)

7.9 S. Francis, A.H. Allison: J. Opt. Soc. Am. **49**, 131 (1959)

7.10 J. Pritchard: *Vibrations of Adsorbed Layers*, Conference Records Series of KFA, ed. by H. Ibach and S. Lehwald (Jülich, Germany 1978) p. 114

7.11 The surface selection rule is valid only in the absence of resonant substrate-molecule interactions. Recent work shows that in the far-infrared the parallel component of the electric field electric field *inside* of the metal surface can resonantly couple to the vibrational motion of the molecule parallel to the surface. This results in anti-absorption bands as discussed in section 7.3.1

7.12 R.G. Greenler: J. Chem. Phys. **44**, 310 (1966)
R.G. Greenler: J. Vac. Sci. Technol. **12**, 1410 (1975)

7.13 J.D.E. McIntyre, D.E. Aspnes: Surf. Sci. **24**, 417 (1971)

7.14 J.R. Stevenson, H. Ellis, R. Bartlett: Appl. Opt. **12**, 2884 (1973)

7.15 W.E. Duncan, G.P. Williams: Appl. Opt. **22**, 2914 (1983)

7.16 G.P. Williams: Int'l J. Infr. and Millimeterwaves **5**, 829 (1984)

7.17 C.A. Brau: Science **239**, 1115 (1988) and references therein

7.18 P. Rabinowitz, B.N. Perry, N. Levinos: IEEE J. QE-**22**, 797 (1986)

7.19 F.M. Hoffmann, N.J. Levinos, B.N. Perry, P. Rabinowitz: Phys. Rev. B. **33**, 4309 (1986) and references therein

7.20 V.M. Bermudez, J.L. Hylden, J.E. Butler: J. Electron Spectrosc. Rel. Phenom. **38**, 143 (1986)

7.21 G.P. Williams, C.J. Hirschmugl, E.M. Kneedler, P. Takacs, M. Shleifer, Y.J. Chabal, F.M. Hoffmann: Phys. Rev. Lett. **62**, 261 (1989)
7.22 J. Pritchard: Surf. Sci. **79**, 231 (1979)
K. Horn, J. Pritchard: Surface Sci. **55**, 701 (1976)
7.23 P. Uvdal, P.A. Karlsson, C. Nyberg, S. Andersson, N.V. Richardson: Surf. Sci. **292**, 167 (1988)
7.24 C.J. Hirschmugl, G.P. Williams, F.M. Hoffmann, Y.J. Chabal: Phys. Rev. Lett. **65**, 480 (1990).
7.25 D. Hoge, M. Teushaus, E. Schweitzer, A.M. Bradshaw: Chem. Phys. Lett. **151**, 230 (1989)
I. Malik, M. Trenary: Surf. Sci. **214**, L237 (1989)
R. Ryberg: Phys. Rev. B. **40**, 8567 (1989)
7.26 W. Akemann, A. Otto: J. Raman Spectrosc. **22**, 797 (1991)
7.27 G. Herzberg: *Molecular Spectra and Molecular Structure, II. Infrared and Raman Spectra of Diatomic Molecules* (Van Norstrand, Princeton, NJ 1945)
7.28 For consistency we continue to use the nomenclature introduced in Fig. 7.2, where E_p denotes the electric field perpendicular to the surface and E_s the field parallel to the surface
7.29 B.N.J. Persson: Surf. Sci. **269/270**, 103 (1992)
7.30 B.N.J. Persson, D. Schumacher, A. Otto: Chem. Phys. Lett. **178**, 204 (1991)
B.N.J. Persson, A.I. Volotkin: Chem. Phys. Letts. **185**, 292 (1991)
B.N.J. Persson: Phys. Rev. B **44**, 3277 (1991)
7.31 C.J. Hirschmugl, P. Dumas, M. Suhren, Y.J. Chabal, F.M. Hoffmann, G.P. Williams: J. Electron Spectrosc. Rel. Phenomena, **64/65**, 67 (1991)
7.32 *Physics and Chemistry of Alkali Adsorption* ed. by H.P. Bonzel, A.M. Bradshaw, G. Ertl (Elsevier, Amsterdam 1989) and references therein
7.33 J. Paul, R.A. de Paola, F.M. Hoffmann: In *Physics and Chemistry of Alkali Adsorption*, ed. by H.P. Bonzel, A.M. Bradshaw, G. Ertl (Elsevier, Amsterdam 1989) p. 213
7.34 F.M. Hoffmann, K.C. Lin, R.G. Tobin, C.J. Hirschmugl, G.P. Williams, P. Dumas: Surf. Sci. **275**, L675 (1992)
7.35 F.M. Hoffmann, M.D. Weisel, W. Eberhardt, Z. Fu: Surf. Sci. **234**, L264 (1990)
7.36 R.R. Smardzewski, L. Andrews: J. Chem. Phys. **57**, 1327 (1972)
7.37 Y. Shao, J. Paul, to be submitted
7.38 R.G. Greenler, R.R. Rahn, J.P. Schwartz: J. Catalysis **23**, 42 (1971)
7.39 P. Dawson, M.M. Hargreave, G.R. Wilkinson: J. Phys. Chem. Solids **34**, 2201 (1973)
7.40 L.H. Dubois: Surf. Sci. **119**, 399 (1982)
7.41 The stability has recently been improved by the installation of a global feedback system at the UV ring at the NSLS [7.42]
7.42 L.H. Yu, R. Biscardi, E. Bozoki, J. Galayada, S. Krinsky, R. Nawroky, O. Singh, G. Vignola: Proc. Particle Accelerator Conf. (Chicago, IL, 1985) p. 1792 (1989)
7.43 G.L. Carr, J. Reichman, D. DiMarzio, M.B. Lee, D. Ederer, K. Miyano, D.R. Muller, A. Vasilakis: J. Semiconductor Sci. and Technol., in press
7.44 Note added in proof: Recently, we have observed strong anti-absorption bands for alkali metals adsorbed on Cu(111). Experimental data show that these bands are *not* due to phonon bands, but can be consistently explained with vibrations parallel to the surface of a pinned charge density wave [7.45]
7.45 F.M. Hoffmann, B.N.J. Persson, G.P. Williams, C.J. Hirschmugl, W. Walter and D.A. King, Phys. Rev. Lett. **72**, 1256 (1994)

Index of Molecules and Adsorbate Systems

Subject Index

Springer-Verlag and the Environment

MIX
Papier aus verantwortungsvollen Quellen
Paper from responsible sources
FSC® C105338

If you have any concerns about our products,
you can contact us on
ProductSafety@springernature.com

In case Publisher is established outside the EU,
the EU authorized representative is:
Springer Nature Customer Service Center GmbH
Europaplatz 3, 69115 Heidelberg, Germany

Printed by Libri Plureos GmbH
in Hamburg, Germany